T0236438

Lecture Notes in Artificial Intelligence   8439

Subseries of Lecture Notes in Computer Science

LNAI Series Editors

Randy Goebel
University of Alberta, Edmonton, Canada
Yuzuru Tanaka
Hokkaido University, Sapporo, Japan
Wolfgang Wahlster
DFKI and Saarland University, Saarbrücken, Germany

LNAI Founding Series Editor

Joerg Siekmann
DFKI and Saarland University, Saarbrücken, Germany

For further volumes:
http://www.springer.com/series/1244

Michael Hanus · Ricardo Rocha (Eds.)

Declarative Programming and Knowledge Management

Declarative Programming Days, KDPD 2013
Unifying INAP, WFLP, and WLP
Kiel, Germany, September 11–13, 2013
Revised Selected Papers

Springer

Editors
Michael Hanus
Universität Kiel
Kiel
Germany

Ricardo Rocha
CRACS & INESC-Porto LA
University of Porto
Porto
Portugal

ISSN 0302-9743   ISSN 1611-3349 (electronic)
ISBN 978-3-319-08908-9   ISBN 978-3-319-08909-6 (eBook)
DOI 10.1007/978-3-319-08909-6
Springer Cham Heidelberg New York Dordrecht London

Library of Congress Control Number: 2014944063

LNCS Sublibrary: SL7 – Artificial Intelligence

© Springer International Publishing Switzerland 2014
This work is subject to copyright. All rights are reserved by the Publisher, whether the whole or part of the material is concerned, specifically the rights of translation, reprinting, reuse of illustrations, recitation, broadcasting, reproduction on microfilms or in any other physical way, and transmission or information storage and retrieval, electronic adaptation, computer software, or by similar or dissimilar methodology now known or hereafter developed. Exempted from this legal reservation are brief excerpts in connection with reviews or scholarly analysis or material supplied specifically for the purpose of being entered and executed on a computer system, for exclusive use by the purchaser of the work. Duplication of this publication or parts thereof is permitted only under the provisions of the Copyright Law of the Publisher's location, in its current version, and permission for use must always be obtained from Springer. Permissions for use may be obtained through RightsLink at the Copyright Clearance Center. Violations are liable to prosecution under the respective Copyright Law.
The use of general descriptive names, registered names, trademarks, service marks, etc. in this publication does not imply, even in the absence of a specific statement, that such names are exempt from the relevant protective laws and regulations and therefore free for general use.
While the advice and information in this book are believed to be true and accurate at the date of publication, neither the authors nor the editors nor the publisher can accept any legal responsibility for any errors or omissions that may be made. The publisher makes no warranty, express or implied, with respect to the material contained herein.

Printed on acid-free paper

Springer is part of Springer Science+Business Media (www.springer.com)

Preface

This volume contains a selection of papers presented at the Kiel Declarative Programming Days 2013, held in Kiel (Germany) during September 11–13, 2013. The Kiel Declarative Programming Days 2013 unified the following events:

- 20th International Conference on Applications of Declarative Programming and Knowledge Management (INAP 2013)
- 22nd International Workshop on Functional and (Constraint) Logic Programming (WFLP 2013)
- 27th Workshop on Logic Programming (WLP 2013)

All these events are centered around *declarative programming*, an advanced paradigm for the modeling and solving of complex problems. These specification and implementation methods attracted increasing attention over the last decades, e.g., in the domains of databases and natural language processing, for modeling and processing combinatorial problems, and for high-level programming of complex, in particular, knowledge-based systems.

The INAP conferences provide a communicative forum for intensive discussion of applications of important technologies around logic programming, constraint problem solving, and closely related computing paradigms. It comprehensively covers the impact of programmable logic solvers in the Internet society, its underlying technologies, and leading-edge applications in industry, commerce, government, and societal services. Previous INAP editions have been held in Japan, Germany, Portugal, and Austria.

The international workshops on functional and logic programming (WFLP) aim at bringing together researchers interested in functional programming, logic programming, as well as the integration of these paradigms. Previous WFLP editions have been held in Germany, France, Spain, Italy, Estonia, Brazil, Denmark, and Japan.

The workshops on (constraint) logic programming (WLP) serve as the scientific forum of the annual meeting of the Society of Logic Programming (GLP e.V.) and bring together researchers interested in logic programming, constraint programming, and related areas like databases, artificial intelligence, and operations research. Previous WLP editions have been held in Germany, Austria, Switzerland, and Egypt.

In 2013 these events were jointly organized under the umbrella of the Kiel Declarative Programming Days in order to promote the cross-fertilizing exchange of ideas and experiences among researchers and students from the different communities interested in the foundations, applications, and combinations of high-level, declarative programming languages and related areas. The technical program of the event included presentations of refereed technical papers and system descriptions. In addition to the selected papers, the scientific program included an invited lecture by Tom Schrijvers (University of Ghent, Belgium).

After the event, the Program Committees invited authors to submit revised versions of their presented papers. Each submission was reviewed by at least three Program Committee members. The meetings of the Program Committees were conducted electronically during July 2013 and February 2014 with the help of the conference management system EasyChair. After careful discussions, the Program Committees decided to accept 15 papers for inclusion in these proceedings.

We would like to thank all authors who submitted papers to this event. We are grateful to the members of the Program Committees and all the additional reviewers for their careful and efficient work in the review process. Finally, we express our gratitude to all members of the local Organizing Committee for their help in organizing a successful event.

April 2014

Michael Hanus
Ricardo Rocha

Conference Organization

WFLP/WLP Conference Chair

Michael Hanus   University of Kiel, Germany

INAP Conference Chair

Ricardo Rocha   University of Porto, Portugal

WFLP/WLP Program Committee

Elvira Albert	Universidad Complutense de Madrid, Spain
Sergio Antoy	Portland State University, USA
François Bry	University of Munich, Germany
Jürgen Dix	Clausthal University of Technology, Germany
Rachid Echahed	CNRS, University of Grenoble, France
Moreno Falaschi	Università di Siena, Italy
Sebastian Fischer	Kiel, Germany
Thom Frühwirth	University of Ulm, Germany
Michael Hanus	University of Kiel, Germany (Chair)
Oleg Kiselyov	Monterey (CA), USA
Herbert Kuchen	University of Münster, Germany
Francisco Javier López-Fraguas	Universidad Complutense de Madrid, Spain
Torsten Schaub	University of Potsdam, Germany
Peter Schneider-Kamp	University of Southern Denmark, Denmark
Dietmar Seipel	University of Würzburg, Germany
Hans Tompits	Vienna University of Technology, Austria
German Vidal	Universidad Politécnica de Valencia, Spain
Janis Voigtländer	University of Bonn, Germany

INAP Program Committee

Salvador Abreu	University of Évora, Portugal
Sergio Alvarez	Boston College, USA
Christoph Beierle	FernUniversität in Hagen, Germany
Philippe Codognet	JFLI/CNRS at University of Tokyo, Japan
Daniel Diaz	University of Paris I, France
Ulrich Geske	University of Potsdam, Germany

Petra Hofstedt	Brandenburg University of Technology Cottbus, Germany
Katsumi Inoue	National Institute of Informatics, Japan
Gabriele Kern-Isberner	University of Dortmund, Germany
Ulrich Neumerkel	Vienna University of Technology, Austria
Vitor Nogueira	University of Évora, Portugal
Enrico Pontelli	New Mexico State University, USA
Ricardo Rocha	University of Porto, Portugal (Chair)
Irene Rodrigues	University of Évora, Portugal
Carolina Ruiz	Worcester Polytechnic Institute, USA
Vítor Santos Costa	University of Porto, Portugal
Dietmar Seipel	University of Würzburg, Germany
Terrance Swift	Universidade Nova de Lisboa, Portugal
Hans Tompits	Vienna University of Technology, Austria
Masanobu Umeda	Kyushu Institute of Technology, Japan
Marina De Vos	University of Bath, UK
Armin Wolf	Fraunhofer FIRST, Berlin, Germany
Osamu Yoshie	Waseda University, Japan

Local Organization

Linda Haberland	University of Kiel, Germany
Michael Hanus	University of Kiel, Germany
Björn Peemöller	University of Kiel, Germany
Fabian Reck	University of Kiel, Germany
Jan Rasmus Tikovsky	University of Kiel, Germany

External Reviewers

Demis Ballis	Tony Ribeiro
Steffen Ernsting	José Miguel Rojas
Raúl Gutiérrez	Javier Romero
Benny Höckner	Miguel A. Salido
Andy Jost	Peter Sauer
Arne König	Luca Torella
Ludwig Ostermayer	Amira Zaki

Contents

Construction of Explanation Graphs
from Extended Dependency Graphs
for Answer Set Programs

Ella Albrecht, Patrick Krümpelmann(✉), and Gabriele Kern-Isberner

Technische Universität Dortmund, Dortmund, Germany
patrick.kruempelmann@cs.tu-dortmund.de

Abstract. Extended dependency graphs are an isomorphic representation form for Answer Set Programs, while explanation graphs give an
explanation for the truth value of a literal contained in an answer set. We
present a method and an algorithm to construct explanation graphs from
a validly colored extended dependency graph. This method exploits the
graph structure of the extended dependency graph to gradually build up
explanation graphs. Moreover, we show interesting properties and relations of the graph structures, such as loops, and we consider both answer
set and well-founded semantics. We also present two different approaches
for the determination of assumptions in an extended dependency graph,
an optimal but exponential and a sub-optimal but linear one.

1 Introduction

Graphs are an excellent tool for the illustration and understanding of nonmonotonic reasoning formalisms, and for the determination and explanation of
models. For answer set programs two graph based representations have recently
been proposed: Extended dependency graphs (EDG) [2] and explanation graphs
(EG) [1]. EDGs are an isomorphic representation of extended logic programs
and use a coloring of the nodes to determine answer sets. Explanation graphs,
on the other hand, provide an explanation for the appearance of a single literal
in an answer set. In [1] it was conjectured that there is a strong relation between
a validly colored extended dependency graph and an explanation graph. In this
work we present a method to construct explanation graphs from a successfully
colored extended dependency graph and prove its correctness. The way of proceeding exploits the structure of the EDG and the fact that explanation graphs
can be built up gradually from smaller sub-explanation graphs.

In [1] assumptions are introduced, which describe literals whose truth value
has to be guessed during the determination process of answer sets, but there is
actually no appropriate method given to find proper assumptions. We present
two systematic approaches which extract assumptions from an EDG. This is the
most difficult part of the construction of EGs, since intra-cyclic as well as inter-
cyclic dependencies between nodes have to be considered. The first approach
makes use of basic properties of assumptions and the graph to reduce the size

M. Hanus and R. Rocha (Eds.): KDPD 2013, LNAI 8439, pp. 1–16, 2014.
DOI: 10.1007/978-3-319-08909-6_1, © Springer International Publishing Switzerland 2014

of assumptions in linear runtime. The second approach exploits cycle structures and their interdependencies to determine the minimal assumptions, which comes with the cost of exponential runtime.

In Sect. 2 we give an introduction to answer set programming and in Sect. 3 we present extended dependency graphs and explanation graphs. Section 4 deals with the construction process of the EGs from a validly colored EDG. The fifth section deals with the different approaches of finding proper assumptions in an EDG.

2   Answer Set Programming

We consider extended logic programs under the answer set semantics [3]. An *extended logic program* P is a set of rules r of the form $r : h \leftarrow a_1, ..., a_n, not\ b_1, ..., not\ b_n$. where $h, a_1, ... a_m, b_1, ..., b_n$ are literals. A *literal* may be of the form x or $\neg x$ where x is a propositional symbol called *atom* and $\neg$ is the classical negation. $head(r) = \{h\}$ denotes the head, $pos(r) = \{a_1, ..., a_m\}$ denotes the positive, and $neg(r) = \{b_1, ..., b_n\}$ the negative body literals of a rule. The *Herbrand base* $\mathcal{H}(P)$ of a logic program P is the set of all grounded literals of P. A literal is *grounded*, if it does not contain a variable. In this work, we assume that the logic programs are grounded, i.e., every literal appearing in the program is grounded.

Let $M \subseteq \mathcal{H}(P)$ be a consistent set of literals, i.e., M does not contain any complementary literals. The Gelfond-Lifschitz reduct of a program P is the program $P^M = \{h \leftarrow a_1, ..., a_n. \mid h \leftarrow a_1, ..., a_n, not\ b_1, ..., not\ b_m. \in P, \{b_1, ..., b_n\} \cap M = \emptyset\}$ that is obtained by removing all rules where a literal $b_i \in M$ appears as a negative body literal, and removing all negative body literals from the remaining rules. M is *closed* under P^M if $head(r) \in M$ whenever $pos(r) \subseteq M$ for every rule $r \in P^M$. M is an *answer set* for P if M is closed under P^M and M is minimal w.r.t. set inclusion.

Answer set semantics may yield multiple models resp. answer sets. Another semantics for logic programs is the well-founded semantics [6]. Its basic idea is that there exist literals which have to be true with certainty and literals which have to be false with certainty. Under the answer set semantics such information gets lost if no answer set exists.

For a logic program P and a logic program P^+ that we get if we remove all rules with negative body literals, the sequence $(K_i, U_i)_{i \geq 0}$ is defined as $K_0 = lfp(T_{P^+, \emptyset})U_0 = lfp(T_{P, K_0})K_i = lfp(T_{P, U_{i-1}})U_i = lfp(T_{P, K_i})$ where $T_{P,V}(S) = \{a \mid \exists r \in P : head(r) = a, pos(r) \subseteq S, neg(r) \cap V = \emptyset\}$. The well-founded model is $WF_P = \langle W^+, W^- \rangle$ where $W^+ = K_j$ is the *well founded* set and $W^- = \mathcal{H}(P) \setminus U_j$ is the *unfounded* set, with j being the first index with $\langle K_j, U_j \rangle = \langle K_{j+1}, U_{j+1} \rangle$. Literals that are neither contained in W^+ nor in W^- are called *undefined*.

3   Graphs for Answer Set Programs

We introduce two types of graphs, the first graph type is the extended dependency graph [2].

Definition 1 (Extended dependency graph). *The* extended dependency graph *(EDG) to a logic program P with its Herbrand base $\mathcal{H}(P)$ is a directed graph $EDG(P) = (V, E)$ with node set $V \subseteq \mathcal{H}(P)$ and edge set $E \subseteq V \times V \times \{+, -\}$, according to the following rules:*

V1 *There is a node a_i^k for every rule $r_k \in P$ where $head(r_k) = a_i$.*

V2 *There is a node a_i^0 to every atom $a_i \in \mathcal{H}(P)$ which does not appear as the head of a rule.*

E1 *There is an edge $(c_j^l, a_i^k, +)$ for every node $c_j^l \in V$ iff there is a rule $r_k \in P$ where $c_j \in pos(r_k)$ and $head(r_k) = a_i$.*

E2 *There is an edge $(c_j^l, a_i^k, -)$ for every node $c_j^l \in V$ iff there is a rule $r_k \in P$ where $c_j \in neg(r_k)$ and $head(r_k) = a_i$.*

To every logic program a unique EDG can be constructed, this means a logical program is isomorphic to its representation as an EDG in the sense that the structure of the program is reflected one-to-one by the structure of the EDG. Properties of a logic program can be obtained from properties of the corresponding EDG and vice versa. Colorings of the graph that comply with the semantics of the edges are called valid colorings and correspond to answer sets of a logic program. A green colored node represents a successfully deduced head of a rule, and a red colored node represents a head of a rule that can not be deduced. A literal is contained in the answer set if there exists a node that represents the literal which is colored green. A literal is not contained in the answer set if all nodes corresponding to the literal are colored red.

Definition 2 (Valid coloring of an EDG). *Let a program P be given. A coloring $\nu : V \rightarrow \{green, red\}$ of the graph $EDG(P) = (V, E)$ is valid, if the following conditions holds:*

1. *$\forall i, k$ where $k \geq 1$, $\nu(a_i^k) = green$ if a_i^k has no incoming edge.*
2. *$\forall i, k$, $\nu(a_i^k) = green$ if the following two conditions are met:*
   (a) *$\forall j, m$ where $(a_j^m, a_i^k, +) \in E$, $\exists a_j^h \in V$ such that $\nu(a_j^h) = green$*
   (b) *$\forall j, m$ where $(a_j^m, a_i^k, -) \in E$, $\nu(a_j^m) = red$*
3. *$\forall i, k$ $\nu(a_i^k) = red$, if at least one of the two following conditions is met:*
   (a) *$\exists j, m$ where $(a_j^m, a_i^k, +) \in E$ and $\forall a_j^h \in V$, $\nu(a_j^h) = red$*
   (b) *$\exists j, m$ where $(a_j^m, a_i^k, -) \in E$ and $\nu(a_j^m) = green$*
4. *For every positive cycle C where $\nu(a_i^k) = green$ for all $a_i^k \in C$ the following condition holds: There is an i and $l \neq k$, such that $\nu(a_i^l) = green$.*

A main feature of EDGs is their representation of cycles and handles for those. A cycle consists of several literals that are dependent in a cyclic way, e.g., given the two rules $r_1 : a \leftarrow b.$ and $r_2 : b \leftarrow a.$, both literals a and b are interdependent in a cyclic way. Generally, cycles can be connected in two different ways to the rest of the program:

- *OR-handle:* Let the rule $r_1 : a \leftarrow \beta.$ be part of a cycle where β may be of the form b or *not* b. If there exists another rule $r_2 : a \leftarrow \delta$ where δ may be of the form d or *not* d, then δ is an OR-handle for the cycle to which r_1 belongs. The OR-handle is called active if δ is true.

$P_1 := \{a \leftarrow not\ b.$
$\quad b \leftarrow not\ a.$
$\quad c \leftarrow \{not\ a\}.$
$\quad c \leftarrow e.$
$\quad d \leftarrow c.$
$\quad e \leftarrow not\ d.$
$\quad f \leftarrow [e],\ not\ f.\}$

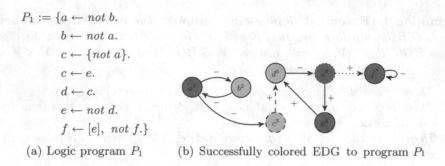

(a) Logic program P_1 (b) Successfully colored EDG to program P_1

Fig. 1. A logic program, the corresponding EDG, and a valid coloring

- *AND-handle*: If a rule r is part of a cycle and has an additional condition γ, that means the rule is of the form $r :\ a \leftarrow \beta, \gamma$ where γ may be of the form c or *not* c, then γ is an AND-handle. The AND-handle is called active if γ is false.

An extended dependency graph extends a normal dependency graph in so far that it distinguishes between AND- and OR-handles.

Example 1. *Figure 1 shows a logic program (a) and the corresponding EDG (b). The EDG has a valid coloring which represents the answer set $\{b, c, d\}$. The AND-handle is marked in the program with [ ] and is dotted in the EDG. The OR-handle is marked in the program with { } and is dashed in the EDG.*

The second graph type is the explanation graph (EG). In contrast to the EDGs, which visualize the structure of a whole logic program, EGs provide an explanation for why a single literal appears or does not appear in an answer set and is always constructed with regard to an answer set and a set of assumptions. Assumptions are literals for which no explanation is needed since their value is assumed. All literals that are qualified for being used as assumptions are called *tentative assumptions* and are formally defined as follows:

Definition 3 (Tentative Assumptions). *Let P be a logic program, M an answer set of P and $WF = \langle WF^+, WF^- \rangle$ the well-founded model of P. Then the set of* tentative assumptions *of P w.r.t. M is*

$$\mathcal{T}\mathcal{A}_P(M) = \{a \mid a \in NANT(P) \text{ and } a \notin M \text{ and } a \notin WF^+ \cup WF^-\}$$

where $NANT(P)$ is the set of all literals appearing in P as a negative body literal: $NANT(P) = \{a \mid \exists r \in P :\ a \in neg(r)\}$.

Given a logic program P and a subset $U \subseteq \mathcal{T}\mathcal{A}_P(M)$ of tentative assumptions, one can obtain the negative reduct $NR(P, U)$ of a program P w.r.t. U by removing all rules where $head(r) \in U$.

Definition 4 (Assumption). *An* assumption *of a program P regarding an answer set M is a set $U \subseteq \mathcal{T}\mathcal{A}_p(M)$ where the well-founded model of the negative reduct corresponds to the answer set M, i. e. $WF_{NR(P,U)} = \langle M, \mathcal{H}(P) \setminus M \rangle$.*

This means that setting all literals of the assumption U to *false* leads to all literals being defined in the well-founded model.

Explanation graphs are based on local consistent explanations (LCE). These are sets of literals which directly influence the truth value of a literal a. For a literal a that is contained in the answer set and a rule where a is the head and all conditions of the rule are fulfilled, i.e., all body literals are true, the LCE consists of all body literals of the rule. Since there may exist several fulfilled rules with a as their head, a can also have various LCEs. For a literal a that is not contained in the answer set, an LCE is a minimal set of literals that together falsify all rules that define a. For this purpose the LCE has to contain one falsified condition from each rule.

Definition 5 (Local Consistent Explanation). *Let a program P be given, let a be a literal, let M be an answer set of P, let U be an assumption and let $S \subseteq \mathcal{H}(P) \cup \{not\ a \mid a \in \mathcal{H}(P)\} \cup \{assume, \top, \bot\}$ be a set of (default-negated) literals and justifying symbols.*

1. *S is an LCE for a^+ w.r.t. (M, U), if $a \in M$ and*
   - *$S = \{assume\}$ or*
   - *$S \cap \mathcal{H}(P) \subseteq M$, $\{c \mid not\ c \in S\} \subseteq (\mathcal{H}(P) \setminus M) \cup U$ and there exists a rule $r \in P$ where $head(r) = a$ and $S = body(r)$. For the case that $body(r) = \emptyset$ one writes $S = \{\top\}$ instead of $S = \emptyset$.*
2. *S is an LCE for a^- w.r.t. (M, U), if $a \in (\mathcal{H}(P) \setminus M) \cup U$ and*
   - *$S = \{assume\}$ or*
   - *$S \cap \mathcal{H}(P) \subseteq (\mathcal{H}(P) \setminus M) \cup U$, $\{c \mid not\ c \in S\} \subseteq M$ and S is a minimal set of literals, such that for every $r \in P$ the following holds: if $head(r) = a$ then $pos(r) \cap S \neq \emptyset$ or $neg(r) \cap \{c \mid not\ c \in S\} \neq \emptyset$. For the case S being the empty set one writes $S = \{\bot\}$.*

In an EDG an edge (a_i^k, a_j^l, s) with $s \in \{+, -\}$ means that the truth value of literal a_j depends on the truth value of literal a_i. In an explanation graph the edges are defined the other way round, so that an edge (a_i, a_j, s) with $s \in \{+, -\}$ means that a_j explains or supports the truth value of a_i. A node in an explanation graph is either annotated with $+$ or $-$, depending on whether the literal is contained in the answer set or not. We define two sets of annotated literals $\mathcal{H}^p = \{a^+ \mid a \in \mathcal{H}(P)\}$ and $\mathcal{H}^n = \{a^- \mid a \in \mathcal{H}(P)\}$. Furthermore we define $literal(a^+) = a$ and $literal(a^-) = a$. The *support* of a node a_i in an EG is the set of all direct successors of a_i in the EG and is formally defined as follows:

Definition 6 (Support). *Let $G = (V, E)$ be a graph with node set $V \subseteq \mathcal{H}^p \cup \mathcal{H}^n \cup \{assume, \top, \bot\}$ and edge set $E \subseteq V \times V \times \{+, -\}$. Then the support of a node $a \in V$ w.r.t. G is defined as:*

- *$support(a, G) = \{literal(c) \mid (a, c, +) \in E\} \cup \{not\ literal(c) \mid (a, c, -) \in E\}$,*
- *$support(a, G) = \{\top\}$ if $(a, \top, +) \in E$,*
- *$support(a, G) = \{\bot\}$ if $(a, \bot, -) \in E$ or*
- *$support(a, G) = \{assume\}$ if $(a, assume, s) \in E$ where $s \in \{+, -\}$.*

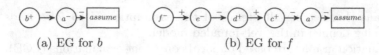

(a) EG for b (b) EG for f

Fig. 2. Explanation graphs for literals b and f in program P_1 w.r.t. answer set $M = \{b, d, c\}$ and assumption $U = \{a\}$

Definition 7 (Explanation graph). *An* explanation graph *for a literal* $a \in \mathcal{H}^p \cup \mathcal{H}^n$ *in a program P w.r.t. an answer set M and an assumption $U \in$ Assumptions(P, M) is a directed graph $G = (V, E)$ with node set $V \subseteq \mathcal{H}^p \cup \mathcal{H}^n \cup \{assume, \top, \bot\}$ and edge set $E \subseteq V \times V \times \{+, -\}$, such that the following holds:*

1. *The only sinks in the graph are assume, $\top$ and $\bot$, where $\top$ is used to explain facts of the program P, $\bot$ is used to explain literals which do not appear as a head of any rule and assume is used to explain literals for which no explanations are needed since their value is assumed to be false.*
2. *If $(c, l, s) \in E$ where $l \in \{assume, \top, \bot\}$ and $s \in \{+, -\}$, then (c, l, s) is the only outgoing edge for every $c \in V$.*
3. *Every node $c \in V$ is reachable from a.*
4. *For every node $c \in V \setminus \{assume, \top, \bot\}$ the support support(c, G) is an LCE for c regarding M and U.*
5. *There exists no $c^+ \in V$, such that $(c^+, assume, s) \in E$ where $s \in \{+, -\}$.*
6. *There exists no $c^- \in V$, such that $(c^-, assume, +) \in E$.*
7. *$(c^-, assume, -) \in E$ iff $c \in U$.*

Example 2. *Figure 2 shows the explanation graphs for literals b and f from the program in Fig. 1a w.r.t. the answer set $M = \{b, d, c\}$ and assumption $= \{a\}$.*

4   Construction of Explanation Graphs

In this section we introduce an approach for the construction of explanation graphs by extracting the required information from a validly colored extended dependency graph. Suppose we are given an extended dependency graph $G = (V, E)$ with a valid coloring $\nu : V \to \{green, red\}$. In the first step, we clean up the EDG by removing irrelevant edges and nodes. Irrelevant edges and nodes are those edges and nodes that do not have influence on the appearance or non-appearance of a literal in the answer set. This means they do not provide an explanation for a literal and hence are not needed for any explanation graph.

Definition 8 (Irrelevant edge, irrelevant node). *An edge (a_i^k, a_j^l, s) is irrelevant if*

- *$\nu(a_i^k) = green$, $\nu(a_j^l) = green$ and $s = -$,*
- *$\nu(a_i^k) = green$, $\nu(a_j^l) = red$ and $s = +$,*

– $\nu(a_i^k) = red$, $\nu(a_j^l) = green$ *and* $s = +$ *or*
– $\nu(a_i^k) = red$, $\nu(a_j^l) = red$ *and* $s = -$.

A node a_i^k is irrelevant if $\nu(a_i^k) = red$ and there exists $l > 0$ where $\nu(a_i^l) = green$.

If an irrelevant node is removed, all its incoming and outgoing edges are also removed. After removing irrelevant edges and nodes we get an EDG $G' = (V', E')$ with $V' \subseteq V$ and $E' \subseteq E$. In the second step, nodes are gradually marked in the EDG. The marking process starts at nodes which have no incoming edges, because the explanation graphs for these nodes do not depend on other nodes. Every time a node is marked, the explanation graphs for the marked node are built. For this purpose five types of transformations are defined. The two first transformations describe the construction of explanation graphs for simple nodes, i.e., nodes which have no incoming edges in the EDG. The third and fourth transformations describe the construction of nodes which are dependent on other nodes, i.e., have incoming edges, distinguished by the color of the nodes. The last transformation is used for the construction of EGs for literals that are used as assumptions.

Transformation 1 (Transformation of fact nodes). *The EG for a node a_i^k which has no incoming edges and satisfies $\nu(a_i^k) = green$ consists of a node a_i^+, a node $\top$ and an edge $(a_i^+, \top, +)$ (Fig. 3a), because such a node corresponds to a fact of the logic program.*

Transformation 2 (Transformation of unfounded nodes). *The EG to a node a_i^k which has no incoming edges and satisfies $\nu(a_i^k) = red$ consists of a node a_i^-, a node $\bot$ and an edge $(a_i^-, \bot, -)$ (Fig. 3b).*

After marking nodes without incoming edges, we can mark nodes in positive cycles (cycles that contain only positive edges) that do not have an active handle, since the corresponding literals do not have a supportive justification and are unfounded in the well-founded model. Since there exists no active handle for the cycle, there is no other explanation for the nodes of the cycle than the one consisting of the cycle itself (with reversed edges). Now we continue marking nodes using the following rules until no more nodes can be marked:

A green node a_i^k can be marked if

– for all a_j^l where $(a_j^l, a_i^k, +) \in E'$, there exists $n \geq 1$ with $a_j^n \in V'$, such that a_j^n is marked, and
– for all nodes a_j^l where $(a_j^l, a_i^k, -) \in E'$, a_j^l is marked.

That means that a green node can be marked, if all its predecessor nodes are marked. For literals which are represented by multiple green nodes, it is sufficient if one of these nodes is marked.

A red node a_i^k can be marked if

– $\exists (a_j^l, a_i^k, -) \in E'$ where a_j^l is marked, or
– $\exists (a_j^l, a_i^k, +) \in E'$ where for all $n \geq 0$, $a_j^n \in V'$ is marked.

That means that a red node can be marked, if at least one of its predecessor nodes is marked. In case that a predecessor literal is represented by multiple red nodes, all these nodes have to be marked.

Lemma 1. *The well-founded set W^+ corresponds to the set of all marked green nodes and the unfounded set W^- corresponds to the set of all marked red nodes.*

Proof sketch. It has to be shown that K_i always contains marked green nodes and $X_i = \mathcal{H}(P) \setminus U_i$ always contains marked red nodes where K_i and U_i are the sets that are generated during the calculation of the well-founded model (see Page 2). For this purpose the fixpoint operator $T_{P,V}$ for the generation of K_i and X_i has to be adjusted to X_i instead of U_i, especially in the adjustment of the operator for X_i positive cycles have to be considered. Then it can be seen, that the resulting operators exactly describe the process of marking green resp. red nodes.

From Lemma 1 we get the following proposition:

Proposition 1. *All unmarked nodes are undefined in the well-founded model.*

Green nodes represent literals that are contained in the answer set. So the local consistent explanation for such a node consists of all direct predecessor nodes (resp. the literals they represent).

Transformation 3 (Transformation of dependent green nodes). *Let $ie(a_i^k)$ be the set of incoming edges of node a_i^k. An EG to a node a_i^k where $\nu(a_i^k) = green$ and $ie(a_i^k) \neq \emptyset$ consists of a node a_i^+ and edges $E_{EG} = \{(a_i^+, EG(a_j), s) \mid (a_j^l, a_i^k, s) \in E'\}$, where $EG(a_j)$ is an explanation graph for a_j (Fig. 3d).*

Red nodes represent literals that are not contained in the answer set. In most cases red nodes have only active edges. The only exception is if a predecessor literal a_j of a red node a_i^k is represented by multiple red nodes, formally $|\{a_j^l \mid (a_j^l, a_i^k, -) \in E'\}| \geq 2$. To get the LCEs for literals represented by a red node, all nodes representing this literal have to be considered. Each node represents a rule where the incoming edges represent the conditions of the rule. An LCE has to contain exactly one violated condition from each rule. Since we have removed all irrelevant edges, every edge represents a violated condition. That means that an LCE contains exactly one incoming edge for every node representing the literal.

Definition 9 (Local consistent explanation in an EDG). *Let $pd(a_i^k) = \{a_j \mid (a_j^l, a_i^k, s) \in E', s \in \{+, -\}\}$ be the set of the predecessor literals of node a_i^k and a_i^1 to a_i^n the nodes representing a literal a_i. We set $L(a_i) = \{\{b_1, ..., b_n\} \mid b_1 \in pd(a_i^1), ..., b_n \in pd(a_i^n)\}$. $L(a_i)$ is an LCE for a_i if $L(a_i)$ is minimal w.r.t. set inclusion.*

Transformation 4 (Transformation of dependent red nodes). *The explanation graph for a node a_i^k where $\nu(a_i^k) = red$ w.r.t. an LCE $L(a_i)$ consists of a node a_i^- and edges*

$$E_{EG} = \{(a_i^-, EG(a_j), s \mid a_j \in L, (a_j^l, a_i^k, s) \in E' \text{ for any } l, k\} \ (Fig. 3e).$$

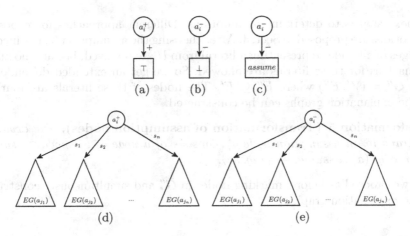

Fig. 3. Templates for constructing an explanation graph for a_i

As mentioned before, every time a node is marked explanation graphs are constructed. It should be remarked that not all explanation graphs for a literal can be created when a node is reached for the first time and marked. This follows from the fact that there can exist multiple nodes for one literal and that a red node can be already marked if one of its predecessors is marked.

Example 3. *In the graph from Fig. 4b for the logic program P_2 (Fig. 4a) we can see that although both nodes are marked, we cannot construct all explanation graphs to the literal a and also e, since e depends on a. The explanation graph for the LCE $\{c, d\}$ of a is missing, because c depends on a cycle where an assumption has to be determined. So, if the node a^1 is reached again by the other edge $(c^3, a^1, +)$ during the marking process, the set of its explanation graphs has to be updated and the information must be propagated to all successor nodes.*

$$P_2 := \{a \leftarrow b, c.$$
$$a \leftarrow d.$$
$$c \leftarrow not\ f.$$
$$e \leftarrow not\ a.$$
$$f \leftarrow not\ c.\}$$

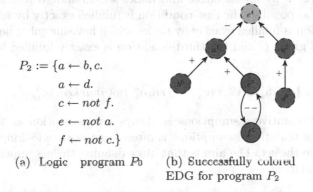

(a) Logic program P_0　　(b) Successfully colored EDG for program P_2

Fig. 4. A logic program and the corresponding EDG

The next step is to determine assumptions. Different approaches for choosing assumptions are proposed in Sect. 4. After choosing the assumption U, the incoming edges of all nodes representing a literal from U are removed, because no other explanations for these literals are allowed. So we get an extended dependency graph $G'' = (V', E'')$ where $E'' \subseteq E'$. The nodes of these literals are marked and the explanation graphs can be constructed.

Transformation 5 (Transformation of assumption nodes). *The explanation graph to an assumption node a_i^k consists of a node a_i^-, a node "assume" and an edge $(a_i^-, assume, -)$ (Fig. 3c).*

Then we proceed as before, marking nodes in G'' and simultaneously constructing the explanation graphs.

5   Choosing Assumptions

In most cases it is desirable to choose as few literals as possible as assumption. Assumptions where no literal can be removed without the set being no assumption anymore, are called *minimal assumptions*. Finding them in an EDG can be very complex, since all dependencies between the unmarked cycles have to be considered. For this purpose two different approaches are presented in this section. The first approach does not consider dependencies between cycles, so that assumptions can be computed in $O(|V| + |E|)$. The disadvantage of this approach is that the determined assumptions are not minimal in most cases. The second approach determines all minimal assumptions of an EDG, but has an exponential complexity.

For the determination of assumptions we first have to determine all tentative assumptions. From the definition of tentative assumptions we know that a tentative assumption has to meet three conditions: (a) it must not be contained in the answer set, (b) it has to appear as negative body literal in a rule and (c) it has to be undefined by the well-founded model. Transferred to a node in an EDG G' where all irrelevant edges and nodes were removed and all nodes are marked as far as possible, the first condition is fulfilled exactly by red nodes, the second condition is fulfilled exactly by nodes which have outgoing negative edges in the original graph G and the third condition is exactly fulfilled by unmarked nodes:

$$\mathcal{TA}(G') = \{a_i \mid a_i^k \in V', \nu(a_i^k) = red, a_i^k \ not \ marked, (a_i^k, a_j^l, -) \in E\}.$$

The set of tentative assumptions is always an assumption as shown in [1]. Since the set of tentative assumptions is often large, we are looking for an approach to reduce the set. The aim is that after defining the assumption, all other nodes can be marked.

Lemma 2. *In a graph without any cycles all nodes can be marked without making assumptions.*

Proof sketch. Since there exist no cycles, the nodes can be ordered into different levels on the graph. Every level contains all nodes from the lower level and nodes whose predecessors are contained in a lower level. Then it can be shown via induction that all nodes can be marked.

When we choose an assumption, the truth value of the literal is fixed to false. Since no further explanations than the one that the literal is an assumption are allowed, all incoming edges of the assumption nodes are removed. We treat cycles here as sets of nodes. Minimal cycles (where minimality is understood w.r.t. set inclusion) will play a crucial role. Choosing one assumption node in each minimal cycle breaks up the minimal cycles. Since all bigger cycles contain a minimal cycle they are also broken up, so there exist no more unmarked cycles. Since it is better to choose as few literals as possible as assumption, we only choose those possible assumptions that are minimal with regard to set inclusion. Then a possible assumption consists of one tentative assumption from each minimal cycle.

Approach 1. *Let $C_1, C_2, ..., C_n$ be all minimal unmarked cycles in G'.*

$$Assumptions(G') = \{\{a_1, a_2, ..., a_n\} \mid a_1 \in C_1, a_2 \in C_2, ..., a_n \in C_n,$$
$$\{a_1, a_2, ... a_n\} \text{ is minimal w.r.t. set inclusion}\}.$$

Of course there still may exist cycles which are marked. But since we know that they can be marked, we can simply replace a marked cycle $C_m = (V_m, E_m)$ by a dummy node c_m with incoming edges $ie(c_m) = \bigcup_{v \in V_m} ie(v)$ and outgoing edges $oe(c_m) = \bigcup_{v \in V_m} oe(v)$. One very simple possibility to determine minimal assumptions is to try all combinations of tentative assumptions. Such a combination is an assumption, if the whole graph can be marked after choosing the assumption. A minimal assumption is then the combination that is successful and minimal with regard to set inclusion. But with an increasing number of tentative assumptions this approach will not be very efficient. For this reason, we will introduce an approach that tries to reduce the number of combinations that have to be checked. Its basis is not to check all tentative assumptions and combinations, but only those literals that are important to determine the value of a so called *critical node*. It is obvious that this approach is only more efficient, if the number of such literals is smaller than the number of tentative assumptions. For the sake of simplicity, the approach is limited to graphs where each literal in a cycle is represented only by one node, i.e., there exist no OR-handles. When OR-handles have to be considered a similar approach can be used, just a more complex case differentiation has to be carried out.

Since marked nodes are irrelevant for the determination of assumptions, we remove all marked nodes and their outgoing and incoming edges from the graph and obtain a sub-graph $G_{unmarked}$. In the next step, we are looking for strongly connected components of $G_{unmarked}$. A strongly connected component is a maximal sub-graph where each node is reachable from each other node. This means that every node has an influence on every other node in the same strongly connected component, so that a strongly connected component behaves like a big cycle. For this reason we call the strongly connected components *linked cycles*.

If a linked cycle consists of several smaller cycles, there exist nodes belonging to multiple cycles. Such a node is *critical*, if its value depends on more than one cycle. Since we have removed all irrelevant edges and have no OR-handles, a red node has only active AND-handles. This means that the truth value of just one predecessor node is sufficient to determine the truth value of the red node. So a red node does not depend on more than one cycle, which means that only green nodes can be critical.

Definition 10 (Critical Node). *Let $LC = (V_{LC}, E_{LC})$ be a linked cycle. A node a_i^k is critical, if $\nu(a_i^k) = green$ and it has at least two incoming edges $(a_j^l, a_i^k, s) \in E_{LC}$ where $s \in \{+, -\}$. The set of all critical nodes of a linked cycle LC will be denoted as $\mathcal{CN}(LC)$.*

If a linked cycle has no critical nodes, a node can be deduced from any other node. Then a minimal assumptions consists of a single literal which is a tentative assumption and is represented by a node of the linked cycle. The set of all minimal assumptions of the linked cycle $LC(V_{LC}, E_{LC})$ in G' is: $\mu Assumptions$ $(LC) = \{\{a_i\} \mid a_i^k \in V_{LC}, a_i \in \mathcal{TA}(G')\}$.

The value of every critical node depends on the value of its predecessor nodes. We will call these nodes pre-conditions.

Definition 11 (Pre-condition). *Let $LC = (V_{LC}, E_{LC})$ be a linked cycle. The pre-conditions for a green critical node a_i^k are:*

$$pre(a_i^k) = \{a_j^l \mid (a_j^l, a_i^k, s) \in E_{LC}, s \in \{+, -\}\}$$

$Preconditions(LC) = \bigcup_{a_i^k \in \mathcal{CN}(LC)} pre(a_i^k)$ *is the set of all pre-conditions in the linked cycle.*

Lemma 3. *Nodes that are not critical can be deduced from at least one critical node.*

Proof sketch. It can be shown by induction that the truth value of a node a_n on a path $c, a_1, a_2, ..., a_{n-1}, a_n$ from a critical node c can be deduced, if $a_1, ... a_n$ are not critical.

So if we can deduce all critical nodes with an assumption, we also can deduce all other literals of the linked cycle with the assumption. $lfp(Succ(\{cn\}))$ calculates the nodes that can be deduced from a critical node $cn \in \mathcal{CN}(LC)$ where $Succ(S) = \{a_j^l \mid (a_i^k, a_j^l, s) \in E_{LC}, s \in \{+, -\}, a_i^k \in S\}$. Then the set of all nodes that can be deduced from a set of nodes S can be calculated with $lfp(T(S))$ where

$$T(S) = \{a_i^k \mid pre(a_i^k) \subseteq S\} \cup \{a_i^k \mid a_i^k \in Succ(a_j^l), a_j^l \in S \cap \mathcal{CN}(LC)\}.$$

Now combinations c of pre-conditions have to be tested for success. A combination c is successful, if all critical nodes can be deduced from them, i.e., $\mathcal{CN}(LC) \subseteq lfp(T(c))$. What we know is that

1. Each combination c has to contain at least one complete pre-condition set $pre(cn)$, $cn \in \mathcal{CN}(LC)$. Otherwise the fix-point operator could not deduce any critical node.
2. Since we are looking for minimal assumption sets, we do not have to check combinations c_1 where we already have found a smaller successful combination c_2, i.e., $c_2 \subseteq c_1$ and $\mathcal{CN}(LC) \subseteq lfp(T(c_2))$.

The way of proceeding is to first test single pre-condition sets $pre(cn) \subset c$, $cn \in \mathcal{CN}(LC)$ for success (exploits fact 1). If a set is successful, we add it to the set of successful combinations C, otherwise it is put to NC. Then we test sets $n \cup \{a_i^k\}$, where $n \in NC$ and $a_i^k \in Preconditions(LC)$. This means we test different combinations of adding one more pre-condition to all sets that have not been successful in the step before (exploits fact 2). Again we add successful combinations to C and set NC to the combinations that were not successful. This is repeated till $NC = \emptyset$ or the set to be tested consists of all pre-conditions. Then C contains all combinations of pre-conditions that suffice to deduce all critical nodes and therefore to deduce also all other nodes in the linked cycle, since they are not critical. For the purpose of determining assumptions, we determine all nodes from which a pre-condition p can be deduced. These nodes lie on paths from critical nodes to the pre-condition.

Definition 12 (Pre-condition paths). *The* pre-condition path *for a pre-condition p from a linked cycle $LC = (V_{LC}, E_{LC})$ can be obtained by $path(p) = lfp(T_{path}(\{p\}))$ where*

$$T_{path}(S) = \{a_i^k \mid (a_i^k, a_j^l, s) \in E_{LC}, s \in \{+, -\}, a_i^k \notin \mathcal{CN}(LC), a_j^l \in S\}.$$

A pre-condition path contains the nodes from which a pre-condition can be deduced. For deducing all nodes of a linked cycle we have to deduce all pre-conditions of a successful combination $c = \{p_1, ..., p_n\}$. This means that we need exactly one node from the path of each pre-condition. Since we want to determine assumptions, the nodes have also to fulfill the other conditions of an assumption.

Definition 13 (Path assumptions). *The set of path assumptions for a path p in a validly colored EDG $G = (V, E)$ is defined by*

$$\mathcal{PA}(p) = \{a_i \mid a_i^k \in p \wedge a_i^k \in \mathcal{TA}(G')\}.$$

Let C be the set of all successful pre-condition combinations $c = \{p_1, ..., p_n\}$. Then the set

$$Assumptions(LC) = \bigcup_{c \in C} \{\{a_1, ..., a_n\} \mid a_1 \in \mathcal{PA}(p_1), ..., a_n \in \mathcal{PA}(p_n)\}$$

is the set of possible minimal assumptions.

Proposition 2. *Minimal assumptions $\mu Assumptions(LC)$ of a linked cycle LC are those sets of $Assumptions(LC)$ that are minimal with regard to set inclusion.*

Proof sketch. It has to be shown that each set $S \in \mu Assumptions(LC)$ is a minimal assumption for the linked cycle LC, i.e., S is an assumption and there exists no set $S' \subset S$, such that S' is an assumption for LC. To show that S is an assumption, it has to be checked, if S meets the conditions of an assumption. To show that S is minimal, one looks at the successful combinations from which S and S' are created and distinguish between different cases. For every case it can be shown by contradiction that S' can not be in $\mu Assumptions(LC)$.

Algorithm 1. Determination of assumptions using Approach 2

Require: Marked graph $G' = (V', E')$, set of linked cycles $independentLCs$
Ensure: All minimal assumptions of G

1: **procedure** FINDASSUMPTIONS(G,$independentLCs$)
2:     **var** graph $G'' = (V'', E'') = G$
3:     **var** set of linked cycles $iLCs$
4:     **for all** $v \in V$ **do**   ▷ Remove marked nodes
5:         **if** v.marked = true **then** G''.removeNode(v) **end if**
6:     **end for**
7:     **if** $V'' = \emptyset$ **then**   ▷ There are no more unmarked nodes
8:         **var** sets of assumptions $\mu Assumptions[independentLCs]$
9:         **for all** $LC \in independentLCs$ **do**
10:             $\mu Assumptions[LC]$ = CALCULATEMINIMALASSUMPTIONSLC(LC)
11:         **end for**
12:         **return** CALCULATEMINIMALASSUMPTIONSGRAPH($\mu Assumptions$)
13:     **else**
14:         $iLCs$ = CALCULATEINDEPENDENTLINKEDCYCLES(G'')
15:         $independentLCs$.add($iLCs$)
16:                              ▷ Replace independent linked cycles by a dummy node
17:         **for all** $LC = (V_{LC}, E_{LC}) \in iLCs$ **do**
18:             **for** $v \in V_{LC}$ **do** G'.removeNode(v) **end for**
19:             G''.addNode($v_{dummy,LC}$)
20:             **for all** $e = (v_s, v_e, s) \in E_{LC}$ **do**
21:                 G''.addEdge($v_{dummy,LC}$,v_e,s)
22:                 G''.removeEdge(e)
23:             **end for**
24:         **end for**
25:         MARKNODES(G'')
26:         **return** FINDASSUMPTIONS(G'',$independentLCs$)
27:     **end if**
28: **end procedure**

Approach 2. *To calculate the minimal assumptions of a graph G Algorithm 1 is used by calling* FINDASSUMPTIONS($G', \emptyset$), *where G' is the graph obtained from G after removing irrelevant edges and nodes and after marking the nodes like described in Sect. 4.* FINDASSUMPTIONS *is a recursive function, which adds further independent linked cycles to the set of all independent linked cycles of G at* *each recursion. A linked cycle is independent if the truth value of its nodes does not depend on the truth value of the nodes in the rest of the graph. Since they are independent of the rest of the graph, they have to contain an assumption. To determine independent linked cycles, first all marked nodes and their incident edges are removed, so we get a graph $G'' = (V'', E'')$ (lines 4–6). If there still exist unmarked nodes in G'', G'' contains independent linked cycles. Independent linked cycles are those linked cycles, which have no incoming edges in G''. A linked cycle $LC = (V_{LC}, E_{LC})$ has no incoming edges if for all $a_i^k \in V_{LC}$ there is no $a_j^l \in V'' \setminus V_{LC}$ and $s \in \{+, -\}$ such that $(a_j^l, a_i^k, s) \in E''$. So all linked cycles without incoming edges are determined (line 15) and added to the set of all independent linked cycles (line 15). Then each independent linked cycle is replaced by a dummy node (lines 17–24).*

In the next step, the graph is marked using the marking rules from Sect. 4 (line 25). Since we have replaced linked cycles without incoming edges by dummy nodes, these nodes have no incoming edges and are marked. With the marking process we can determine which linked cycles are dependent of the independent linked cycles, that we already have determined. The procedure is repeated until all independent linked cycles of the graph are found. This is the case, if there exist no unmarked nodes in the graph (line 7).

Then for each linked cycle the minimal assumptions are calculated. The function CALCULATEMINIMALASSUMPTIONSLC *(line 13) contains following steps: the critical nodes of the linked cycle LC and their pre-conditions are specified and the pre-condition paths are calculated, then successful combinations of pre-conditions are looked for. The path assumptions are determined and used to calculate the minimal assumptions of the linked cycle.*

After determining the minimal assumptions of all independent linked cycles, the minimal assumptions of the graph are calculated with the function CALCU-LATEMINIMALASSUMPTIONSGRAPH *(line 12) which exploits Proposition 3.*

Proposition 3. *We can obtain the minimal assumptions of the EDG by taking one minimal assumption of each independent linked cycle:*

$$\mu Assumptions(G') = \{\{a_1 \cup \cdots \cup a_n\} \mid a_1 \in \mu Assumptions(LC_1), \ldots$$
$$a_n \in \mu Assumptions(LC_n)\}$$

Proof sketch. For every assumption $a \in \mu Assumptions(G')$ two things have to be shown: (a) a is an assumption for the EDG and (b) a is minimal. (a) can be directly shown by the stopping condition of the algorithm. For the proof of (b) a literal is removed from a and it can be shown that then not all nodes in the EDG can be marked, because of the definition of the minimal assumption in an EDG and the definition of independent linked cycles.

6 Conclusion

We presented an approach to construct explanation graphs from validly colored Extended Dependency Graphs. We exploited that the logic program is already present in graph form. In EDGs the nodes may differ in number of incoming edges and coloring, so different types of transformations were defined to transfer a node from the EDG to an EG. For the determination of assumptions it was necessary to determine the well-founded model. A strong relationship between well-founded models and the marking process of an EDG was observed, which means that an unmarked node represents a literal which is undefined in the well-founded model. We presented two different approaches for the determination of assumptions. While the first approach determines non-minimal assumptions in $O(|V| + |E|)$, the determination of minimal assumptions has an exponential complexity.

Acknowledgement. This work has been supported by the German Research Foundation DFG, Collaborative Research Center SFB876, Project A5. (http://sfb876. tu-dortmund.de)

References

1. Pontelli, E., Son, T.C., Elkhatib, O.: Justifications for logic programs under answer set semantics. Theor. Pract. Logic Program. **9**(1), 1–56 (2009)
2. Constantini, S., Provetti, A.: Graph representations of consistency and truth-dependencies in logic programs with answer set semantics. In: The 2nd International IJCAI Workshop on Graph Structures for Knowledge Representation and Reasoning (GKR 2011), vol. 1 (2011)
3. Gelfond, M., Lifschitz, V.: Classical negation in logic programs and disjunctive data-bases. New Gener. Comput. **9**(3–4), 365–385 (1991)
4. Lloyd, J.W.: Foundations of Logic Programming, 2nd edn. Springer, Heidelberg (1987)
5. Brignoli, G., Costantini, S., Provetti, A.: Characterizing and computing stable models of logic programs: the non-stratified case. In: Proceedings of the Conference on Information Technology, Bhubaneswar, India. AAAI Press (1999)
6. Van Gelder, A., Ross, K.A., Schlipf, J.S.: The well-founded semantics for general logic programs. J. ACM **38**(3), 620–650 (1991)
7. Dimopoulos, Y., Torres, A.: Graph theoretical structures in logic programs and default theories. Theor. Comput. Sci. **170**, 209–244 (1996)

Sharing and Exchanging Data

Rana Awada[1]($\boxtimes$), Pablo Barceló[2], and Iluju Kiringa[1]

[1] EECS, University of Ottawa, Ottawa, Canada
{rawad049,Iluju.kiringa}@uottawa.ca
[2] Department of Computer Science, University of Chile, Santiago, Chile
pbarcelo@dcc.uchile.cl

Abstract. Exchanging and integrating data that uses different vocabularies are two prominent problems in the database literature. These problems have been, so far, solved separately, and never been addressed together in a unified setting. In this paper, we propose a class of mappings - called *DSE*, for *data sharing and exchange* - that represents this unified setting. We introduce a DSE setting with particular interpretation of related data where ordinary data exchange or data integration cannot be applied. We define the class of *DSE* solutions in a DSE setting, that allow to store a part of explicit data and a set of inference rules used to generate the complete set of exchanged data. We identify among those a particular DSE solution with good properties; namely, one that contains a minimal amount of explicit data. Finally, we define the set of certain answers to conjunctive queries.

Keywords: Data exchange · Data coordination · Knowledge exchange

1 Introduction

Different problems of accessing and integrating data residing in independent sources have received wide attention in the literature, and different systems were introduced to solve these problems, e.g. distributed databases, federated databases, data exchange settings, and (peer-to-peer) data coordination settings.

Data exchange [7] defines the problem of moving data residing in independent applications and accessing it through a new target schema. This process of exchange only allows to move data from a source into a target that uses the same set of vocabularies, and thus, transformation occurs to the structure of the data, and not to the data itself. All data integration and coordination systems [2,14,15], on the other hand, use different query re-writing methods to allow access to data residing in independent sources, that possibly use different vocabularies, without having to exchange it and while maintaining autonomy.

We show in what follow that a collaborative process – including coordination tools for managing different vocabularies of different sources and exchange tools – would yield interoperability capabilities that are beyond the ones that can be offered today by any of the two tasks separately.

M. Hanus and R. Rocha (Eds.): KDPD 2013, LNAI 8439, pp. 17–32, 2014.
DOI: 10.1007/978-3-319-08909-6_2, © Springer International Publishing Switzerland 2014

Recall that a data exchange (DE) setting [7] $\mathfrak{S}$ consists of a source schema $\mathbf{S}$, a target schema $\mathbf{T}$, and a set Σ_{st} of database dependencies – the so-called source-to-target dependencies – that describes structural changes made to data as we move it from source to target. This exchange solution supports exchanging information between two applications that refer to the same object using the same instance value. However, there exist cases where objects in the source are named and referred to differently in the target. A motivating example of such an exchange scenario is exchanging data about students applying for program transfers from one university to a different one. Indeed, different universities can offer different courses and a course in one university can possess one or more equivalent courses in the second. Given what we mentioned so far about a DE setting, we can easily deduce that DE does not support such type of exchange.

Unlike data exchange, data coordination (DC) settings [1] solved the problem of integrating information of different sources that possess different yet related domains of constants by using the *mapping table* construct [9] in their query rewriting mechanisms. A mapping table specifies for each source value the set of *related* (or *corresponding*) target values. DC settings have been studied mainly in peer-to-peer networks, where sources – called peers – possess equal capabilities and responsibilities in terms of propagating changes and retrieving related information. A DC setting $\mathfrak{S}$ consists of two schemas $\mathbf{S}_1$ and $\mathbf{S}_2$, and a set of mapping tables $\{\mathcal{M}\}$. We give in the following example a data coordination instance that integrates data from two different universities with different domains of constants, and we show that query re-writing techniques still miss to return results that are *inferred* by certain semantics of mapping tables.

Example 1. Let $\mathfrak{S}$ be a DC setting. Suppose that $\mathbf{S}_1$ in $\mathfrak{S}$ is a schema for the University of Ottawa (UOO) and $\mathbf{S}_2$ in $\mathfrak{S}$ is the schema of the University of Carleton (UOC).

Suppose $\mathbf{S}_1$ has the relations: *Student(Sname, Sage)*, *Course(Cid, Cname, Pname)*, and *Enroll(Sname, Cid, Cgrade)*. Also, let $\mathbf{S}_2$ consist of the relation symbols *St(Sname, Sage, Saddress)*, *Cr(Cid, Cname, Pname)* and *Take(Sname, Cid, Cgrade)*.

Relation *Student (St)* stores students name and age (and address) information. Relation *Course (Cr)* stores courses ids and names information, in addition to the program name which provides each course. Finally, relation *Enroll (Take)* stores the set of courses that each student completed.

Further, assume that $\mathbf{S}_1$ and $\mathbf{S}_2$ are connected by a mapping table $\mathcal{M}$ that consists of the following pairs $\{(CSI1390, ECOR1606), (CSI1390, COMP1005), (CS, CS), (ENG, ENG)\}$.

Let I be an instance of $\mathbf{S}_1$ and $J = \{St\ (Alex, 18, Ottawa),\ Cr\ (ECOR1606,$ *Problem Solving and Computers, ENG)*, $Cr\ (COMP1005,$ *Introduction to Computer Science I, CS)*, *Take (Alex, ECOR1606, 80))*$\}$ be an instance of $\mathbf{S}_2$.

According to [2], posing a query q to I that computes the list of students considered to have finished CS courses in UOO, will re-write q to a query q' to retrieve a similar list from UOC following the semantics of $\mathcal{M}$. A query q' can be the following: q': Select *Sname* From *Cr, Take* Where $Cr.Cid = Take.Cid$ And $Cr.Pname = 'CS'$. In this case, the answer of posing q' to J is $\emptyset$. □

Assume that UOO accredits a 'CS' course to a student doing program transfer from UOC only if this student finishes an equivalent 'CS' course, according to $\mathcal{M}$, in UOC. In Example 1, *Alex* is not considered as finished a 'CS' course at UOC. Therefore, if *Alex* does a transfer to the CS program in UOO, he will not be credited the $CSI1390$ course. However, if the semantics of the mapping table $\mathcal{M}$ in this example specify that course $CSI1390$ in UOO is equivalent to the ENG course $ECOR1606$ in UOC, and course $CSI1390$ in UOO is equivalent to the CS course $COMP1005$ in UOC, then it can be deduced that courses $ECOR1606$ and $COMP1005$ are considered equivalent with respect to UOO according to $\mathcal{M}$. Therefore, given the fact that $Take\ (Alex, ECOR1606, 80) \in J$ in Example 1 along with the equivalence semantics in $\mathcal{M}$, *Alex* is considered to have finished the equivalent CS course $COMP1005$ in UOC and he should be credited the 'CS' course $CSI1390$ with a grade 80 if he did a transfer to UOO.

To solve such a problem, we introduce a new class of settings, called *data sharing and exchange* (DSE) settings, where exchange occurs between a source and a target that use different sets of vocabularies. Despite the importance of the topic, the fundamentals of this process have not been laid out to date. In this paper, we embark on the theoretical foundations of such problem, that is, exchanging data between two independent applications with different sets of domains of constants. DSE settings extend DE settings with a *mapping table* $\mathcal{M}$, introduced in [9], to allow collaboration at the instance level. In addition, the set of source-to-target dependencies Σ_{st} in DSE refers to such mapping table so that coordination of distinct vocabularies between applications takes place together with the exchange.

From what we have mentioned so far about DSE, one would think that all DSE instances can be reduced to a usual DE instance where the source schema is extended with a mapping table $\mathcal{M}$. However, we argue in this paper that there exist DSE settings with particular interpretation of related data in mapping tables where DSE is different than a DE setting (as we show later in Example 2). One such particular interpretation of related data that we consider in this paper is: a source element is mapped to a target element only if both are considered to be equivalent (i.e. denote the same object). In this DSE scenario, DSE and DE are different because source and target data can be incomplete with respect to the "implicit" information provided by the semantics of mapping tables. To formalize this idea we use techniques developed by Arenas et al. in [5], where authors introduced a *knowledge exchange* framework for exchanging knowledge bases. It turns out that this framework suits our requirements, and in particular, allows us to define the exchange of both explicit and implicit data from source to target. Our main contributions in this work are the following:

(1) Universal DSE Solutions. We formally define the semantics of a DSE setting and introduce the class of universal DSE solutions, that can be seen as a natural generalization of the class of universal data exchange solutions [7] to the DSE scenario, and thus, as "good" solutions. A universal DSE solution consists of a subset of explicit data that is necessary to infer the remaining implicit information using a given set Σ_t of rules in the target.

(2) Minimal Universal DSE Solutions. We define the class of minimal universal DSE solutions which are considered as "best" solutions. A minimal universal DSE solution contains the minimal amount of explicit data required to compute the complete set of explicit and implicit data using a set of target rules Σ_t. We show that there exists an algorithm to generate a *canonical minimal universal* DSE solution, with a well-behaved set Σ_t of target rules, in LOGSPACE.

(3) Query Answering. We formally define the set of DSE certain answers for conjunctive queries. We also show how to compute those efficiently using canonical minimal universal DSE solutions.

2   Preliminaries

A *schema* **R** is a finite set $\{R_1, \ldots, R_k\}$ of relation symbols, with each R_i having a fixed arity $n_i > 0$. Let D be a countably infinite domain. An *instance* I of **R** assigns to each relation symbol R_i of **R** a finite n_i-ary relation $R_i^I \subseteq \mathsf{D}^{n_i}$. Sometimes we write $R_i(\bar{t}) \in I$ instead of $\bar{t} \in R_i^I$, and call $R_i(\bar{t})$ a *fact* of I. The *domain dom*(I) of instance I is the set of all elements that occur in any of the relations R_i^I. We often define instances by simply listing the facts that belong to them. Further, every time that we have two disjoint schemas **R** and **S**, an instance I of **R** and an instance J of **S**, we define (I, J) as the instance K of schema **R** $\cup$ **S** such that $R^K = R^I$, for each $R \in \mathbf{R}$, and $S^K = S^J$, for each $S \in \mathbf{S}$.

Data Exchange Settings. As is customary in the data exchange literature [7,8], we consider instances with two types of values: constants and nulls.[1] More precisely, let Const and Var be infinite and disjoint sets of constants and nulls, respectively, and assume that D = Const $\cup$ Var. If we refer to a schema **S** as a *source* schema, then we assume that for an instance I of **S**, it holds that $dom(I) \subseteq$ Const; that is, source instances are assumed to be "complete", as they do not contain missing data in the form of nulls. On the other hand, if we refer to a schema **T** as a *target* schema, then for every instance J of **T**, it holds that $dom(J) \subseteq$ Const$\cup$Var; that is, target instances are allowed to contain null values.

A *data exchange* (DE) *setting* is a tuple $\mathfrak{S} = (\mathbf{S}, \mathbf{T}, \Sigma_{st})$, where **S** is a source schema, **T** is a target schema, **S** and **T** do not have predicate symbols in common, and Σ_{st} consists of a set of *source-to-target tuple-generating* dependencies (st-tgds) that establish the relationship between source and target schemas. An st-tgd is a FO-sentence of the form:

$$\forall \bar{x} \forall \bar{y} \, (\phi(\bar{x}, \bar{y}) \rightarrow \exists \bar{z} \, \psi(\bar{x}, \bar{z})), \tag{1}$$

---

[1] We usually denote constants by lowercase letters $a, b, c, \ldots$, and nulls by symbols $\bot, \bot', \bot_1, \ldots$

where $\phi(\bar{x}, \bar{y})$ is a conjunction of relational atoms over $\mathbf{S}$ and $\psi(\bar{x}, \bar{z})$ is a conjunction of relational atoms over $\mathbf{T}$.[2] A *source* (resp. *target*) instance K for $\mathfrak{S}$ is an instance of $\mathbf{S}$ (resp. $\mathbf{T}$). We usually denote source instances by $I, I', I_1, \ldots$, and target instances by $J, J', J_1, \ldots$.

An instance J of $\mathbf{T}$ is a *solution* for an instance I under $\mathfrak{S} = (\mathbf{S}, \mathbf{T}, \Sigma_{st})$, if the instance (I, J) of $\mathbf{S} \cup \mathbf{T}$ satisfies every st-tgd in Σ_{st}. If $\mathfrak{S}$ is clear from the context, we say that J is a solution for I.

The data exchange literature has identified a class of preferred solutions, called the *universal* solutions, that in a precise way represents all other solutions. In order to define these solutions, we need to introduce the notion of homomorphism between instances. Let K_1 and K_2 be instances of the same schema $\mathbf{R}$. A *homomorphism* h from K_1 to K_2 is a function $h : dom(K_1) \rightarrow dom(K_2)$ such that: (1) $h(c) = c$ for every $c \in \mathsf{Const} \cap dom(K_1)$, and (2) for every $R \in \mathbf{R}$ and tuple $\bar{a} = (a_1, \ldots, a_k) \in R^{K_1}$, it holds that $h(\bar{a}) = (h(a_1), \ldots, h(a_k)) \in R^{K_2}$. Let $\mathfrak{S}$ be a DE setting, I a source instance and J a solution for I under $\mathfrak{S}$. Then J is a *universal* solution for I under $\mathfrak{S}$, if for every solution J' for I under $\mathfrak{S}$, there exists a homomorphism from J to J'.

For the class of data exchange settings that we referred to in this paper, every source instance has a universal solution [7]. Further, given a DE setting $\mathfrak{S}$, there is a procedure (based on the *chase* [6]) that computes a universal solution for each source instance I under $\mathfrak{S}$. In the case when $\mathfrak{S}$ is fixed such procedure works in LOGSPACE. Assuming $\mathfrak{S}$ to be fixed is a usual and reasonable assumption in data exchange [7], as mappings are often much smaller than instances. We stick to this assumption for the rest of the paper.

Mapping Tables. Coordination can be incorporated at the *data* level, through the use of *mapping* tables [9]. These mechanisms were introduced in data coordination settings [2] to establish the correspondence of related information in different domains. In its simplest form, mapping tables are just binary tables containing pairs of corresponding identifiers from two different sources. Formally, given two domains D_1 and D_2, not necessarily disjoint, a mapping table over $(\mathsf{D}_1, \mathsf{D}_2)$ is nothing else than a subset of $\mathsf{D}_1 \times \mathsf{D}_2$. Intuitively, the fact that a pair (d_1, d_2) belongs to the mapping table implies that value $d_1 \in \mathsf{D}_1$ *corresponds* to value $d_2 \in \mathsf{D}_2$. Notice that the exact meaning of "correspondence" between values is unspecified and depends on the application.

In this paper we deal with a very particular interpretation of the notion of correspondence in mapping tables. We assume that the fact that a pair (a, b) is in a mapping table implies that a and b are equivalent objects. We are aware of the fact that generally mapping tables do not interpret related data in this way. However, we argue that this particular case is, at the same time, practically relevant (e.g. in peer-to-peer settings [9]) and theoretically interesting (as we will see along the paper).

This particular interpretation of mapping tables implies that they may contain implicit information that is not explicitly listed in their extension. For instance,

---

[2] We usually omit universal quantification in front of st-tgds and express them simply as $\phi(\bar{x}, \bar{y}) \rightarrow \exists \bar{z}\, \psi(\bar{x}, \bar{z})$.

assume that $\mathcal{M}$ is a mapping table that consists of the pairs $\{(a, c), (b, c), (b, d)\}$. Since a and c are equivalent, and the same is true about b and c, we can *infer* that a and b are equivalent. Also, we can infer using the same reasoning that c and d are equivalent. Such implicit information is, of course, valuable, and cannot be discarded at the moment of using the mapping table as a coordination tool. In particular, we will use this view of mapping tables as being incomplete with respect to its implicit data when defining the semantics of DSE settings.

3   Data Sharing and Exchange Settings

We formally define in this section DSE settings that extend DE settings to allow collaboration via mapping tables.

Definition 1 (DSE setting). *A data sharing and exchange (DSE) setting is a tuple $\mathfrak{S} = (\mathbf{S}, \mathbf{T}, \mathcal{M}, \Sigma_{st})$, where: (1) $\mathbf{S}$ and $\mathbf{T}$ are a source and a target schema, respectively; (2) $\mathcal{M}$ is a binary relation symbol that appears neither in $\mathbf{S}$ nor in $\mathbf{T}$, and that is called a* source-to-target mapping *(we call the first attribute of $\mathcal{M}$ the source attribute and the second one the target attribute); and (3) Σ_{st} consists of a set of* mapping st-tgds, *which are FO sentences of the form*

$$\forall \bar{x} \forall \bar{y} \forall \bar{z} \, (\phi(\bar{x}, \bar{y}) \wedge \mu(\bar{x}, \bar{z}) \rightarrow \exists \bar{w} \, \psi(\bar{z}, \bar{w})), \tag{2}$$

where (i) $\phi(\bar{x}, \bar{y})$ and $\psi(\bar{z}, \bar{w})$ are conjunctions of relational atoms over $\mathbf{S}$ and $\mathbf{T}$, resp., (ii) $\mu(\bar{x}, \bar{z})$ is a conjunction of atomic formulas that only use the relation symbol $\mathcal{M}$, (iii) $\bar{x}$ is the tuple of variables that appear in $\mu(\bar{x}, \bar{z})$ in the positions of source attributes of $\mathcal{M}$, and (iv) $\bar{z}$ is the tuple of variables that appear in $\mu(\bar{x}, \bar{z})$ in the positions of target attributes of $\mathcal{M}$.

We provide some terminology and notations before explaining the intuition behind the different components of a DSE setting. As before, instances of $\mathbf{S}$ (resp. $\mathbf{T}$) are called source (resp. target) instances, and we denote source instances by $I, I', I_1, \ldots$ and target instances by $J, J', J_1, \ldots$. Instances of $\mathcal{M}$ are called *source-to-target* mapping tables (st-mapping tables). By slightly abusing notation, we denote st-mapping tables also by $\mathcal{M}$.

Let $\mathfrak{S} = (\mathbf{S}, \mathbf{T}, \mathcal{M}, \Sigma_{st})$ be a DSE setting. We distinguish between the set of source constants, denoted by $\mathsf{Const}^{\mathbf{S}}$, and the set of target constants, denoted by $\mathsf{Const}^{\mathbf{T}}$, since applications that collaborate on data usually have different data domains. As in the case of usual data exchange, we also assume the existence of a countably infinite set Var of labelled nulls (that is disjoint from both $\mathsf{Const}^{\mathbf{S}}$ and $\mathsf{Const}^{\mathbf{T}}$). Also, in a DSE the domain of a source instance I is contained in $\mathsf{Const}^{\mathbf{S}}$, while the domain of a target instance J belongs to $\mathsf{Const}^{\mathbf{T}} \cup \mathsf{Var}$. On the other hand, the domain of the st-mapping table $\mathcal{M}$ is a subset of $\mathsf{Const}^{\mathbf{S}} \times \mathsf{Const}^{\mathbf{T}}$. Thus, coordination between the source and the target at the data level occurs when $\mathcal{M}$ identifies which source and target constants denote the same object. The intuition behind usual st-tgds is that they specify how source data has to be transformed to conform to the target schema (that is, coordination at the schema level). However, since in the DSE scenario we are interested in transferring data based on the source instance as well as on the correspondence between source and target constants given by the st-mapping table that interprets $\mathcal{M}$, the mapping st-tgds extend usual st-tgds with a conjunction μ that filters the target data that is related via $\mathcal{M}$ with the corresponding source data.

More formally, given a source instance I and an st-mapping table $\mathcal{M}$, the mapping st-tgd $\phi(\bar{x},\bar{y}) \wedge \mu(\bar{x},\bar{z}) \rightarrow \exists \bar{w}\psi(\bar{z},\bar{w})$ enforces the following: whenever $I \models \phi(\bar{a},\bar{b})$, for a tuple $(\bar{a},\bar{b})$ of constants in $\mathsf{Const}^{\mathbf{S}} \cap dom(I)$, and the tuple $\bar{c}$ of constants in $\mathsf{Const}^{\mathbf{T}}$ is related to $\bar{a}$ via μ (that is, $\mathcal{M} \models \mu(\bar{a},\bar{c})$), then it must be the case that $J \models \psi(\bar{c},\bar{d})$, for some tuple $\bar{d}$ of elements in $dom(J) \cap (\mathsf{Const}^{\mathbf{T}} \cup \mathsf{Var})$, where J is the materialized target instance. In usual DE terms, we should say that J is a solution for I and $\mathcal{M}$ under $\mathfrak{S}$, i.e. $(((I \cup \{\mathcal{M}\}), J) \models \Sigma_{st})$. However, as we see in the next section, solutions have to be defined differently in DSE. Therefore, to avoid confusions, we say J is a *pre-solution* for I and $\mathcal{M}$ under $\mathfrak{S}$.

Example 2. Let $\mathfrak{S} = (\mathbf{S},\mathbf{T},\mathcal{M},\Sigma_{st})$ be a DSE setting. In reference to Example 1, assume that $\mathbf{S}$ in $\mathfrak{S}$ is the schema of UOC and $\mathbf{T}$ in $\mathfrak{S}$ is the schema of UOO.

Suppose that $\mathcal{M}$ in $\mathfrak{S}$ consists of the following pairs $\{(ECOR1606, CSI1390),$ $(COMP1005, CSI1390)$, $(COMP1005, CSI1790)$, (CS,CS), (ENG,ENG), $(Alex,Alex), (18,18)\}$. Finally, let Σ_{st} consist of the following st-mapping dependencies:

(a) $St(x,y,z) \wedge Take(x,w,u) \wedge Cr(w,v,\text{'CS'}) \wedge \mathcal{M}(x,x') \wedge \mathcal{M}(y,y')$
$$\rightarrow Student(x',y').$$
(b) $St(x,y,z) \wedge Take(x,w,u) \wedge Cr(w,v,\text{'CS'}) \wedge \mathcal{M}(x,x') \wedge \mathcal{M}(w,w') \wedge \mathcal{M}(u,u')$
$$\rightarrow Enroll(x',w',u').$$

It is clear that this DSE instance is exchanging information of UOC students that have taken 'CS' courses with the list of courses they have finished. Also, $\mathcal{M}$ specifies that the *Introduction to Computers* course with $Cid = CSI1390$ in UOO has a French version course *Introduction aux Ordinateurs* with $Cid = CSI1790$ provided at UOO. Let $I = \{St(Alex,18,Ottawa), Cr(ECOR1606, Problem\ Solving\ and\ Computers, ENG), Cr(COMP1005, Introduction\ to\ Computer\ Science\ I, CS), Take\ (Alex, ECOR1606, 80)\}$ be an instance of $\mathbf{S}$. Then, $J = \emptyset$ is a pre-solution for I and $\mathcal{M}$ under $\mathfrak{S}$. $\qquad\square$

We can see in Example 2 that in the pre-solution J, *Alex* is not considered as have finished a 'CS' course. However, if the st-mapping table $\mathcal{M}$ follows the semantics we adopt in this paper, then *Alex* should be considered to have completed the 'CS' course *Introduction to Computer Science I*. Therefore, we can easily deduce that in a DSE setting $\mathfrak{S}$, we cannot identify solutions with pre-solutions. One reason is that a source instance I in $\mathfrak{S}$ can be *incomplete* with respect to the semantics of $\mathcal{M}$ as the case in Example 2 above. A second reason is that data mappings in an st-mapping table $\mathcal{M}$ in $\mathfrak{S}$ can also be *incomplete* with respect to the semantics of $\mathcal{M}$ as we shall show in Sect. 4. Data mappings in an st-mapping table $\mathcal{M}$ are usually specified by domain specialists. However, $\mathcal{M}$

should record not only the associations suggested by the domain specialists, but also the ones inferred by its semantics. Therefore, to capture the real semantics of the DSE problem, we came up with a more sophisticated notion of a solution that we introduce in the following section.

4   DSE and Knowledge Exchange

From now on we use the equivalence relation $\sim$ as $a \sim b$ to intuitively denote that a and b, where $\{a, b\} \subseteq \mathsf{Const}^{\mathbf{S}}$ (or $\{a, b\} \subseteq \mathsf{Const}^{\mathbf{T}}$) are *inferred* by the semantics of an st-mapping table $\mathcal{M}$ as equivalent objects. Let us revisit Example 2. There are two ways in which the data in $\mathfrak{S}$ is incomplete: First of all, since $\mathcal{M}(ECOR1606, CSI1390)$ holds in $\mathfrak{S}$, then UOC course $ECOR1606$ is equivalent to the UOO course $CSI1390$. Also, since $\mathcal{M}(COMP1005, CSI1390)$ holds, then UOC course $COMP1005$ is equivalent to the UOO course $CSI1390$. Therefore, we can deduce that $ECOR1606 \sim COMP1005$ with respect to the target UOO. This means, according to semantics of $\sim$, the source instance I is incomplete, since I should include the tuple $Take(Alex, COMP1005, 80)$ in order to be complete with respect to $\mathcal{M}$.

Second, since $\mathcal{M}(COMP1005, CSI1390)$ holds in $\mathfrak{S}$, then the UOC course $COMP1005$ is equivalent to the UOO course $CSI1390$ according to the semantics of $\mathcal{M}$. Also, since $\mathcal{M}(COMP1005, CSI1790)$ holds in $\mathfrak{S}$, then course $COMP$ 1005 is equivalent to the UOO course $CSI1790$. Therefore, we can deduce that $CSI1390 \sim CSI1790$, according to the semantics of $\mathcal{M}$. This implies that $\mathcal{M}$ is incomplete, since the fact that $\{(ECOR1606, CSI1390), (COMP1005, CSI1390), (COMP1005, CSI1790)\} \subseteq \mathcal{M}$ entails from the semantics of $\sim$ the fact that $(ECOR1606, CSI1790) \in \mathcal{M}$. Therefore, we say I and $\mathcal{M}$ are incomplete in the sense that they do not contain all the data that is implied by the semantics of $\mathcal{M}$. Further, it is not hard to see that the completion process we just sketched can become recursive in more complex DSE instances.

Given the above reasoning, again one would think to solve the DSE problem by reducing it to a DE setting $\mathfrak{S}$ with a set of dependencies defined over a combined schema $(\mathbf{S} \cup \{\mathcal{M}\})$, which complete the source instance I and the st-mapping table $\mathcal{M}$ under $\mathfrak{S}$ with the additional data entailed by the semantics of $\mathcal{M}$. However, usually the real reason behind defining dependencies over a databases schema is to ensure that data stored in the extension of this schema or new coming tuples follow the structure of it [16]. Also, such constraints are not treated as additional *implicit* data that represents specific semantics in st-mapping tables, and whose purpose is to entail new facts in addition to the stored ones. Therefore, to apply the intuition explained in Example 2 and to generate "good" solutions in a DSE setting, it is assumed to be fundamental that both explicit data stored in I and implicit data entailed by the semantics of $\mathcal{M}$ are exchanged to the target. Contrarily, we show in what follow that semantics of a DSE setting vary from those of DE setting since there exist solutions which consist of a portion of the fully exchanged set of data and yet are still good solutions that proved in Sect. 7 to be efficient for conjunctive query answering.

From what we explained so far, we conclude that the real semantics of a DSE setting is based on the explicit data contained in I and $\mathcal{M}$, in addition to the implicit data obtained by following a completion process for the source, the target, and $\mathcal{M}$.

We define below a set of FO sentences, of type full tgds[3], over a schema $\mathbf{S} \cup \mathcal{M}$ ($\mathbf{T} \cup \mathcal{M}$) extended with a fresh binary relation symbol EQUAL that appears neither in $\mathbf{S}$ nor in $\mathbf{T}$ and that captures the semantics of $\sim$ in a recursive scenario, which formally defines this completion process:

Definition 2 (Source and Target Completion). *Let $\mathfrak{S} = (\mathbf{S}, \mathbf{T}, \mathcal{M}, \Sigma_{st})$ be a DSE setting. The* source completion of $\mathfrak{S}$, *denoted by Σ_s^c, is the conjunction of the following FO sentences over the schema $\mathbf{S} \cup \{\mathcal{M}, \text{EQUAL}\}$:*

1. *For each $S \in \mathbf{S} \cup \{\mathcal{M}\}$ of arity n and $1 \leq i \leq n$:* $\forall x_1 \cdots \forall x_n (S(x_1, \ldots, x_i, \ldots, x_n) \to \text{EQUAL}(x_i, x_i))$.
2. $\forall x \forall y (\text{EQUAL}(y, x) \to \text{EQUAL}(x, y))$.
3. $\forall x \forall y \forall z (\text{EQUAL}(x, z) \wedge \text{EQUAL}(z, y) \to \text{EQUAL}(x, y))$.
4. $\forall x \forall y \forall z (\mathcal{M}(x, z) \wedge \mathcal{M}(y, z) \to \text{EQUAL}(x, y))$.
5. $\forall x \forall y \forall z \forall w (\mathcal{M}(x, z) \wedge \text{EQUAL}(x, y) \wedge \text{EQUAL}(z, w) \to \mathcal{M}(y, w))$.
6. *For each $S \in \mathbf{S}$ of arity n:* $\forall x_1, y_1 \cdots \forall x_n, y_n (S(x_1, \ldots, x_n) \wedge \bigwedge_{i=1}^{n} \text{EQUAL}(x_i, y_i) \to S(y_1, \ldots, y_n))$.

The target completion of $\mathfrak{S}$, *denoted Σ_t^c, is defined analogously by simply replacing the role of $\mathbf{S}$ by $\mathbf{T}$ in Σ_s^c, and then adding the rule 7. $\forall x \forall y \forall z (\mathcal{M}(z, x) \wedge \mathcal{M}(z, y) \to \text{EQUAL}(x, y))$ that defines the completion of $\mathcal{M}$ over the target.*

Notice that the first 3 rules of Σ_s^c make sure that EQUAL is an equivalence relation on the domain of the source instance. The fourth rule detects which source elements have to be declared equal by the implicit knowledge contained in the st-mapping table. The last two rules allow to complete the interpretation of $\mathcal{M}$ and the symbols of $\mathbf{S}$, by adding elements declared to be equal in EQUAL. The intuition for Σ_t^c is analogous.

Summing up, data in a DSE scenario always consists of two modules: (1) The explicit data stored in the source instance I and the st-mapping table $\mathcal{M}$, and (2) the implicit data formalized in Σ_s^c and Σ_t^c. This naturally calls for a definition in terms of *knowledge exchange* [5], as defined next. A knowledge base (KB) over schema $\mathbf{R}$ is a pair (K, Σ), where K is an instance of $\mathbf{R}$ (the explicit data) and Σ is a set of logical sentences over $\mathbf{R}$ (the implicit data). The knowledge base representation has been used to represent various types of data including ontologies in the semantic web, which are expressed using different types of formalisms including *Description Logic* (DL) [12].

The set of *models* of (K, Σ) [5], denoted by $\mathsf{Mod}(K, \Sigma)$, is defined as the set of instances of $\mathbf{R}$ that contain the explicit data in K and satisfy the implicit data in Σ; that is, $\mathsf{Mod}(K, \Sigma)$ corresponds to the set $\{K' \mid K'$ is an instance of $\mathbf{R}$, $K \subseteq K'$ and $K' \models \Sigma\}$. In DSE, we consider source KBs of the form $((I \cup \{\mathcal{M}\}), \Sigma_s^c)$, which intuitively correspond to completions of the source instance

---

[3] Full tgds are tgds that do not use existential quantication.

I with respect to the implicit data in $\mathcal{M}$, and, analogously, target KBs of the form $((J \cup \{\mathcal{M}\}), \Sigma_t^c)$.

A good bulk of work has recently tackled the problem of exchange of KBs that are defined using different DL languages [4]. we formalize the notion of (universal) DSE solution to extend the KB (universal) solution introduced in [5]. The main difference is that in DSE solutions we need to coordinate the source and target information provided by $\mathcal{M}$, as opposed to KB solutions that require no data coordination at all. This is done by establishing precise relationships in a (universal) DSE solution between the interpretation of $\mathcal{M}$ in $\mathbf{S}$ and $\mathbf{T}$, respectively. KB exchange in DL showed that target KB (universal) solutions [5] present several limitations since these can miss some semantics of the source KB [4]. Universal DSE solutions, on the other hand, do not have those limitations and they reflect the semantics in the source and the st-mapping table accurately.

From now on, $K_{\mathbf{R'}}$ denotes the restriction of instance K to a subset $\mathbf{R'}$ of its schema $\mathbf{R}$. Let $\mathfrak{S} = (\mathbf{S}, \mathbf{T}, \mathcal{M}, \Sigma_{st})$ be a DSE setting, I a source instance, $\mathcal{M}$ an st-mapping table, J a target instance. Recall that Σ_s^c, Σ_t^c are the source and target completions of $\mathfrak{S}$, respectively. Then:

1. J is a *DSE solution* for I and $\mathcal{M}$ under $\mathfrak{S}$, if for every $K \in \mathrm{Mod}((J \cup \{\mathcal{M}\}), \Sigma_t^c)$ there is $K' \in \mathrm{Mod}((I \cup \{\mathcal{M}\}), \Sigma_s^c)$ such that the following hold: (a) $K'_{\mathcal{M}} \subseteq K_{\mathcal{M}}$, and (b) $K_{\mathbf{T}}$ is a pre-solution for $K'_{\mathbf{S}}$ and $K'_{\mathcal{M}}$ under $\mathfrak{S}$.
2. In addition, J is a *universal* DSE solution for I and $\mathcal{M}$ under $\mathfrak{S}$, if J is a DSE solution, and for every $K' \in \mathrm{Mod}((I \cup \{\mathcal{M}\}), \Sigma_s^c)$ there is $K \in \mathrm{Mod}((J \cup \{\mathcal{M}\}), \Sigma_t^c)$ such that (a) $K_{\mathcal{M}} \subseteq K'_{\mathcal{M}}$, and (b) $K_{\mathbf{T}}$ is a pre-solution for $K'_{\mathbf{S}}$ and $K'_{\mathcal{M}}$ under $\mathfrak{S}$.

In Example 2, $J = \{Student(Alex, 18), Enroll\ (Alex, CSI1390, 80), Enroll\ (Alex, CSI1790, 80)\}$ is a universal DSE solution for I and $\mathcal{M}$ under $\mathfrak{S}$. We define below a simple procedure $\mathtt{CompUnivDSESol}_{\mathfrak{S}}$ that, given a DSE setting $\mathfrak{S} = (\mathbf{S}, \mathbf{T}, \mathcal{M}, \Sigma_{st})$ and a source instance I and an st-mapping table $\mathcal{M}$, it generates a universal DSE solution J for I and $\mathcal{M}$ under $\mathfrak{S}$.
$\mathtt{CompUnivDSESol}_{\mathfrak{S}}$:
Input: A source instance I, an st-mapping table $\mathcal{M}$, and a set Σ_{st} of st-tgds.
Output: A Canonical Universal DSE solution J for I and $\mathcal{M}$ under $\mathfrak{S}$.

1. Apply the source completion process, Σ_s^c, to I and $\mathcal{M}$, and generate $\hat{I}$ and $\hat{\mathcal{M}}$ respectively.
2. Apply a procedure (based on the *chase* [6]) to the instance $(\hat{I} \cup \{\hat{\mathcal{M}}\})$, and generate a canonical universal pre-solution J for $\hat{I}$ and $\hat{\mathcal{M}}$.

The procedure $\mathtt{CompUnivDSESol}_{\mathfrak{S}}$ works as follows: step 1 applies the source completion process Σ_s^c, given in Definition 2, to I and $\mathcal{M}$, and returns as outcome the source instance $\hat{I}$ and the st-mapping table $\hat{\mathcal{M}}$ that are complete with respect to the implicit data in $\mathcal{M}$. Next, step 2 generates a canonical universal pre-solution J for $\hat{I}$ and $\hat{\mathcal{M}}$ such that $((\hat{I} \cup \hat{\mathcal{M}}), J) \vDash \Sigma_{st}$.

We can combine the fact that universal solutions in fixed data exchange settings $\mathfrak{S} = (\mathbf{S}, \mathbf{T}, \Sigma_{st})$ can be computed in LOGSPACE [3] with some deep results in the computation of symmetrical binary relations [10], to show that universal DSE solutions can be computed in LOGSPACE:

Proposition 1. *Let* $\mathfrak{S} = (\mathbf{S}, \mathbf{T}, \mathcal{M}, \Sigma_{st})$ *be a fixed DSE setting. Then computing a universal DSE solution J for a source instance I and an st-mapping table* $\mathcal{M}$ *is in* LOGSPACE.

Proof. To prove that the CompUnivDSESol$_{\mathfrak{S}}$ procedure computes a universal DSE solution for I and $\mathcal{M}$ under $\mathfrak{S}$ in LOGSPACE, we rely on the following facts: (1) In Σ_s^c (in step 1 to step 4), we compute the transitive closures of the *symmetrical* binary table EQUAL. Computing the transitive closure of symmetrical binary relations is solvable in LOGSPACE [10]; (2) steps 5 and 6 of Σ_s^c compute the complete instances $\hat{I}$ and the st-mapping table $\hat{\mathcal{M}}$ using EQUAL. This step can be computed by applying the naive chase procedure to I and $\mathcal{M}$ using rules 5 and 6 of Σ_s^c. Following the result in [7] that a naive chase procedure runs in LOGSPACE, we can deduce that generating $\hat{I}$ and $\hat{\mathcal{M}}$ using steps 5 and 6 of Σ_s^c is in LOGSPACE; (3) step 2 of CompUnivDSESol$_{\mathfrak{S}}$ procedure can be reduced to the problem of computing a universal solution for $(\hat{I} \cup \{\hat{\mathcal{M}}\})$ in a fixed DE setting $\mathfrak{S}$. Consequently, similar to steps 5 and 6 in Σ_s^c, this process works in LOGSPACE in a fixed DSE setting; (4) finally, since LOGSPACE is closed under composition [11], we conclude that CompUnivDSESol$_{\mathfrak{S}}$ is in LOGSPACE.

5   Minimal Universal DSE Solutions

In the context of ordinary data exchange, "best" solutions – called cores – are universal solutions with minimal size. In knowledge exchange, on the other hand, "best" solutions are cores that materialize a minimal amount of explicit data. Intuitively, a *minimal* universal DSE (MUDSE) solution is a core universal DSE solution J that contains a minimal amount of explicit data in J with respect to Σ_t^c, and such that no universal DSE solution with strictly less constants is also a universal DSE solution with respect to Σ_t^c.

We define this formally: Let $\mathfrak{S}$ be a DSE setting, I be a source instance, $\mathcal{M}$ an st-mapping table, and J a universal DSE solution for I and $\mathcal{M}$ under $\mathfrak{S}$. Then J is a MUDSE solution for I and $\mathcal{M}$ under $\mathfrak{S}$, if: (1) There is no proper subset J' of J such that J' is a universal DSE solution for I and $\mathcal{M}$ under $\mathfrak{S}$, and; (2) There is no universal DSE solution J' such that $dom(J') \cap \mathsf{Const}^{\mathbf{T}}$ is properly contained in $dom(J) \cap \mathsf{Const}^T$.

So, in Example 2, $J = \{Student(Alex, 18), Enroll\,(Alex, CSI1390, 80)\}$ is a MUDSE solution for I and $\mathcal{M}$ under $\mathfrak{S}$. Note that the DSE solution $J' = \{Student(Alex, 18), Enroll\,(Alex, CSI1390, 80), Enroll\,(Alex, CSI1790, 80)\}$ is a core universal DSE solution, however it is not the most compact one. Condition (2) in the definition of MUDSE solutions is not part of the original definition of minimal solutions in knowledge exchange [5]. However, this condition is necessary as we see below.

Assume that the universal DSE solution in Example 2 includes the following two facts $\{Teach(Anna, CSI1390), Teach(Anna, CSI1790)\}$, where $\mathbf{T}$

is extended with the relation $Teach(Tid, Cid)$ which specifies the teachers and the list of courses they teach. Then, the DSE solution $J = \{Student(Alex, 18),$ $Enroll\,(Alex, CSI1390, 80), Teach(Anna, CSI1790)\}$ does not satisfy condition (2) and provides us with redundant information with respect to I and $\mathcal{M}$, since we can conclude that $CSI1390$ and $CSI1790$ are equivalent courses. A MUDSE solution however would be $J = \{Student(Alex, 18), Enroll(Alex, CSI1390, 80),$ $Teach(Anna, CSI1390)\}$.

We define below a procedure $\texttt{CompMUDSEsol}_{\mathfrak{S}}$, that given a DSE setting $\mathfrak{S}$, a source instance I, and an st-mapping table $\mathcal{M}$, it computes a MUDSE solution J^* for I and $\mathcal{M}$ under $\mathfrak{S}$ in LOGSPACE. This procedure works as follows:
$\texttt{CompMUDSEsol}_{\mathfrak{S}}$:
Input: A source instance I, an st-mapping table $\mathcal{M}$, and a set Σ_{st} of st-tgds.
Output: A Minimal Universal DSE solution J^* for I and $\mathcal{M}$ under $\mathfrak{S}$.

1. Apply the source completion process, Σ_s^c, to I and $\mathcal{M}$, and generate $\hat{I}$ and $\hat{\mathcal{M}}$ respectively.
2. Define an equivalence relation $\sim$ on $dom(\hat{\mathcal{M}}) \cap \mathsf{Const}^{\mathbf{T}}$ as follows: $c_1 \sim c_2$ iff there exists a source constant a such that $\hat{\mathcal{M}}(a, c_1)$ and $\hat{\mathcal{M}}(a, c_2)$ hold.
3. Compute equivalence classes $\{C_1, \ldots, C_m\}$ for $\sim$ over $dom(\hat{\mathcal{M}}) \cap \mathsf{Const}^{\mathbf{T}}$ such that c_1 and c_2 exist in C_i only if $c_1 \sim c_2$.
4. Choose a set of witnesses $\{w_1, \ldots, w_m\}$ where $w_i \in C_i$, for each $1 \leq i \leq m$.
5. Compute from $\hat{\mathcal{M}}$ the instance $\mathcal{M}_1 := \texttt{replace}(\hat{\mathcal{M}}, w_1, \ldots, w_m)$ by replacing each target constant $c \in C_i \cap dom(\hat{\mathcal{M}})$ $(1 \leq i \leq m)$ with $w_i \in C_i$.
6. Apply a procedure (based on the *chase* [6]) to the instance $(\hat{I} \cup \{\mathcal{M}_1\})$, and generate a canonical universal pre-solution J for $\hat{I}$ and $\mathcal{M}_1$.
7. Apply a procedure (based on the core [8]) to the target instance J and generate the target instance J^* that is the core of J.

We prove the correctness of $\texttt{CompMUDSEsol}_{\mathfrak{S}}$ in the following Theorem.

Theorem 1. *Let $\mathfrak{S}$ be a DSE setting, I a source instance, and $\mathcal{M}$ an st-mapping table. Suppose that J^* is an arbitrary result for $\texttt{CompMUDSEsol}_{\mathfrak{S}}(I, \mathcal{M})$. Then, J^* is a minimal universal DSE solution for I and $\mathcal{M}$ under $\mathfrak{S}$.*

In data exchange, the smallest universal solutions are known as *cores* and can be computed in LOGSPACE [8]. With the help of such result we can prove that MUDSE solutions can be computed in LOGSPACE too. Also, in this context MUDSE solutions are unique up to isomorphism:

Theorem 2. *Let $\mathfrak{S}$ be a fixed DSE setting. There is a LOGSPACE procedure that computes, for a source instance I and an st-mapping table $\mathcal{M}$, a MUDSE solution J for I and $\mathcal{M}$ under $\mathfrak{S}$. Also, for any two MUDSE solutions J_1 and J_2 for I and $\mathcal{M}$ under $\mathfrak{S}$, it is the case that J_1 and J_2 are isomorphic.*

Proof. The proof of the first part of this theorem is very similar to the proof of Proposition 1, with the difference that steps 2, 3, and 4 in $\texttt{CompMUDSEsol}_{\mathfrak{S}}$ seem to be non-deterministic since they involve choosing a set of witnesses $\{w_1, \ldots, w_m\}$ for $\{C_1, \ldots, C_m\}$. Clearly, different sets of witnesses may yield different target instances. However, each possible choice of witnesses leads to a minimal universal DSE solution. In addition, according to [7,8], generating cores can be computed in LOGSPACE by applying the *naive* chase and the simple *Greedy* algorithm [8]. Finally, since LOGSPACE is closed under composition [11], we can deduce that the procedure CompMUDSEsol$_\mathfrak{S}$ is computed in LOGSPACE.

6   Query Answering

In data exchange, one is typically interested in the *certain answers* of a query Q, that is, the answers of Q that hold in each possible solution [7]. For the case of DSE we need to adapt this definition to solutions that are knowledge bases. Formally, let $\mathfrak{S}$ be a DSE setting, I a source instance, $\mathcal{M}$ an st-mapping table, and Q a FO conjunctive query over $\mathbf{T}$. The set of certain answers of Q over I and $\mathcal{M}$ and under $\mathfrak{S}$, denoted certain$_\mathfrak{S}((I \cup \{\mathcal{M}\}), Q)$, corresponds to the set of tuples that belong to the evaluation of Q over $K_\mathbf{T}$, for each DSE solution J for I and $\mathcal{M}$ and $K \in \mathsf{Mod}((J \cup \{\mathcal{M}\}), \Sigma_t^c)$.

Example 3. We refer to the DSE setting given in Example 2. Let $Q(x, y, z) = Enroll(x, y, z)$. Then, certain$_\mathfrak{S}((I \cup \{\mathcal{M}\}), Q) = \{Enroll(Alex, CSI1390, 80),$ $Enroll(Alex, CSI1790, 80)\}$. □

In DE, certain answers of unions of CQs can be evaluated in LOGSPACE by directly posing them over a universal solution [7], and then discarding tuples with null values. The same complexity bound holds in DSE by applying a slightly different algorithm. In fact, certain answers cannot be simply obtained by posing Q on a universal DSE solution J, since J might be incomplete with respect to the implicit data in Σ_t^c.

One possible solution would be to apply the target completion program Σ_t^c to a universal DSE solution J (denoted as $\Sigma_t^c(J)$) as a first step, then apply Q to $\Sigma_t^c(J)$. A second method is to compute certain answers of Q using a MUDSE solution. A MUDSE solution J in DSE possesses an interesting property, that is, applying Q to J returns a set of certain answers U that minimally represents the set of certain answers U' returned when Q is applied to $\Sigma_t^c(J)$. We can compute certain$_\mathfrak{S}((I \cup \{\mathcal{M}\}), Q)$ directly using J, by first applying rules in Σ_t^c, excluding rule 6, to generate the binary table EQUAL. Then complete the evaluation of Q on J, $Q(x_1, \ldots, x_n)$, and return $\hat{Q}(y_1, \ldots, y_n) = Q(x_1, \ldots, x_n) \wedge \bigwedge_{i=1}^n \mathrm{EQUAL}(x_i, y_i)$.

Adopting the second method to compute certain answers using MUDSE solutions and EQUAL proved in Sect. 7 to exhibit a much better performance in run times than the first method. These results make MUDSE solutions distinguished for their compactness and for their performance in query answering. We also obtain the following result:

Proposition 2. *Let* $\mathfrak{S} = (\mathbf{S}, \mathbf{T}, \mathcal{M}, \Sigma_{st})$ *be a fixed DSE setting,* I *a source instance,* $\mathcal{M}$ *an st-mapping table,* J *a MUDSE solution, and* Q *a fixed CQ over* $\mathbf{T}$*. Then,* certain$_\mathfrak{S}((I \cup \{\mathcal{M}\}), Q) = \hat{Q}(J)$ *where* $\hat{Q}(y_1, \ldots, y_n) = Q(x_1, \ldots, x_n) \wedge$ $\bigwedge_{i=1}^n \mathrm{EQUAL}(x_i, y_i)$

In addition, we prove in the following proposition that we can still compute the set of certain answers of a conjunctive query Q in LOGSPACE.

Proposition 3. *Let $\mathfrak{S} = (\mathbf{S}, \mathbf{T}, \mathcal{M}, \Sigma_{st})$ be a fixed DSE setting and Q a fixed union of CQs. There is a* LOGSPACE *procedure that computes* certain$_\mathfrak{S}$$((I \cup \{\mathcal{M}\}), Q)$, *given a source instance I and an st-mapping table $\mathcal{M}$.*

Proof. Let Q be a fixed union of conjunctive queries. The fact that computing the set of certain answers of Q, certain$_\mathfrak{S}$$((I \cup \{\mathcal{M}\}), Q)$ is in LOGSPACE, is based on the following facts: (1) following Proposition 1, generating a universal DSE solution is in LOGSPACE; and (2) following Theorem 2, generating a MUDSE solution is in LOGSPACE; and finally (3) it is known from [11] that the data complexity of any FO formula is in LOGSPACE, and thus checking if a fixed conjunctive query is satisfied in a database instance is in LOGSPACE.

7   Experiments

We implement the knowledge exchange semantics we introduced in this paper in a DSE prototype system. This system effectively generates universal DSE and MUDSE solutions that can be used to compute certain answers for CQs using the two methods introduced in Sect. 6. We used the DSE scenario of Example 2 extended with the st-tgd: $Cr(x, y, z) \wedge \mathcal{M}(x, x') \wedge \mathcal{M}(y, y') \wedge \mathcal{M}(z, z') \to Course$ (x', y', z'). Due to the lack of a benchmark that enforces recursion of the $\sim$ equivalence relation in the st-mapping table $\mathcal{M}$ and due to size restrictions, we synthesized the data in our experiments.

We show in our experiments that as the percentage of recursion increases in an st-mapping table, the run time to generate a universal DSE solution exceeds the time to generate a MUDSE solution. We also show that computing certain answers using a MUDSE solution is more effective than using a universal DSE solution. The experiments were conducted on a Lenovo workstation with a Dual-Core Intel(R) 1.80 GHz processor running Windows 7, and equipped with 4GB of memory and a 297 GB hard disk. We used Python (v2.7) to write the code and PostgreSQL (v9.2) database system.

DSE and MUDSE Solutions Computing Times. We used in this experiment a source instance I of 4,500 tuples, and 500 of those were courses information. The DSE system leveraged the work done in the state of the art ++Spicy system [13] to generate MUDSE solutions. We mapped courses data in the source to common target courses in $\mathcal{M}$, with different $\sim$ equivalence percentages (to enforce a recursive $\sim$ relation). The remaining set of source data was mapped to itself in $\mathcal{M}$. Figure 1 shows that as the percentage of recursion in $\sim$ equivalence relation over $\mathcal{M}$ increases, the run times to generate universal DSE and MUDSE solutions increase. The reason is, as the $\sim$ percentage increases, the number of source values (and target values) inferred to be $\sim$ increases, and thus the size of EQUAL created in Σ_s^c and Σ_t^c increases. Also, since target instances are usually larger than $\mathcal{M}$, the run time of completing the former to generate DSE solutions exceeds the time of completing the later when generating MUDSE solutions.

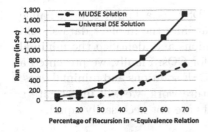

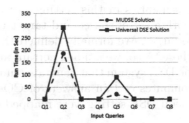

Fig. 1. MUDSE and Universal DSE solutions generation times

Fig. 2. Queries run times against a core of a universal DSE solution and a MUDSE solution

Table 1. List of queries

Q1	Fetch all the students names and the name of courses they have taken
Q2	Fetch the list of pairs of students ids and names that took the same course
Q3	Fetch all the students names and the grades they have received
Q4	Fetch the list of pairs of courses names that belong to the same program
Q5	Fetch for each student id the pair of courses that he has finished with the same grade
Q6	Fetch all the courses ids and their names
Q7	Fetch all the students ids and their names
Q8	Fetch the list of pairs of students ids that possess the same address

Conjunctive Queries Computing Times. We have selected a set of 8 queries to compare the performance of computing certain answers using a universal DSE solution (following the first method in Sect. 6) versus a MUDSE solution (following the second method in Sect. 6). We list the queries in Table 1.

We applied the list of input queries to a DSE instance where the $\sim$ percentage is 40 % and a course in the source is mapped to a maximum of two courses in the target. We chose a universal DSE solution, with a property of being a core of itself, that had around 18,000 records, and a MUDSE solution that contained around 4,900 records. Figure 2 shows that computing the sets of certain answers for the input conjunctive queries using a MUDSE solution take less run times than when computing these using a DSE solution. In addition, the deterioration in performance of query execution against the DSE solution appeared more in queries $Q2$ and $Q5$ than the remaining queries, is because both queries apply join operations to the *Enroll* table that involves a lot of elements which are inferred to be *equivalent* by $\mathcal{M}$.

8 Concluding Remarks

We introduced a DSE setting which exchanges data between two applications that have distinct schemas and distinct yet related sets of vocabularies. To capture the semantics of this setting, we defined DSE as a knowledge exchange system with a set of source and target rules that infer the implicit data should be in the target. We formally defined DSE solutions and identified the minimal among those. Also, we studied certain answers for CQs. Finally, we presented a prototype DSE system that generates universal DSE solutions and minimal ones, and it computes certain answers of CQs. In future work, we will investigate a more general DSE setting were mapped elements are not necessarily equal.

Acknowledgments. We thank NSERC for providing us the grants.

References

1. Bernstein, P., Giunchiglia, F., Kementsietsidis, A., Mylopoulos, J., Serani, L., Zaihrayeu, I.: Data management for Peer-to-Peer computing: a vision. In: Proceedings of the Workshop on the Web and Databases (WebDB'02) (2002)
2. Arenas, M., Kantere, V., Kementsietsidis, A., Kiringa, I., Miller, R.J., Mylopoulos, J.: The hyperion project: from data integration to data coordination. In: ACM SIGMOD Record, pp. 53–58 (2003)
3. Arenas, M., Barceló, P., Libkin, L., Murlak, F.: Relational and XML Data Exchange. Morgan and Claypool Publishers, New York (2010)
4. Arenas, M., Botoeva, E., Calvanese, D.: Knowledge base exchange. In: Proceedings of Description Logics (2011)
5. Arenas, M., Perez, J., Reutter, J.L.: Data exchange beyond complete data. In: Proceedings of PODS, pp. 83–94 (2011)
6. Beeri, C., Vardi, M.Y.: A proof procedure for data dependencies. J. ACM **71**(4), 718–741 (1984)
7. Fagin, R., Kolaitis, P.G., Miller, R.J., Popa, L.: Data exchange: semantics and query answering. Theor. Comput. Sci **336**(1), 89–124 (2005)
8. Fagin, R., Kolaitis, P.G., Popa, L.: Data exchange: getting to the core. ACM Trans. Database Syst **30**(1), 174–210 (2005)
9. Kementsietsidis, A., Arenas, M., Miller, R.J.: Mapping data in peer-to-peer systems: semantics and algorithmic issues. In: Proceedings of the 2003 ACM SIGMOD International Conference on Management of Data, pp. 325–336 (2003)
10. Reinghold, O.: Undirected connectivity in log-space. J. ACM **55**(4), 1–24 (2008)
11. Arenas, M., Reutter, J., Barceló, P.: Query languages for data exchange: beyond unions of conjunctive queries. In: Proceedings of the 12th International Conference on Database Theory, pp. 73–83 (2009)
12. Baader, F., Calvanese, D., McGuinness, D.L., Nardi, D., Patel-Schneider, P.F.: The Description Logic Handbook. Cambridge University Press, Cambridge (2003)
13. Marnette, B., Mecca, G., Papotti, P.: ++Spicy: an open-source tool for second-generation schema mapping and data exchange. In: Proceedings of the VLDB, pp. 1438–1441 (2011)
14. Levy, A.Y., Rajaraman, A., Ordille, J.: Querying heterogeneous information sources using source descriptions. In: Proceedings of VLDB, pp. 251–262 (1996)
15. Larson, J.A., Sheth, A.P.: Federated database systems for managing distributed, heterogeneous, and autonomous databases. ACM Comput. Surv **22**(3), 183–236 (1990)
16. Motik, B., Horrocks, I., Sattler, U.: Bridging the gap between OWL and relational databases. J. Web Semant. **7**(2), 74–89 (2009)

Propositional Encoding of Constraints over Tree-Shaped Data

Alexander Bau and Johannes Waldmann[✉]

HTWK Leipzig, Fakultät IMN, 04277 Leipzig, Germany
waldmann@imn.htwk-leipzig.de

Abstract. We present a functional programming language for specifying constraints over tree-shaped data. The language allows for Haskell-like algebraic data types and pattern matching. Our constraint compiler CO4 translates these programs into satisfiability problems in propositional logic. We present an application from the area of automated analysis of termination of rewrite systems, and also relate CO4 to Curry.

1 Motivation

The paper presents a high-level declarative language CO4 for describing constraint systems. The language includes user-defined algebraic data types and recursive functions defined by pattern matching, as well as higher-order and polymorphic types. This language comes with a compiler that transforms a high-level constraint system into a satisfiability problem in propositional logic. This is motivated by the following.

Constraint solvers for propositional logic (SAT solvers) like Minisat [ES03] are based on the Davis-Putnam-Logemann-Loveland (DPLL) [DLL62] algorithm and extended with conflict-driven clause learning (CDCL) [SS96] and preprocessing. They are able to find satisfying assignments for conjunctive normal forms with 10^6 and more clauses in a lot of cases quickly. SAT solvers are used in industrial-grade verification of hardware and software.

With the availability of powerful SAT solvers, *propositional encoding* is a promising method to solve constraint systems that originate in different domains. In particular, this approach had been used for automatically analyzing (non-)termination of rewriting [KK04,ZSHM10,CGSKT12] successfully, as can be seen from the results of International Termination Competitions (most of the participants use propositional encodings).

So far, these encodings are written manually: the programmer has to construct explicitly a formula in propositional logic that encodes the desired properties. Such a construction is similar to programming in assembly language: the advantage is that it allows for clever optimizations, but the drawbacks are that the process is inflexible and error-prone.

This author is supported by an ESF grant.

M. Hanus and R. Rocha (Eds.): KDPD 2013, LNAI 8439, pp. 33–47, 2014.
DOI: 10.1007/978-3-319-08909-6_3, © Springer International Publishing Switzerland 2014

This is especially so if the data domain for the constraint system is remote from the "sequence of bits" domain that naturally fits propositional logic. In typical applications, data is not a flat but hierarchical (e.g., using lists and trees), and one wants to write constraints on such data in a direct way.

Therefore, we introduce a constraint language CO4 that comes with a compiler to propositional logic. Syntactically, CO4 is a subset of Haskell [Jon03], including data declarations, case expressions, higher order functions, polymorphism (but no type classes). The advantages of re-using a high level declarative language for expressing constraint systems are: the programmer can rely on established syntax and semantics, does not have to learn a new language, can re-use his experience and intuition, and can re-use actual code. For instance, the (Haskell) function that describes the application of a rewrite rule at some position in some string or term can be directly used in a constraint system that describes a rewrite sequence with a certain property.

A constraint programming language needs some way of parameterizing the constraint system to data that is not available when writing the program. For instance, a constraint program for finding looping derivations for a rewrite system R, will not contain a fixed system R, but will get R as run-time input.

A formal specification of compilation is given in Sect. 2, and a concrete realization of compilation of first-order programs using algebraic data types and pattern matching is given in Sect. 3. In these sections, we assume that data types are finite (e.g., composed from Bool, Maybe, Either), and programs are total. We then extend this in Sect. 4 to handle infinite (that is, recursive) data types (e.g., lists, trees), and partial functions. Note that a propositional encoding can only represent a finite subset of values of any type, e.g., lists of Booleans with at most 5 elements, so partial functions come into play naturally.

We then treat in Sect. 5 briefly some ideas that serve to improve writing and executing CO4 programs. These are higher-order functions and polymorphism, as well as hash-consing, memoization, and built-in binary numbers.

Next, we give an application of CO4 in the termination analysis of rewrite systems: In Sect. 6 we describe a constraint system for looping derivations in string rewriting. We compare this to a hand-written propositional encoding [ZSHM10], and evaluate performance. The subject of Sect. 7 is the comparison of CO4 to Curry [Han13], using the standard N-Queens-Problem as a test case.

Our constraint language and compiler had been announced in short workshop contributions at *HaL 8* (Leipzig, 21 June 13), and *Haskell and Rewriting Techniques* (Eindhoven, 26 June 13). The current paper is extended and revised from our contribution to *Workshop on Functional and Logic Programming* (Kiel, 11 September 13). Because of space restrictions, we still leave out some technicalities in Sects. 2 and 3, and instead refer to the full version [BW13].

2   Semantics of Propositional Encodings

In this section, we introduce CO4 syntax and semantics, and give the specification for compilation of CO4 expressions, in the form of an invariant (it should hold for all sub-expressions). When applied to the full input program, the specification implies that the compiler works as expected: a solution for the constraint system can be found via the external SAT solver. We defer discussion of our implementation of this specification to Sect. 3, and give here a more formal, but still high-level view of the CO4 language and compiler.

Evaluations on Concrete Data. We denote by $\mathbb{P}$ the set of expressions in the input language. It is a first-order functional language with algebraic data types, pattern matching, and global and local function definitions (using `let`) that may be recursive. The concrete syntax is a subset of Haskell. We give examples— which may appear unrealistically simple but at this point we cannot use higher-order or polymorphic features. These will be discussed in Sect. 5.

E.g., `f p u` is an expression of $\mathbb{P}$, containing three variables `f`, `p` and `u`. We allow only *simple* patterns (a constructor followed by variables), and we require that pattern matches are *complete* (there is exactly one pattern for each constructor of the respective type). Nested patterns can be translated to this form.

Evaluation of expressions is defined in the standard way: The domain of *concrete values* $\mathbb{C}$ is the set of data terms. For instance, `Just False` $\in \mathbb{C}$. A *concrete environment* is a mapping from program variables to $\mathbb{C}$. A *concrete evaluation function* concrete-value : $E_{\mathbb{C}} \times \mathbb{P} \to \mathbb{C}$ computes the value of a concrete expression $p \in \mathbb{P}$ in a concrete environment $e_{\mathbb{C}}$. Evaluation of function and constructor arguments is strict.

Evaluations on Abstract Data. The CO4 compiler transforms an input program that operates on concrete values, to an *abstract program* that operates on *abstract values*. An abstract value contains propositional logic formulas that may contain free propositional variables. An abstract value represents a set of concrete values. Each assignment of the propositional values produces a concrete value.

We formalize this in the following way: the domain of abstract values is called $\mathbb{A}$. The set of assignments (mappings from propositional variables to truth values $\mathbb{B} = \{0, 1\}$) is called Σ, and there is a function decode : $\mathbb{A} \times \Sigma \to \mathbb{C}$.

We now specify abstract evaluation. (The implementation is given in Sect. 3.) We use *abstract environments* $E_{\mathbb{A}}$ that map program variables to abstract values, and an *abstract evaluation function* abstract-value : $E_{\mathbb{A}} \times \mathbb{P} \to \mathbb{A}$.

Allocators. As explained in the introduction, the constraint program receives known and unknown arguments. The compiled program operates on abstract values.

The abstract value that represents a (finite) set of concrete values of an unknown argument is obtained from an *allocator*. For a property $q : \mathbb{C} \to \mathbb{B}$ of concrete values, a *q-allocator* constructs an object $a \in \mathbb{A}$ that represents all concrete objects that satisfy q:

$$\forall c \in \mathbb{C} : q(c) \longleftrightarrow \exists v \in \Sigma . c = \mathrm{decode}(a, v).$$

We use allocators to specify that c uses constructors that belong to a specific type. Later (with recursive types, see Sect. 4) we also specify a size bound for c. An example is an allocator for lists of Booleans of length ≤ 4.

As a special case, an allocator for a singleton set is used for encoding a known concrete value. This *constant allocator* is given by a function $\mathsf{encode} : \mathbb{C} \to \mathbb{A}$ with the property that $\forall c \in \mathbb{C}, \sigma \in \Sigma : \mathsf{decode}(\mathsf{encode}(c), \sigma) = c$.

Correctness of Constraint Compilation. The semantical relation between an expression p (a concrete program) and its compiled version $\mathsf{compile}(p)$ (an abstract program) is given by the following relation between concrete and abstract evaluation:

Definition 1. *We say that* $p \in \mathbb{P}$ *is compiled* correctly *if*

$$\forall e \in E_\mathbb{A} \; \forall \sigma \in \Sigma : \mathsf{decode}(\mathsf{abstract\text{-}value}(e, \mathsf{compile}(p)), \sigma)$$
$$= \mathsf{concrete\text{-}value}(\mathsf{decode}(e, \sigma), p) \tag{1}$$

Here we used $\mathsf{decode}(e, \sigma)$ as notation for lifting the decoding function to environments, defined element-wise by

$$\forall e \in E_\mathbb{A} \; \forall v \in \mathrm{dom}(e) \; \forall \sigma \in \Sigma : \mathsf{decode}(e, \sigma)(v) = \mathsf{decode}(e(v), \sigma).$$

Application of the Correctness Property. We are now in a position to show how the stages of CO4 compilation and execution fit together.

The top-level parametric constraint is given by a declaration `constraint k u = b` (cf. Fig. 1) where `b` (the *body*, a concrete program) is of type `Bool`. It will be processed in the following stages:

1. *compilation* produces an abstract program $\mathsf{compile}(b)$,
2. *abstract computation* takes a concrete parameter value $p \in \mathbb{C}$ and a q-allocator $a \in \mathbb{A}$, and computes the abstract value

$$V = \mathsf{abstract\text{-}value}(\{k \mapsto \mathsf{encode}(p), u \mapsto a\}, \mathsf{compile}(b))$$

```
data Bool        = False  | True
data Maybe_Bool = Nothing | Just Bool

and2 :: Bool -> Bool -> Bool
and2 x y = case x of { False -> False ; True -> y }

f :: Maybe_Bool -> Maybe_Bool -> Maybe_Bool
f p q = case p of
    Nothing -> Nothing
    Just x -> case q of Nothing -> Nothing
                        Just y  -> Just (and2 x y)

constraint :: Maybe_Bool -> Maybe_Bool -> Bool
constraint p u = case f p u of Nothing -> False
                          Just r    -> r
```

Fig. 1. Exemplary constraint-system in CO4

3. *solving* calls the back-end SAT solver to determine $\sigma \in \Sigma$ with $\mathsf{decode}(V, \sigma) = \mathrm{TRUE}$. If this was successful,
4. *decoding* produces a concrete value $s = \mathsf{decode}(a, \sigma)$,
5. and optionally, *testing* checks $\mathsf{concrete\text{-}value}(\{k \mapsto p, u \mapsto s\}, b) = \mathrm{TRUE}$.

The last step is just for reassurance against implementation errors, since the invariant implies that the test returns True. This highlights another advantage of re-using Haskell for constraint programming: one can easily check the correctness of a solution candidate.

3   Implementation of a Propositional Encoding

In this section, we give a realization for abstract values, and show how compilation creates programs that operate correctly on those values, as specified in Definition 1.

Encoding and Decoding of Abstract Values. The central idea is to represent an abstract value as a tree, where each node contains an encoding for a symbol (a constructor) at the corresponding position, and the list of concrete children of the node is a prefix of the list of abstract children (the length of the prefix is the arity of the constructor).

The encoding of constructors is by a sequence of formulas that represent the number of the constructor in binary notation.

We denote by F the set of propositional logic formulas. At this point, we do not prescribe a concrete representation. For efficiency reasons, we will allow some form of sharing. Our implementation[1] assigns names to subformulas by doing the Tseitin transform [Tse83] on-the-fly, creating a fresh propositional literal for each subformula.

Definition 2. *The set of abstract values* $\mathbb{A}$ *is the smallest set with* $\mathbb{A} = \mathrm{F}^* \times \mathbb{A}^*$. *An element* $a \in \mathbb{A}$ *thus has shape* $(\overrightarrow{f}, \overrightarrow{a})$ *where* $\overrightarrow{f}$ *is a sequence of formulas, called the* flags *of* a, *and* $\overrightarrow{a}$ *is a sequence of abstract values, called the* arguments *of* a.
We introduce notation

- flags : $\mathbb{A} \to \mathrm{F}^*$ *gives the flags of an abstract value*
- flags$_i$: $\mathbb{A} \to \mathrm{F}$ *gives the* i*-th flag of an abstract value*
- arguments : $\mathbb{A} \to \mathbb{A}^*$ *gives the arguments of an abstract value,*
- argument$_i$: $\mathbb{A} \to \mathbb{A}$ *gives the* i*-th argument of an abstract value*

The sequence of flags of an abstract value encodes the number of its constructor. We use the following variant of a binary encoding: For each data type T with c constructors, we use as flags a set of sequences $S_c \subseteq \{0,1\}^*$ with $|S_c| = c$ and such that each long enough $w \in \{0,1\}^*$ does have exactly one prefix in S_c:

$$S_1 = \{\epsilon\}; \qquad \text{for } n > 1: \quad S_n = 0 \cdot S_{\lceil n/2 \rceil} \cup 1 \cdot S_{\lfloor n/2 \rfloor}$$

---

[1] https://github.com/apunktbau/satchmo-core

Note that $\forall c : S_c \subset F^c$, i.e. each sequence of flags represents a sequence of constant propositional formulas.

Example 1. $S_2 = \{0, 1\}$, $S_3 = \{00, 01, 1\}$, $S_5 = \{000, 001, 01, 10, 11\}$.

The lexicographic order of S_c induces a bijection $\mathsf{numeric}_c : S_c \to \{1, \ldots, c\}$ to map sequence of flags to constructor indices.

The encoding function (from concrete to abstract values) is defined by

$$\mathsf{encode}_T(C(v_1, \ldots)) = (\mathsf{numeric}_c^-(i), [\mathsf{encode}_{T_1}(v_1), \ldots])$$

where C is the i-th constructor of type T, and T_j is the type of the j-th argument of C. Note that here, $\mathsf{numeric}_c^-(i)$ denotes a sequence of constant flags (formulas) that represents the corresponding binary string.

For decoding, we need to take care of extra flags and arguments that may have been created by the function merge (Definition 4) that is used in the compilation of $\mathtt{case}$ expressions. Therefore, we extend the mapping $\mathsf{numeric}_c$ to longer strings by $\mathsf{numeric}_c(u \cdot v) := \mathsf{numeric}_c(u)$ for each $u \in S_c, v \in \{0, 1\}^*$. This is possible by the unique-prefix condition. For example, $\mathsf{numeric}_5(10) = 4$ and thus $\mathsf{numeric}_5(101) = 4$.

Example 2. Given the type declaration $\mathtt{data\ Bool = False\ |\ True}$ the concrete value $\mathtt{True}$ can be represented by the abstract value $a_1 = \mathsf{encode}_{\mathsf{Bool}}(\mathtt{True}) = ([x], [])$ and assignment $\{x \mapsto 1\}$, since $\mathtt{True}$ is the second of two constructors, and $\mathsf{numeric}_2(1) = 2$. The same concrete value $\mathtt{True}$ can also be represented by the abstract value $a_2 = ([x, y], [a_1])$ and assignment $\{x \mapsto 1, y \mapsto 0\}$, since $\mathsf{numeric}_2(10) = 2$. This shows that extra flags and extra arguments are ignored in decoding.

We give a formal definition: for a type T with c constructors, $\mathsf{decode}_T((f, a), \sigma)$ is the concrete value $v = C_i(v_1, \ldots)$ where $i = \mathsf{numeric}_c(f\sigma)$, and C_i is the i-th constructor of T, and $v_j = \mathsf{decode}_{T_j}(a_j, \sigma)$ where T_j is the type of the j-th argument of C_i.

As stated, this is a partial function, since any of f, a may be too short. For this section, we assume that abstract values always have enough flags and arguments for decoding, and we defer a discussion of partial decodings to Sect. 4.

Allocators for Abstract Values. Since we consider (in this section) finite types only, we restrict to *complete* allocators: for a type T, a complete allocator is an abstract value $a \in \mathbb{A}$ that can represent each element of T: for each $e \in T$, there is some σ such that $\mathsf{decode}_T(a, \sigma) = e$.

Example 3.

Type	Complete allocator
$\mathtt{data\ Bool\ =\ False\ |\ True}$	$a_1 = ([x_1], [])$
$\mathtt{data\ \ Ordering\ =\ LT\ |\ EQ\ |\ GT}$	$a_2 = ([x_1, x_2], [])$
$\mathtt{data\ \ EBO\ =\ Left\ Bool\ |\ Right\ Ordering}$	$a_3 = ([x_1], [([x_2, x_3], [])])$

where x_i are (Boolean) variables. We compute $\mathsf{decode}_{\mathrm{EBO}}(a_3, \sigma)$ for $\sigma = \{x_1 = 0, x_2 = 1, x_3 = 0\}$): Since $\mathsf{numeric}_2(0) = 1$, the top constructor is $\mathtt{Left}$. It has one argument, obtained as $\mathsf{decode}_{\mathrm{Bool}}(([x_2, x_3], []), \sigma)$. For this we compute $\mathsf{numeric}_2(10) = 2$, denoting the second constructor ($\mathtt{True}$) of $\mathtt{Bool}$. Thus, $\mathsf{decode}_{\mathrm{EBO}}(a_3, \sigma) = \mathtt{Left\ True}$.

Compilation of Programs. In the following we illustrate the actual transformation of the input program (that operates on concrete values) to an abstract program (operating on abstract values).

Generally, compilation keeps structure and names of the program intact. For instance, if the original program defines functions f and g, and the implementation of g calls f, then the transformed program also defines functions f and g, and the implementation of g calls f.

Compilation of variables, bindings, and function calls is straightforward, and we omit details.

We deal now with pattern matches. They appear naturally in the input program, since we operate on algebraic data types. The basic plan is that *compilation removes pattern matches*. This is motivated as follows. Concrete evaluation of a pattern match (in the input program) consists of choosing a branch according to a concrete value (of the discriminant expression). Abstract evaluation cannot access this concrete value (since it will only be available after the SAT solver determines an assignment). This means that we cannot abstractly evaluate pattern matches. Therefore, they are transformed into a linear program by compilation.

We restrict to pattern matches where patterns are *simple* (a constructor followed by variables) and *complete* (one branch for each constructor of the type).

Definition 3 (Compilation, pattern match). *Consider a pattern match expression e of shape* $\mathtt{case}\ d\ \mathtt{of}\ \{\ldots\}$, *for a discriminant expression d of type T with c constructors.*

We have $\mathsf{compile}(e) = \mathtt{let}\ x = \mathsf{compile}(d)\ \mathtt{in}\ \mathsf{merge}_c(\mathsf{flags}(x), b_1, \ldots)$ *where x is a fresh variable, and b_i represents the compilation of the i-th branch.*

Each such branch is of shape $C\ v_1 \ldots v_n \rightarrow e_i$, where C is the i-th constructor of the type T.

Then b_i is obtained as $\mathtt{let}\ \{v_1 = \mathsf{argument}_1(x); \ldots\}\ \mathtt{in}\ \mathsf{compile}(e_i)$.

Example 4. The following listing shows the abstract counterpart of function $\mathtt{constraint}$ from example 1:

```
constraint :: A -> A -> A
constraint p u =
  let _128 = f p u
  in
    let _133 = encode_Bool(False)
        _134 = argument_1(_128)
    in
      merge(flags(_128),(_133,_134))
```

The abstract value of the pattern match's discriminant f p u is bound to variable _128. The result of evaluating all compiled branches are bound to fresh variables _133 and _134. Finally, the resulting value is computed by mergeing _133 and _134.

The auxiliary function merge combines the abstract values from branches of pattern matches, according to the flags of the discriminant.

Definition 4 (Combining function). merge : $F^* \times \mathbb{A}^c \to \mathbb{A}$ *combines abstract values so that* merge$(\vec{f}, a_1, \ldots, a_c)$ *is an abstract value* $(\vec{g}, z_1, \ldots, z_n)$, *where*

- *number of arguments:* $n = \max(|\,\mathsf{arguments}(a_1)|, \ldots, |\,\mathsf{arguments}(a_c)|)$
- *number of flags:* $|\vec{g}| = \max(|\,\mathsf{flags}(a_1)|, \ldots, |\,\mathsf{flags}(a_c)|)$
- *combining the flags:*

$$\text{for } 1 \le i \le |\vec{g}|, \qquad g_i \leftrightarrow \bigwedge_{1 \le j \le c} (\mathsf{numeric}_c(\vec{f}) = j \to \mathsf{flags}_i(a_j)) \qquad (2)$$

- *combining the arguments recursively:*

$$\text{for each } 1 \le i \le n, z_i = \mathsf{merge}(\vec{f}, \mathsf{argument}_i(a_1), \ldots, \mathsf{argument}_i(a_c)).$$

Example 5. Consider the expression case e of False -> u; True -> v, where e,u,v are of type Bool, represented by abstract values $([f_e], [])$, $([f_u], [])$, $([f_v], [])$. The case expression is compiled into an abstract value $([f_r], [])$ where

$$f_r = \mathsf{merge}_2([f_e], ([f_u], []), ([f_v], []))$$
$$= (\mathsf{numeric}_2(f_e) = 1 \to f_u) \wedge (\mathsf{numeric}_2(f_e) = 2 \to f_v)$$
$$= (\overline{f_e} \to f_u) \wedge (f_e \to f_v)$$

We refer to [BW13] for the full specification of compilation, and proofs of correctness.

We mention already here one way of optimization: if all flags of the discriminant are constant (i.e., known during abstract evaluation, before running the SAT solver) then abstract evaluation will evaluate only the branch specified by the flags, instead of evaluating all, and merging the results. Typically, flags will be constant while evaluating expressions that only depend on the input parameter, and not on the unknown.

4   Partial Encoding of Infinite Types

We discuss the compilation and abstract evaluation for constraints over infinite types, like lists and trees. Consider declarations

```
data N = Z | S N
double :: N -> N
double x = case x of {Z -> Z ; S x' -> S (S (double x'))}
```

Assume we have an abstract value a to represent x. It consists of a flag (to distinguish between Z and S), and of one child (the argument for S), which is another abstract value. At some depth, recursion must stop, since the abstract value is finite (it can only contain a finite number of flags). Therefore, there is a child with no arguments, and it must have its flag set to [FALSE] (it must represent Z).

There is another option: if we leave the flag open (it can take on values FALSE or TRUE), then we have an abstract value with (possibly) a constructor argument missing. When evaluating the concrete program, the result of accessing a non-existing component gives a bottom value. This corresponds to the Haskell semantics where each data type contains bottom, and values like S (S ⊥) are valid. To represent these values, we extend our previous definition to:

Definition 5. *The set of abstract values* $\mathbb{A}_\perp$ *is the smallest set with* $\mathbb{A}_\perp = F^* \times \mathbb{A}_\perp^* \times F$, *i.e. an abstract value is a triple of flags and arguments (cf. definition 2) extended by an additional* definedness *constraint.*

We write def : $\mathbb{A}_\perp \to F$ *to give the definedness constraint of an abstract value, and keep* flags *and* argument *notation of Definition 2.*

The decoding function is modified accordingly: $\text{decode}_T(a, \sigma)$ for a type T with c constructors is $\perp$ if $\text{def}(a)\sigma = \text{FALSE}$, or $\text{numeric}_c(\text{flags}(a)\sigma)$ is undefined (because of "missing" flags), or $|\,\text{arguments}(a)|$ is less than the number of arguments of the decoded constructor.

The correctness invariant for compilation (Eq. 1) is still the same, but we now interpret it in the domain $\mathbb{C}_\perp$, so the equality says that if one side is $\perp$, then both must be. Consequently, for the application of the invariant, we now require that the abstract value of the top-level constraint under the assignment *is defined* and TRUE. Abstract evaluation is extended to $\mathbb{A}_\perp$ by the following:

- *explicit bottoms*: a source expression **undefined** results in an abstract value ([], [], 0) (flags and arguments are empty, definedness is False)
- *constructors are lazy*: the abstract value created by a constructor application has its definedness flag set to True
- *pattern matches are strict*: the definedness flag of the abstract value constructed for a pattern match is the conjunction of the definedness of the discriminant with the definedness of the results of the branches, combined by merge.

5 Extensions for Expressiveness and Efficiency

We briefly present some enhancements of the basic CO4 language. To increase expressiveness, we introduce higher order functions and polymorphism. To improve efficiency, we use hash-consing and memoization, as well as built-in binary numbers.

More Haskell Features in CO4. For formulating the constraints, expressiveness in the language is welcome. Since we base our design on Haskell, it is natural to include some of its features that go beyond first-order programs: higher order functions and polymorphic types.

Our program semantics is first-order: we cannot (easily) include functions as result values or in environments, since we have no corresponding abstract values for functions. Therefore, we instantiate all higher-order functions in a standard preprocessing step, starting from the main program.

Polymorphic types do not change the compilation process. The important information is the same as with monomorphic typing: the total number of constructors of a type, and the number (the encoding) of one constructor.

In all, we can use in CO4 a large part of the Haskell Prelude functions. CO4 just compiles their "natural" definition, e.g.,

```
and xs = foldl (&&) True xs ; a ++ b = foldr (:) b a
```

Memoization. We describe another optimization: in the abstract program, we use memoization for all subprograms. That is, during execution of the abstract program, we keep a map from (function name, argument tuple) to result. Note that arguments and result are abstract values. This allows to write "natural" specifications and still get a reasonable implementation.

For instance, the lexicographic path order $>_{lpo}$ (cf. [BN98]) defines an order over terms according to some precedence over symbols. Its textbook definition is recursive, and leads to an exponential time algorithm, if implemented literally. For evaluating $s >_{lpo} t$ the algorithm still does only compare subterms of s and t, and in total, there are $|s| \cdot |t|$ pairs of subterms, and this is also the cost of the textbook algorithm with a memoizing implementation.

For memoization we frequently need table lookups. For fast lookups we need fast equality tests (for abstract values). We get these by *hash-consing*: abstract constructor calls are memoized as well, so that abstract nodes are globally unique, and structural equality is equivalent to pointer equality.

Memoization is awkward in Haskell, since it transforms pure functions into state-changing operations. This is not a problem for CO4 since this change of types only applies to the abstract program, and thus is invisible on the source level.

Built-in Data Types and Operations. Consider the following natural definition:

```
not a   = case a of {False -> True ; True -> False}
```

The abstract value for a contains one flag each (and no arguments). CO4 will compile not in such a way that a fresh propositional variable is allocated for the result, and then emit two CNF clauses by Definition 4. This fresh result variable is actually not necessary since we can invert the polarity of the input literal directly. To achieve this, Booleans and (some of) their operations are handled specially by CO4.

Similarly, we can model binary numbers as lists of bits:

```
data [] a = [] | a : [a] ; data Nat = Nat [Bool]
```

An abstract value for a k-bit number then is a tree of depth k. At each level, we need one flag for the list constructor (Nil or Cons), and one flag for the list element (False or True). Instead of this, we provide built-in data types Nat_k that represent a k-bit number as one abstract node with k flags, and no arguments. These types come with standard arithmetical and relational operations.

We remark that a binary propositional encoding for numbers is related to the "sets-of-intervals" representation that a finite domain (FD) constraint solver would typically use. A partially assigned binary number, e.g., $[*, 0, 1, *, *]$, also represents a union of intervals, here, $[4..7] \cup [20..23]$. Assigning variables can be thought of as splitting intervals. See Sect. 7 an application of CO4 to a typical FD problem.

6   Case Study: Loops in String Rewriting

We use CO4 for compiling constraint systems that describe looping derivations in rewriting. We make essential use of CO4's ability to encode (programs over) unknown objects of algebraic data types, in particular, of lists of unknown lengths, and with unknown elements.

The application is motivated by automated analysis of programs. A *loop* is an infinite computation, which may be unwanted behaviour, indicating an error in the program's design. In general, it is undecidable whether a rewriting system admits a loop. Loops can be found by enumerating finite derivations.

Our approach is to write the predicate "the derivation d conforms to a rewrite system R and d is looping" as a Haskell function, and solve the resulting constraint system, after putting bounds on the sizes of the terms that are involved.

Previous work uses several heuristics for enumerations resp. hand-written propositional encodings for finding loops in string rewriting systems [ZSHM10].

We compare this to a propositional encoding via CO4. We give here the type declarations and some code examples. Full source code is available[2].

In the following, we show the data declarations we use, and give code examples.

- We represent symbols as binary numbers of flexible width, since we do not know (at compile-time) the size of the alphabet: `type Symbol = [ Bool ]`.
- We have words: `type Word = [Symbol]`, rules: `type Rule = (Word, Word)`, and rewrite systems `type SRS = [Rule]`.
- A rewrite step $(p \mathbin{+\!\!+} l \mathbin{+\!\!+} s) \rightarrow_R (p \mathbin{+\!\!+} r \mathbin{+\!\!+} s)$, where rule (l, r) is applied with left context p and right context s, is represented by `Step p (l,r) s` where

  `data Step = Step Word Rule Word`

- a derivation is a list of steps: `type Derivation = [Step]`, where each step uses a rule from the rewrite system, and consecutive steps fit each other:

  `conformant :: SRS -> Derivation -> Bool`

---

[2] https://github.com/apunktbau/co4/blob/master/CO4/Example/Loop/Standalone.hs

Table 1. Finding looping derivations in rewrite systems.

	Gebhardt/03		Gebhardt/08		Zantema_04/z042		Zantema_06/loop1	
	CO4	TTT2	CO4	TTT2	CO4	TTT2	CO4	TTT2
s#vars	132232	23759	132168	23696	248990	32180	132024	21880
#clauses	448543	39541	448351	39445	854949	50150	447935	35842
Solving	97 s	8 s	6 s	20 s	5 s	1 s	4 s	1 s

- a derivation is looping if the output of the last step is a subword of the input of the first step

```
constraint :: SRS -> Looping_Derivation -> Bool
constraint srs (Looping_Derivation pre d suf) =
  conformant srs d && eqWord (pre ++ start d ++ suf) (result d)
```

This is the top-level constraint. The rewrite system `srs` is given at run-time. The derivation is unknown. An allocator represents a set of derivations with given maximal length (number of steps) and width (length of words).

Overall, the complete CO4 code consists of roughly 100 lines of code. The code snippets above indicate that the constraint system literally follows the textbook definitions. E.g., note the list-append (++) operators in `constraint`.

In contrast, Tyrolean Termination Tool 2 (TTT2, version 1.13)[3] contains a hand-written propositional encoding for (roughly) the same constraint[4] consisting of roughly 300 lines of (non-boilerplate) code. The TTT2 implementation explicitly allocates propositional variables (this is implicit in CO4), and explicitly manipulates indices (again, this is implicit in our ++).

Table 1 compares the performance of our implementation to that of TTT2 on some string rewriting systems of the Termination Problems Data Base[5] collection. We restrict the search space in both tools to derivations of length 16 and words of length 16. All test were run on a Intel Xeon CPU with 3 GHz and 12 GB RAM. CO4's test results can be replicated by running `cabal test --test-options="loop-srs"`.

We note that CO4 generates larger formulas, for which, in general, MiniSat-2.2.0 needs more time to solve. There are rare cases where CO4's formula is solved faster.

7   A Comparison to Curry

We compare the CO4 language and implementation to that of the functional logic programming language Curry [Han13], and its PAKCS-1.11.1 implementation (using the SICSTUS-4.2.3 Prolog system).

---

[3] http://colo6-c703.uibk.ac.at/ttt2/
[4] ttt2/src/processors/src/nontermination/loopSat.ml
[5] http://termination-portal.org/wiki/TPDB

CO4 source code	Curry source code

```
constraint n l =                    import CLPFD
   all (\ x -> le (nat8 1) x
           && le x n ) l            queens options n l =
  && all_safe l                        gen_vars n =:= l &
                                       domain l 1 (length l) &
all_safe l = case l of                 all_safe l &
   []   -> True                        labeling options l
   q:qs -> safe q qs (nat8 1)
          && all_safe qs          all_safe [] = success
                                  all_safe (q:qs) = safe q qs 1
safe q qs p = case qs of                   & all_safe qs
   []        -> True
   q1 : qs' ->                     safe _ [] _ = success
        no_attack q q1 p           safe q (q1:qs) p = no_attack q q1 p
    && safe q qs' (increment p)            & safe q qs (p+#1)

no_attack q1 q2 p =               no_attack q1 q2 p = q1 /=# q2
   neq q1 q2 && neq (add q1 p) q2           & q1 /=# q2+#p
         && neq q1 (add q2 p)               & q1 /=# q2-p

le         = leNat8               gen_vars n = if n==0
neq a b    = not (eqNat8 a b)        then []
add        = plusNat8                else var : gen_vars (n-1)
increment x = add x (nat8 1)             where var free
```

Fig. 2. Two approaches to solve the n queens problem

A common theme is that both languages are based on Haskell (syntax and typing), and extend this by some form of non-determinism, so the implementation has to realize some form of search.

In Curry, nondeterminism is created lazily (while searching for a solution). In CO4, nondeterminism is represented by additional Boolean decision variables that are created beforehand (in compilation).

The connection from CO4 to Curry is easy: a CO4 constraint program with top-level constraint main :: Known − > Unknown − > Bool is equivalent to a Curry program (query) main k u =:= True where u free (Fig. 2).

In the other direction, it is not possible to translate a Curry program to a CO4 program since it may contain locally free variables, a concept that is not supported in CO4. All free variables are globally defined by the allocator of the unknown parameter of the top-level constraint. For doing the comparison, we restrict to CO4 programs.

Example 6. We give an example where the CO4 strategy seems superior: the n queens problem.

We compare our approach to a Curry formulation (taken from the PAKCS online examples collection) that uses the CLPFD library for finite-domain

Table 2. Time for finding one solution of the n queens problem

n	8	12	16	20	24	32	64	128
CO4	0.08 s	0.16 s	0.31 s	0.57 s	0.73	1.59 s	10.8 s	53.1 s
Curry/PAKCS	0.02 s	0.13 s	0.43 s	8.54 s	> 10 m	> 10 m	> 10 m	> 10 m

constraint programming. Our CO4 formulation uses built-in 8-bit binary numbers (Sect. 5) but otherwise is a direct translation. Note that with 8 bit numbers we can handle board sizes up to 2^7: we add co-ordinates when checking for diagonal attacks.

Table 2 shows the run-times on several instances of the n queens problem. CO4's runtime is the runtime of the abstract program in addition to the runtime of the SAT-solver. The run-times for PAKCS were measured using the `:set +time` flag after compiling the Curry program in the PAKCS evaluator. Tests were done on a Intel Core 2 Duo CPU with 2.20 GHz and 4 GB RAM.

The PAKCS software also includes an implementation of the n queens problem that does not use the CLPFD library. As this implementation already needs 6 seconds to solve a $n = 8$ instance, we omit it in the previous comparison.

8   Discussion

In this paper we described the CO4 constraint language and compiler that allows to write constraints on tree-shaped data in a natural way, and to solve them via propositional encoding.

We presented the basic ideas for encoding data and translating programs, and gave an outline of a correctness proof for our implementation.

We gave an example where CO4 is used to solve an application problem from the area of termination analysis. This example shows that SAT compilation has advantages w.r.t. manual encodings.

We also gave an experimental comparison between CO4 and Curry, showing that propositional encoding is an interesting option for solving finite domain (FD) constraint problems. Curry provides lazy nondeterminism (creating choice points on-the-fly). CO4 does not provide this, since choice points are allocated before abstract evaluation.

Work on CO4 is ongoing. Our immediate goals are, on the one hand, to reduce the size of the formulas that are built during abstract evaluation, and on the other hand, to extend the source language with more Haskell features.

References

[BN98] Baader, F., Nipkow, T.: Term Rewriting and all That. Cambridge University Press, New York (1998)

[BW13] Bau, A., Waldmann, J.: Propositional encoding of constraints over tree-shaped data. CoRR, abs/1305.4957 (2013)

[CGSKT12] Codish, M., Giesl, J., Schneider-Kamp, P., Thiemann, R.: Sat solving for termination proofs with recursive path orders and dependency pairs. J. Autom. Reasoning **49**(1), 53–93 (2012)

[DLL62] Davis, M., Logemann, G., Loveland, D.W.: A machine program for theorem-proving. Commun. ACM **5**(7), 394–397 (1962)

[ES03] Eén, N., Sörensson, N.: An extensible SAT-solver. In: Giunchiglia, E., Tacchella, A. (eds.) SAT 2003. LNCS, vol. 2919, pp. 502–518. Springer, Heidelberg (2004)

[Han13] Hanus, M.: Functional logic programming: from theory to curry. In: Voronkov, A., Weidenbach, C. (eds.) Programming Logics. LNCS, vol. 7797, pp. 123–168. Springer, Heidelberg (2013)

[Jon03] Peyton Jones, S. (ed.): Haskell 98 Language and Libraries, The Revised Report. Cambridge University Press, Cambridge (2003)

[KK04] Kurihara, M., Kondo, H.: Efficient BDD encodings for partial order constraints with application to expert systems in software verification. In: Orchard, B., Yang, C., Ali, M. (eds.) IEA/AIE 2004. LNCS (LNAI), vol. 3029, pp. 827–837. Springer, Heidelberg (2004)

[SS96] Marques Silva, J.P., Sakallah, K.A.: Grasp - a new search algorithm for satisfiability. In: ICCAD, pp. 220–227 (1996)

[Tse83] Tseitin, G.S.: On the complexity of derivation in propositional calculus. In: Siekmann, J., Wrightson, G. (eds.) Automation of Reasoning. Symbolic Computation, pp. 466–483. Springer, Heidelberg (1983)

[ZSHM10] Zankl, H., Sternagel, C., Hofbauer, D., Middeldorp, A.: Finding and certifying loops. In: van Leeuwen, J., Muscholl, A., Peleg, D., Pokorný, J., Rumpe, B. (eds.) SOFSEM 2010. LNCS, vol. 5901, pp. 755–766. Springer, Heidelberg (2010)

On a High-Level Approach to Implementing Exact Real Arithmetic in the Functional Logic Programming Language Curry

Christoph Beierle[✉] and Udo Lelitko

Department of Computer Science,
FernUniversität in Hagen, 58084 Hagen, Germany
`beierle@fernuni-hagen.de`

Abstract. While many functions on the real numbers are not exactly computable, the theory of exact real arithmetic investigates the computation of such functions with respect to any given precision. In this paper, we present an approach to implementing exact real arithmetic based on *Type-2 Theory of Effectivity* in the functional logic language Curry. It is demonstrated how the specific features of Curry can be used to obtain a high-level realisation that is close to the underlying theoretical concepts. The new Curry data type Real and its corresponding functions can easily be used in other function definitions.

1 Introduction

The set $\mathbb{R}$ of real numbers is not countable, and many functions and relations on $\mathbb{R}$ are not exactly computable [14]. In computer systems, the real numbers are usually approximated by a subset of the rational numbers. While for this representation highly efficient floating point arithmetic is available, in general the computed results may deviate from the correct values, and in unfavourable situations, e.g. rounding errors may accumulate to significantly corrupted results.

The subject of computable analysis deals with computing functions and relations on $\mathbb{R}$. Within exact real arithmetic, one can provide an arbitrary precision $\epsilon > 0$ such that the maximal deviation of the computed result from the correct result is guaranteed to be less than ϵ. While there are packages available providing exact real arithmetic in different programming languages, mostly in C and C++, but also others (see e.g. [2,4,8,9]), in this paper we present an approach to realise exact real arithmetic in the functional logic language Curry [1,5]. The main objective of this approach that has been implemented in [6] is to demonstrate how the specific features of Curry can conveniently be used to develop a high-level implementation that is rather close to the underlying theoretical concepts and notations as given by the Type-2 Theory of Effectivity (TTE) [14].

After briefly recalling the background of TTE as far as needed here (Sect. 2), we start with an abstract view of the new Curry data type Real (Sect. 3), describe some auxiliary types and functions (Sect. 4), and present the implementation

M. Hanus and R. Rocha (Eds.): KDPD 2013, LNAI 8439, pp. 48–64, 2014.
DOI: 10.1007/978-3-319-08909-6_4, © Springer International Publishing Switzerland 2014

of Real on the basis of quickly converging Cauchy sequences of rational numbers (Sect. 5). Various derived functions are introduced in Sect. 6, and functions for obtaining unique results and for deciding properties are given in Sect. 7. In Sect. 8, we conclude and point out further work.

2 Background and Related Work

For functions on $\mathbb{N}$ (or on *finite* words), there are well-established concepts of effectively computable functions, e.g. Turing machines. While there are different concepts, basically, they are all equivalent. On the other hand, for functions on $\mathbb{R}$ (or on *infinite* words), there are also different approaches to computable analysis, but these approaches are not all equivalent. There are differences in content and in technical details [14]. In this paper, we will focus on exact real arithmetic based on *Type-2 Theory of Effectivity (TTE)* [14].

In traditional (Type-1) Computability Theory, functions $f : \Sigma^* \to \Sigma^*$, possibly being partial, over *finite* words are considered, and a computable function is given by a Turing machine. Computability on other sets S, e.g. on rational numbers, arrays, or trees, can be defined by using words as names for elements of S and by interpreting words computed by a Turing machine as elements of S.

This notion of computability can not be applied to functions on real numbers since real numbers can not be represented by finite words. For instance, $\sqrt{2}$ has only an infinite decimal representation. Type-2 Theory of Effectivity extends Type-1 computability to *infinite* words by taking (possibly partial) functions $f : \Sigma^\omega \to \Sigma^\omega$ over infinite words into account. A computable function is then given by a *Type-2 machine* [14] transforming infinite sequences to infinite sequences.

Such a Type-2 Machine is a Turing machine M with k *one-way*, read-only input tapes, finitely many, two-way work tapes, and a single *one-way*, write-only output tape. The function f_M computed by M is specified by the following two cases, where $y_1, \ldots, y_k \in \Sigma^* \cup \Sigma^\omega$ are the words on the k input tapes [14]:

Case 1:
$$f_M(y_1, \ldots, y_k) = y_0 \in \Sigma^*$$

iff M *halts* on input $y_1, \ldots, y_k$ with y_0 on the output tape
Case 2:
$$f_M(y_1, \ldots, y_k) = y_0 \in \Sigma^\omega$$

iff M *computes forever* on input $y_1, \ldots, y_k$ and *writes* y_0 on the output tape.

In TTE, for $Y_i \in \{\Sigma^*, \Sigma^\omega\}$ with $i \in \{0, \ldots, k\}$, a partial function

$$f : Y_1 \times \ldots \times Y_k \to Y_0$$

over finite and infinite words is computable iff there is a Type-2 machine M such that $f = f_M$. Note that $f_M(y_1, \ldots, y_k)$ is *undefined* if M computes forever, but writes only finitely many symbols on the output tape. Thus, whenever the result of $f_M(y_1, \ldots, y_k)$ is an infinite word, M must keep printing out the next symbol after a finite amount of time, and since the output tape is write-only, this symbol can not be erased afterwards.

Of course, infinite computations can not be finished in reality, but finite computations on *finite initial parts* of inputs producing *finite initial parts* of outputs can be realised up to any arbitrary precision. Increasing the precision of a computation then means to extend the initial part of the input that is used for the computation, i.e., to take more symbols from the input into account.

While the decimal representation is often used for real numbers, this representation can not be used for exact real arithmetic within TTE. To illustrate this, we show that addition on $\mathbb{R}$ can not be computed by a Type-2 machine M when using the decimal representation. Suppose we want to add the two given inputs $y_1 = 0.6666666666\ldots$ and $y_2 = 0.3333333333\ldots$. After reading finitely many input symbols, M must write either "0." or "1." on the output tape. However, this may be wrong, e.g., if the next input symbols on the input tapes are 8 and 3, or 6 and 2. Thus, there is *no* Type-2 machine computing addition on $\mathbb{R}$ and using decimal representation.

Instead of decimals, we can use Cauchy sequences for representing real numbers, and using this representation, addition on $\mathbb{R}$ can easily be defined by a Type-2 machine. Every real number $x \in \mathbb{R}$ is given by the limit value of some quickly converging Cauchy sequence $r_0, r_1, r_2, \ldots$ of rational numbers [12]. Such a Cauchy sequence can be seen as a function

$$f : \mathbb{N} \to \mathbb{Q} \quad \text{with} \quad \lim_{k \to \infty} f(k) = x$$

where the condition *quickly converging* ensures that

$$|f(k) - x| \leq 2^{-k} \tag{1}$$

for all $k \in \mathbb{N}$; note that the condition of quickly converging is essential here, and just using Cauchy sequences would not be sufficient [14]. If $r_0, r_1, r_2, \ldots$ and $r_0', r_1', r_2', \ldots$ are Cauchy sequence representing x and x', respectively, then

$$r_1 + r_1', \ r_2 + r_2', \ r_3 + r_3', \ \ldots \tag{2}$$

represents $x + x'$. Note that the first component from the original sequences has been dropped in (2). Similarly, other operations on $\mathbb{R}$ can be achieved by componentwise operations on the respective Cauchy sequences where the number of leading elements to be dropped, called the *look-ahead*, depends on the operation and on its arguments [14]. While the look-ahead for addition of real numbers represented by quickly converging Cauchy sequences is always 1, independently of the arguments to be added, the look-ahead for multiplication also depends on its arguments. Given the two Cauchy sequences above,

$$r_m \times r_m', \ r_{m+1} \times r_{m+1}', \ r_{m+2} \times r_{m+2}', \ \ldots$$

is a Cauchy sequence representing the product $x \times x'$ where the look-ahead m is determined as follows (cf. [14, Theorem 4.3.2]): It is the smallest natural number m such that

$$|r_0| + 2 \leq 2^{m-1} \quad \text{and} \quad |r_0'| + 2 \leq 2^{m-1} \tag{3}$$

which, by multiplying both sides by 2 and taking the multiplicative inverse, holds iff

$$\left(\frac{1}{2}\right)^m \le \frac{1}{2 \times (\max\{|r_0|, |r_0'|\} + 2)}. \tag{4}$$

For instance, the sequence $11, 10.5, 10, 25, 10.125, \ldots$ (i.e., $r_i = 10 + 2^{-i}$) is a representation of $x = 10$. The look-ahead m for the multiplication of $r_0, r_1, r_2, \ldots$ with itself is thus determined by $11 + 2 \le 2^{m-1}$, i.e. $m = 5$. Hence, the first element of the sequence representing the result of the multiplication is $r_5 \times r_5 = 100 + 20 \times 2^{-5} + 2^{-10}$. Note that for this element, $|r_5 \times r_5 - 100| \le 2^0$ holds as required by (1).

When comparing two real numbers, one has to take into account that the relations $=, \le, <$ between two real numbers are not exactly computable; similarly, many other relations and functions on the real numbers are not computable [12]. Therefore, in TTE the crucial notion of a *multi-function* is used. For instance, the equality on real numbers corresponds to a multi-function

$$eq : \mathbb{R} \times \mathbb{R} \rightrightarrows \{true, false\}$$

that may yield *true*, *false* or both or no result at all [12, 14].

In this paper, we will present a high-level implementation of exact real arithmetic based on TTE and using quickly converging Cauchy sequences of rational numbers as representation of real numbers. For realising exact real arithmetic, many other representations can be used. For example, each $x \in \mathbb{R}$ can be represented by a sequence of nested intervals (l, u) containing x with l, u being rational numbers (e.g. [3]), by continued fractions (e.g. [13]), or by a signed digit representation employing negative digits as used in e.g. [10]; for a comprehensive discussion of different representations and the relationships among them see [14]. Our choice of using Cauchy sequences of rational numbers for a high-level implementation is motivated by the observation that this approach is quite intuitive and close to the underlying concepts. For instance, it uses rational numbers at the core of the representation, functions on real numbers can often be defined via componentwise application of the corresponding function on rational numbers (as illustrated in the addition and multiplication examples given above), and for many real-valued functions there are well-established definitions via limits of sequences of function applications on rational numbers.

Functional programming languages are a natural choice for a high-level implementation of exact real arithmetic, and there are various approaches using Haskell (e.g. [7, 10]). Besides infinite sequences, also the concept of a multi-function is essential in the TTE framework; any function depending on the computation of a real value might be a multi-function. Therefore, when using Curry instead of Haskell, also the multi-functions of TTE can be modelled naturally, namely by Curry's non-determinism which is not available in Haskell. Simulating multi-functions in Haskell would require significant additional effort, and the resulting solution would be not as close to the abstract specification in the TTE framework as is the case for our Curry approach.

3   An Abstract View on the Data Type Real

We start with introducing a new data type Real. As the rational numbers are a proper subset of the real numbers, there is a function embedding rational numbers into reals

```
realq ::   Rat -> Real
```

where Rat is the type for rational numbers. For illustrating basic computation functions on real numbers, we will consider for instance

```
add    ::  Real -> Real -> Real
neg    ::  Real -> Real
mul    ::  Real -> Real -> Real
dvd    ::  Real -> Real -> Real
power ::   Int -> Real ->Real
exp    ::  Real -> Real
```

realising addition, additive inverse, multiplication, division, power, and the exponential function. Further prominent examples of functions on real numbers that we have implemented are the transcendental functions like logarithm and the trigonometric functions.

The concept of a multi-function can be modelled nicely by Curry's notion of a non-deterministic function. The crucial requirement is that for any arbitrary prescribed precision that can be given as an additional parameter, the correct result is among the returned results. In order to add a precision parameter when modelling e.g. TTE's multi-function $eq : \mathbb{R} \times \mathbb{R} \rightrightarrows \{true, false\}$, we introduce a new data type Fuzzybool taking a rational number as precision parameter into account:

```
data Fuzzybool = Fuzzy (Rat -> Bool)
```

The equality relation on real numbers is then modelled by

```
eq ::  Real -> Real -> Fuzzybool
```

where for eq x y = Fuzzy f the function f is a non-deterministic function mapping rational numbers to Booleans. Evaluating eq x y with respect to precision r is done by the non-deterministic function:

```
defuzzy ::  Rat -> Fuzzybool -> Bool
defuzzy  r  (Fuzzy  f) = f  r
```

It will always be guaranteed that for any r, the correct result is among the results returned by defuzzy r (eq x y); thus, if defuzzy r (eq x y) returns a unique result, then this result is the unique correct result. Furthermore, if x and y are Cauchy sequences representing different real numbers $\tilde{x}$ and $\tilde{y}$, then there is a precision $r \in \mathbb{Q}$ with $0 < r < |\tilde{x} - \tilde{y}|$ such that defuzzy r (eq x y) returns the unique result *false*. If x and y represent the same real number, then defuzzy r (eq x y) may return both *true* and *false* for any precision r > 0. Note that this does not mean that two reals can be both equal and unequal at the same time; it is just not possible to refine defuzzy to a deterministic function since at least in the TTE framework, equality on $\mathbb{R}$ is undecidable, not just when using Cauchy sequences, but for any representation [14, Theorem 4.1.16]. Similarly, the functions

```
le       :: Real -> Real -> Fuzzybool
leq      :: Real -> Real -> Fuzzybool
isPositive :: Real -> Fuzzybool
isZero   :: Real -> Fuzzybool
```

realise the predicates *less*, *less or equal*, *is a positive number*, and *is zero* on $\mathbb{R}$ that are also not exactly computable.

Before developing a concrete Curry representation of Real and defining functions and predicates on Real, we first sketch some auxiliary types and functions.

4 Auxiliary Types and Functions

Module Structure. In our implementation of Real, we will reuse function names for different types, especially for arithmetic operations. For instance, there will be functions add, mul, le, leq, etc. on rational and real numbers. In order to handle this overloading of names, we use the following imports in the implementation and also in this paper since we want to present the actual Curry source code (which can be found at www.fernuni-hagen.de/wbs/data/realcurry.tar.gz):

```
import qualified rationals as q
import intervals as i
import fuzzybool
import qualified utils as u
```

Integers are as in Curry. The module utils provides some utility functions; those that are used in this paper will be explained when they are referenced. The other three auxiliary modules for rational numbers, fuzzybool, and intervals are presented in the following.

Rational Numbers. We introduce rational numbers as quotients of integers such that the denominator is positive:

```
data Rat = Rat Int Int

ratn :: Int -> Rat
ratn n = Rat n 1

ratf :: Int -> Int -> Rat
ratf n d | d > 0 = Rat n d
         | d < 0 = Rat (-n) (-d)
```

Addition, subtraction, multiplication, division, and additive and multiplicative inverse are given by the functions add, sub, mul, dvd, neg, and inv defined as expected, e.g.:

```
mul :: Rat -> Rat -> Rat
mul (Rat n1 d1) (Rat n2 d2) = Rat (n1*n2) (d1*d2)

inv :: Rat -> Rat
inv (Rat n d) | n /= 0 = ratf d n
```

Comparing rational numbers is achieved by the functions:

```
eq  :: Rat -> Rat -> Bool
le  :: Rat -> Rat -> Bool
leq :: Rat -> Rat -> Bool
```

For computing the look-ahead (cf. Sect. 2), we will use the auxiliary function

```
minexp :: Rat -> Rat -> Int
minexp x b | leq (ratn 1) x = 0
           | otherwise = (minexp (dvd x b) b) + 1
```

The function call minexp x b returns the smallest natural number n such that $x \geq b^n$ holds. For instance, minexp (Rat 1 1000) (Rat 1 2) = 10 since $1/1000 \not\geq (1/2)^9$ and $1/1000 \geq (1/2)^{10}$.

Fuzzybool. The data type Fuzzybool and the function defuzzy have already been given in Sect. 3; additionally, we will need logical operators:

```
andf :: Fuzzybool -> Fuzzybool -> Fuzzybool
andf a b = Fuzzy (\r -> (defuzzy r a) && (defuzzy r b))

orf :: Fuzzybool -> Fuzzybool -> Fuzzybool
orf a b = Fuzzy (\r -> (defuzzy r a) || (defuzzy r b))

notf :: Fuzzybool -> Fuzzybool
notf a = Fuzzy (\r -> not (defuzzy r a))
```

Thus, these operators on Fuzzybool are defined to correspond to the usual definitions. While defuzzy r b might return true, false, or both Boolean values, the correct value is always among the returned results; note that for any precision r, this property is preserved by the logical operations on Fuzzybool, e.g., if b is a conjunction like andf b1 b2.

Intervals. Intervals of rational numbers are used for checking the required precision when computing and comparing real numbers and thus provide a basis for realising not exactly computable functions. An interval is valid only if its lower bound is less or equal to its upper bound.

```
data Interval = Interval q.Rat q.Rat

isValid :: Interval -> Bool
isValid (Interval a b) = q.leq a b
```

The computation of relations on real numbers will be reduced to three non-deterministic functions on intervals. The function isZero yields *true* if 0 is in the interval, and it yields *false* if the interval contains a number that is not equal to 0. The function isPositive yields *true* if the given interval contains a positive number, and it yields *false* if the interval contains a number that is not positive:

```
isZero :: Interval -> Bool
isZero (Interval a b) | q.leq a (q.ratn 0) && q.leq (q.ratn 0)
    b = True
isZero (Interval a b) | q.le a (q.ratn 0)  || q.le (q.ratn 0)
    b = False

isPositive :: Interval -> Bool
isPositive (Interval _ b) | q.le (q.ratn 0) b = True
isPositive (Interval a _) | q.leq a (q.ratn 0) = False

isNegative :: Interval -> Bool
isNegative (Interval a _) | q.le a (q.ratn 0) = True
isNegative (Interval _ b) | q.leq (q.ratn 0) b = False
```

5 Representing Real Numbers in Curry

The approach of representing real numbers as Cauchy sequences can immediately be realised in Curry by defining

```
data Real :: Cauchy (Int -> q.Rat)
```

and the embedding of Rat into Real is then given by:

```
realq :: q.Rat -> Real
realq a = (Cauchy (\_ -> a))
```

For addition, we employ componentwise operation, observing that the look-ahead is 1 (cf. Sect. 2); subtraction and additive inverse are also easily defined:

```
add :: Real -> Real -> Real
add a b = Cauchy(\k -> let m=k+1 in q.add (get a m) (get b m))

sub :: Real -> Real -> Real
sub a b = add a (neg b)

neg :: Real -> Real
neg a = Cauchy(\k -> q.neg (get a k))

get :: Real -> Int -> q.Rat
get (Cauchy x) k = x k
```

Using the auxiliary function minexp (Sect. 4), the computation of the look-ahead m for multiplication in Curry mirrors exactly the condition given by (4):

```
mul :: Real -> Real -> Real
mul a b = let x = q.max (q.abs (get a 0)) (q.abs (get b 0))
          in let m = lahmul x
                in Cauchy(\k -> q.mul (get a (k+m)) (get b (k+m)))

lahmul :: q.Rat -> Int
lahmul arg = q.minexp (q.dvd (q.ratn 1)(q.mul (q.ratn 2)
      (q.add arg (q.ratn 2)))) (q.ratf 1 2)
```

When computing the multiplicative inverse of a real number, we can use
the function for the multiplicative inverse on rational numbers, but there is
a particular subtlety to be taken into account. In the function definition for
rational numbers, we can easily check that $x \neq 0$ for an argument x, and leave
the function application undefined for $x = 0$; cf. the definition of q.inv in Sect. 4.
However, for a real value $x \in \mathbb{R}$, the relation $x \neq 0$ is not exactly computable.
Before applying q.inv to the elements of a Cauchy sequence representing x, we
therefore have to determine an appropriate look-ahead m:

```
inv :: Real -> Real
inv arg =
  let n = snd (until
                   (\x -> q.leq (fst x) (q.abs (get arg (snd x))))
                   (\x -> (q.dvd (fst x) (q.ratn 2),(snd x)+1))
                   ((q.ratn 3),0)
               )
  in let m=2*n
      in Cauchy(\k-> q.inv (get arg (k+m)))
```

A call until f1 f2 z of the function until provided by Curry repeatedly applies f2
to z (maybe zero times) until f1 applied to the obtained result yields true. Thus,
if inv is applied to $r_0, r_2, r_2, \ldots$, then n determined in inv is the smallest natural
number n such that $3 \times 2^{-n} \leq |r_n|$, and the look-ahead for inv is $m = 2n$ (cf. [14,
Theorem 4.3.2]). Note that inv is a partial function: If for $x \in \mathbb{R}$ represented by
$r_0, r_2, r_2, \ldots$ the relationship $x \neq 0$ holds, then the condition in the call of until
in inv will eventually be true, causing the call of until as well as the call of inv to
terminate and to yield the correct result. If $x = 0$ holds, then the call of inv will
not terminate since equality on reals is undecidable [14].

Using inv, it is now easy to define division on Real since $a/b = ab^{-1}$ holds:

```
dvd :: Real -> Real -> Real
dvd a b = mul a (inv b)
```

The basic relations on $\mathbb{R}$ discussed in Sect. 3 are implemented by reducing
them to the question of checking whether a number is 0 or positive:

```
eq :: Real -> Real -> Fuzzybool
eq x y = isZero (sub y x)

le :: Real -> Real -> Fuzzybool
le x y = isPositive (sub y x)
```

```
leq :: Real -> Real -> Fuzzybool
leq x y = (notf . isPositive) (sub x y)
```

The two functions isZero and isPositive are in turn reduced to the corresponding functions on intervals. To do so, the resulting object of type Fuzzybool contains a function that for any precision uses an interval realising this precision with respect to the given x of type Real.

```
isPositive :: Real -> Fuzzybool
isPositive x = Fuzzy(\r -> i.isPositive(toInterval r x))

isZero :: Real -> Fuzzybool
isZero x = Fuzzy(\r -> i.isZero(toInterval r x))

isNegative :: Real -> Fuzzybool
isNegative x = Fuzzy(\r -> i.isNegative (toInterval r x))
```

Given numbers r of type Rat and x of type Real, the auxiliary function toInterval determines an interval containing the real number in $\mathbb{R}$ represented by x and approximating that number with precision r.

```
toInterval :: q.Rat -> Real -> Interval
toInterval r x = let r2 = q.mul (q.ratf 1 2) r
                 in let y = approx r2 x
                    in Interval (q.sub y r2) (q.add y r2)

approx :: q.Rat -> Real -> q.Rat
approx r x = get x (prec r)

prec :: q.Rat -> Int
prec r | q.le (q.ratn 0) r = q.minexp r (q.ratf 1 2)
```

The approximation given by approx determines the smallest natural number n such that $(1/2)^n \leq r$ holds in order to determine the position in the Cauchy sequence to be used for obtaining the bounds of the interval.

6 Derived Functions

In the previous sections, we presented the complete Curry code for a core set of types and functions providing an implementation of exact real arithmetic where the data type Real provides a representation for the real numbers in the sense that any given precision can be obtained. On the basis of this core set of functions, in [6] a series of further functions are defined and realised in Curry, e.g. a general root function or the trigonometric functions. Here, we will give some examples of how derived functions on Real can be defined in Curry.

Consider the *sign* function on $\mathbb{R}$

$$sign(x) = \begin{cases} 1 & \text{if } x > 0 \\ 0 & \text{if } x = 0 \\ -1 & \text{if } x < 0 \end{cases}$$

which is not exactly computable. Thus, *sign* corresponds to a multi-function that is realised by a non-deterministic function in Curry. Since we want to determine *sign* with a given precision, the Curry function sgn gets this precision value as an additional parameter:

```
sgn :: q.Rat -> Real -> Int
sgn r x | defuzzy r (isPositive x) == True = 1
sgn r x | defuzzy r (isZero x) == True = 0
sgn r x | defuzzy r (isNegative x) == True = -1
```

As an example for constructing an irrational number, we use the well-known Newton's method for computing $\sqrt{2}$. The sequence $x_0, x_1, x_2, \ldots$

$$x_{k+1} = \frac{1}{2}\left(x_k + \frac{2}{x_k}\right) \tag{5}$$

with $x_0 > 0$ has the limit $\lim_{k\to\infty} x_k = \sqrt{2}$. Analysing the sequence $e_0, e_1, e_2, \ldots$ with $e_k = |x_k - \sqrt{2}|$ yields $e_{k+1} = |\frac{1}{2}\frac{e_k^2}{x_k}|$ (e.g. [11, p. 5]). When choosing the start value $x_0 = 2$, induction on k shows that $e_k \leq 2^{-k}$ so that $x_0, x_1, x_2, \ldots$ is a quickly converging Cauchy sequence. Using Real, this method of computing $\sqrt{2}$ can be implemented in Curry in an iterative manner by the function

```
sqrt2Newton :: Real
sqrt2Newton = Cauchy(\k -> sqrt2Newtonsub (q.ratn 2) k)

sqrt2Newtonsub :: q.Rat -> Int -> q.Rat
sqrt2Newtonsub x k =
   if k==0 then x
           else sqrt2Newtonsub (q.mul (q.ratf 1 2) (q.add x
   (q.dvd (q.ratn 2) x))) (k-1)
```

Note that the typical use of sqrt2Newton is not the enumeration of $x_0, x_1, x_2, \ldots$, but rather to compute $\sqrt{2}$ with a given precision, say 2^{-k}. For this, we can use sqrt2Newton to compute x_k efficiently from x_0 by k applications of the recurrence equation (5). However, with increasing k, the precision given by x_k is much higher than the precision 2^{-k} as required by (1) [11]. In order to reduce the unnecessarily high precision and thus unnecessarily large integers in the underlying representation of rationals, we can alternatively use (5) to generate a quickly converging Cauchy sequence $x_0', x_1', x_2', \ldots$ where for x_k' we apply the recurrence equation (5) only so many times that we can ensure that x_k' satisfies the condition given in (1). For instance, since the Banach fixed-point theorem tells us that $|x_k - \sqrt{2}| \leq |x_k - x_{k-1}|$, it suffices to ensure $|\frac{1}{2}(x_k' + \frac{2}{x_k'}) - x_k'| \leq 2^{-k}$, yielding:

```
sqrt2 :: Real
sqrt2 = Cauchy (\k -> if k == 0 then (q.ratn 2) else
          sqrt2sub (q.ratn 2) (q.power k (q.ratf 1 2)))

sqrt2sub :: q.Rat -> q.Rat -> q.Rat
sqrt2sub x r =
```

```
let x2 = q.mul (q.ratf 1 2) (q.add x (q.dvd (q.ratn 2) x))
in if q.leq (q.abs (q.sub x x2)) r then x2
   else sqrt2sub x2 r
```

For an element x of type Real, let $\tilde{x} \in \mathbb{R}$ denote its intended real value, i.e. the limit of the underlying Cauchy sequence. Then for any precision r, x correctly represents $\tilde{x}$ in the sense that the interval determined by toInterval x r contains $\tilde{x}$. We will demonstrate this behaviour of our Curry implementation using sqrt2. For printing a result, we use the function

```
dec :: Real -> Int -> String
```

that takes an element x of type Real and a natural number k. It returns the value of x as a string containing k decimal places, such that the returned decimal representation is correct with precision 10^{-k}, but without any rounding being applied. For instance, taking sqrt2 and k = 20 or k = 100 we get:

```
real> dec sqrt2 20
Result: "1.41421356237309504880"
More solutions? [Y(es)/n(o)/a(ll)]
No more solutions.

real> dec sqrt2 100
Result: "1.41421356237309504880168872420969807856967187537694 8
    07317667973799073247846210703885038753432764157 27"
More solutions? [Y(es)/n(o)/a(ll)]
No more solutions.
```

The decimal representation obtained by dec for the multiplication of sqrt2 with itself and for k = 100 yields

```
real> dec (mul sqrt2 sqrt2) 100
Result: "1.9999999999999999999999999999999999999999999999999 9
    99999999999999999999999999999999999999999999999 9"
More solutions? [Y(es)/n(o)/a(ll)]
Result: "2.0000000000000000000000000000000000000000000000000 0
    00000000000000000000000000000000000000000000000 0"
More solutions? [Y(es)/n(o)/a(ll)]
No more solutions.
```

which are finite prefixes of the exactly two different infinite decimal representations of the real value 2, illustrating that also dec is a multi-function.

Similarly as for addition or multiplication, computing the nth power x^n of a real number x and an integer n can be done by componentwise computations on the representing Cauchy sequence. For negative exponents, the equality $x^{-n} = (1/x)^n$ is used, and for $n > 1$, the required look-ahead is $(n-1)$-times the look-ahead for multiplication of x with itself:

```
power :: Int -> Real -> Real
power n arg | n==0 = realq (q.ratn 1)
            | n==1 = arg
            | n<0 = power (-n) (inv arg)
            | n>1 = let lah = (n-1)*lahmul (q.abs (get arg 0))
                    in Cauchy (\k-> q.power n (get arg (lah+k)))
```

In many cases, real numbers or functions on real numbers are defined using a power series. For instance, for the exponential function we have

$$e^x = \sum_{k=0}^{\infty} \frac{1}{k!} x^k. \tag{6}$$

In general, a function given by a power series

$$\sum_{k=0}^{\infty} a_k x^k \tag{7}$$

is well-defined for any $|x| < R_K$ where $R_K = (\limsup_{n \to \infty} \sqrt[n]{|a_n|})^{-1}$ is the radius of convergence of (7). The computation of (7) for $x \in \mathbb{R}$ can be approximated by a finite sum $a_0 y^0 + \ldots a_N y^N$ with $y \in \mathbb{Q}$. The error of this approximation depends on the coefficients a_k, on the number of summands N, and on $|x - y|$. In order to eliminate the dependence on the a_k, we will require $|a_k| \leq 1$ for all $k \geq 0$. This restriction is satisfied for many power series expressions defining functions like exp, sin, cos, etc., and since it implies $R_K \geq 1$, for any $r \in \mathbb{Q}$ with $0 < r < 1$, (7) converges for all $x \in \mathbb{R}$ with $|x| \leq r$. If a is a sequence $(a_k)_{k \geq 0}$ of rational numbers with $|a_k| \leq 1$ and r is a rational number with $0 < r < 1$, then a call powerser a r x of the auxiliary function

```
powerser :: (Int -> q.Rat) -> q.Rat -> Real -> Real
```

computes (7) for $\tilde{x}$ if x represents $\tilde{x} \in \mathbb{R}$ and $|\tilde{x}| \leq r$. The argument r is used in the estimation of the error of the approximation and in the determination of the number of summands needed in order to meet condition (1) (cf. [14, Chap. 4.3], [7]); a smaller r means fewer summands, but also a smaller function domain. Thus, by choosing r = 1/2, the function

```
exp1 :: Real -> Real
exp1 = powerser (\k -> q.ratf 1 (u.fact k)) (q.ratf 1 2)
```

can be used for computing the exponential function (6) for any values $|x| \leq 1/2$. A straightforward method to extend the computation of e^x to values $|x| > 1/2$ which is also used in e.g. [7] is to exploit the property

$$e^x = \left(e^{\frac{x}{m}}\right)^m \tag{8}$$

and to choose m to be an integer such that $|x/m| \leq 1/2$, allowing us to use exp1 for the inner and power for the outer exponentiation in (8):

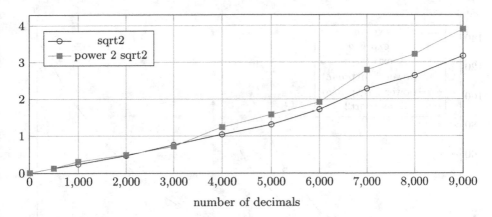

Fig. 1. Execution times (given in sec) for computing k decimals of $\sqrt{2}$ and $(\sqrt{2})^2$ with dec sqrt2 k and dec (power 2 sqrt2) k

```
exp arg = let m=2*(estimate (abs arg))+1
  in power m (expl (mul (realq (q.ratf 1 m)) arg))

estimate :: Real -> Int
estimate arg = q.floor (q.add (get arg 2) (q.ratf 1 2))

abs :: Real -> Real
abs arg = Cauchy (\k -> q.abs (get arg k))
```

For a rational number x, the auxiliary function floor returns the largest integer less or equal to x.

Although in our Curry implementation we neglected any efficiency issues, in Figs. 1 and 2 we present runtimes for evaluating various function calls. Figure 1 suggests that sqrt2 can be used for computing up to several thousands decimals of $\sqrt{2}$ and $(\sqrt{2})^2$ in several seconds, while computing real numbers using exp is more expensive (Fig. 2). When using sqrt2Newton instead of sqrt2, the respective computation times are much higher. However, as shown in [14, Chapter 7], it is not reasonable to use Cauchy representations for complexity investigations; instead, other representations, e.g. signed digits, should be used.

7   Decision Functions and Obtaining Unique Results

Sometimes, one might be interested in computing a function or relation on real numbers not just with a given precision, but one would like to get a *unique* result. In order to try to achieve this, one could start with some precision value and sharpen it until the given function yields a unique result. The function decide realises this approach.

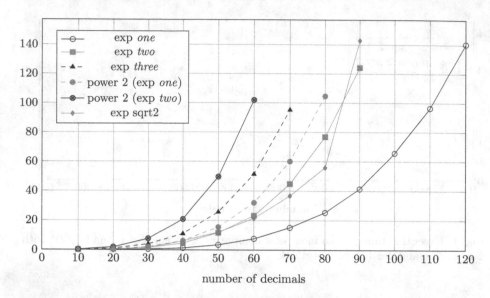

Fig. 2. Execution times (given in sec) for computing k decimals with dec fc k for different function calls fc; in the legend above, *one* stands for (realq (q.ratn 1)), *two* for (realq (q.ratn 2)), and *three* for (realq (q.ratn 3))

```
decide :: (q.Rat -> a -> b) -> a -> b
decide f x = let fkt r | not (null res) =
                            if u.eqset [head res] res
                            then head res
                            else fkt (q.mul (q.ratf 1 2 ) r)
                        where res = u.results2 f r x
             in fkt (q.ratn 1)
```

A call results2 f r x of the utility function results2 collects all results of the in general non-deterministic computation f r x and returns them in a list, and the utility function eqset yields *true* if the two lists given as arguments contain the same set of elements.

Of course, there is a pitfall in using decide. While it is guaranteed that the result returned by decide f x is the correct and unique value of f x, the computation of decide f x may not terminate.

Analogously, one might not be satisfied with obtaining a Boolean value just with a given precision, but one wants to have a certification that the given precision is sufficient to guarantee that no other result is possible. The function sure can ensure this:

```
sure :: q.Rat -> Fuzzybool -> Bool
sure p b = u.eqset (u.results2 defuzzy p b) [True]
```

If sure p b yields *true*, then the corresponding property is guaranteed to hold. If sure p b yields *false*, it may be the case that the property does not hold, but it may also be the case that the property holds, but the given precision p is not sufficient to determine this. Likewise, the function possible tries to determine whether it is possible that a property holds with respect to a given precision, and the function impossible refutes a property regarding a given precision:

```
possible    :: q.Rat -> Fuzzybool -> Bool
impossible :: q.Rat -> Fuzzybool -> Bool
```

Using decide, we can easily define a function for proving or refuting a property where no precision has to be given:

```
proof :: Fuzzybool -> Bool
proof = decide defuzzy
```

Again, due to the underlying characteristics of exact real arithmetic, the execution of proof b may not terminate. However, if proof b terminates, its result is guaranteed to be both unique and correct.

8 Conclusions and Further Work

Using the specific features of Curry, we presented the core types and functions needed for a high-level implementation of exact real arithmetic based on TTE [14] in Curry. In this approach, the non-deterministic functions of Curry are crucial for realising the concept of multi-functions required in real arithmetic as developed in TTE as even a standard test like $x \leq y$ is not exactly computable if x and y are real numbers.

While the use of Curry's non-determinism is essential for our approach, the representation of real numbers does not allow the instantiation of unknown variables of type Real, although this would be an interesting feature.

The new Curry type Real can easily be used in other function definitions. By taking the presented abstract view of Real, the programmer is freed from the possible burden of having to deal with the details of the underlying Real implementation. If a particular computation involves a not exactly computable function, Curry might yield more than one result, but in any case, the correct result will be among the obtained results. For a not exactly computable function, a precision argument can be given so that the correct result can be computed up to any given precision. Since the objective of the approach presented here was to achieve a high-level, declarative solution being as close as possible to the respective theoretical concepts, we did not consider efficiency here; in future work, we plan to address this issue.

Acknowledgements. We would like to thank the anonymous reviewers of this article for their detailed and helpful comments.

References

1. Antoy, S., Hanus, M.: Functional logic programming. Commun. ACM **53**(4), 74–85 (2010)
2. Briggs, K.: Implementing exact real arithmetic in python, C++ and C. Theor. Comput. Sci. **351**(1), 74–81 (2006)
3. Escardó, M.H.: PCF extended with real numbers. Theor. Comput. Sci. **162**(1), 79–115 (1996)
4. Gowland, P., Lester, D.R.: A survey of exact arithmetic implementations. In: Blank, J., Brattka, V., Hertling, P. (eds.) CCA 2000. LNCS, vol. 2064, pp. 30–47. Springer, Heidelberg (2001)
5. Hanus, M.: Functional logic programming: from theory to Curry. In: Voronkov, A., Weidenbach, C. (eds.) Programming Logics - Essays in Memory of Harald Ganzinger. LNCS, vol. 7797, pp. 123–168. Springer, Heidelberg (2013)
6. Lelitko, U.: Realisation of exact real arithmetic in the functional logic programming language Curry. Diploma thesis, Department of Computer Science, FernUniversität in Hagen, Germany (2013) (in German)
7. Lester, D.R., Gowland, P.: Using PVS to validate the algorithms of an exact arithmetic. Theor. Comput. Sci. **291**(2), 203–218 (2003)
8. Marcial-Romero, J.R., Escardó, M.H.: Semantics of a sequential language for exact real-number computation. Theor. Comput. Sci. **379**(1-2), 120–141 (2007)
9. Müller, N.T.: The iRRAM: exact arithmetic in C++. In: Blank, J., Brattka, V., Hertling, P. (eds.) CCA 2000. LNCS, vol. 2064, pp. 222–252. Springer, Heidelberg (2001)
10. Plume, D.B.: A calculator for exact real number computation. University of Edinburgh (1998). http://www.dcs.ed.ac.uk/home/mhe/plume/
11. Scott, L.R.: Numerical Analysis. Princeton Univ. Press, Princeton (2011)
12. Tavana, N., Weihrauch, K.: Turing machines on represented sets a model of computation for analysis. Log. Methods Comput. Sci. **7**(2), 1–21 (2011)
13. Vuillemin, J.: Exact real computer arithmetic with continued fractions. IEEE Trans. Comput. **39**(8), 1087–1105 (1990)
14. Weihrauch, K.: Computable Analysis. Springer, Berlin (2000)

On Axiomatic Rejection
for the Description Logic $\mathcal{ALC}$

Gerald Berger and Hans Tompits[✉]

Institut Für Informationssysteme 184/3, Technische Universität Wien,
Favoritenstraße 9-11, 1040 Vienna, Austria
{berger,tompits}@kr.tuwien.ac.at

Abstract. Traditional proof calculi are mainly studied for formalising the notion of *valid inference*, i.e., they axiomatise the valid sentences of a logic. In contrast, the notion of *invalid inference* received less attention. Logical calculi which axiomatise invalid sentences are commonly referred to as *complementary calculi* or *rejection systems*. Such calculi provide a proof-theoretic account for deriving non-theorems from other non-theorems and are applied, in particular, for specifying proof systems for nonmonotonic logics. In this paper, we present a sound and complete sequent-type rejection system which axiomatises *concept non-subsumption* for the description logic $\mathcal{ALC}$. Description logics are well-known knowledge-representation languages formalising ontological reasoning and provide the logical underpinning for semantic-web reasoning. We also discuss the relation of our calculus to a well-known tableau procedure for $\mathcal{ALC}$. Although usually tableau calculi are syntactic variants of standard sequent-type systems, for $\mathcal{ALC}$ it turns out that tableaux are rather syntactic counterparts of complementary sequent-type systems. As a consequence, counter models for witnessing concept non-subsumption can easily be obtained from a rejection proof. Finally, by the well-known relationship between $\mathcal{ALC}$ and multi-modal logic **K**, we also obtain a complementary sequent-type system for the latter logic, generalising a similar calculus for standard **K** as introduced by Goranko.

1 Introduction and Overview

Research on proof theory is usually guided by the semantic concept of *validity*, finding appropriate (i.e., sound and complete) proof calculi for various types of logics. This is reasonable insofar as logical methods have been devised since their very beginning for characterising the valid sentences by virtue of their form rather than their semantic denotations. However, the complementary notion of validity, that is, *invalidity*, has rarely been studied by syntactic means. From a proof-theoretic point of view, the invalidity of sentences is largely established by the exhaustive search for counter models.

Proof systems which axiomatise the invalid sentences of a logic are commonly coined under the terms *complementary calculi* or *rejection systems*. Such calculi formalise *proofs for invalidity*, i.e., with the existence of a *sound* and *complete*

M. Hanus and R. Rocha (Eds.): KDPD 2013, LNAI 8439, pp. 65–82, 2014.
DOI: 10.1007/978-3-319-08909-6_5, © Springer International Publishing Switzerland 2014

rejection system of a logic under consideration, one is able to check for the invalidity of a sentence by syntactic deduction. Another way to characterise this notion is that a proof in such a complementary calculus witnesses the non-existence of a proof in a corresponding (sound and complete) assertional proof system.

To the best of our knowledge, a first systematic theory of rejection was established by Jan Łukasiewicz in his work on describing Aristotelian syllogistic in terms of modern logic [1]. Indeed, already Aristotle, the forefather of modern logic, recognised that showing the invalidity of a syllogistic form is not only possible by providing a counterexample, but also by employing some form of axiomatic reasoning. This notion was put into formal, axiomatic terms by Łukasiewicz.

Up to now, rejection systems have been studied for different families of logics including classical logic [2,3], intuitionistic logic [4,5], modal logics [6,7], and many-valued logics [8] (for an overview, cf., e.g., Wybraniec-Skardowska [9] and Caferra and Peltier [10]). Many of them are analytic sequent-type systems, which proved fruitful in axiomatising invalidity without explicitly referring to validity. In contrast, the fundamental rule of rejection in the system of Łukasiewicz, stating that

$$\text{if } \varphi \supset \psi \text{ is asserted and } \psi \text{ is rejected, then } \varphi \text{ is rejected,}$$

makes reference to an accepting proof system too.

Besides a general proof-theoretic interest in such calculi, they received also attention in research on proof theory for nonmonotonic logics. In particular, Bonatti and Olivetti [11] employed complementary sequent-type systems when devising proof systems for default logic [12], autoepistemic logic [13], and propositional circumscription [14]. Furthermore, in logics in which the validity of a formula φ is tantamount to checking unsatisfiability of the negation of φ, a complementary calculus provides a proof-theoretic account of satisfiability checking as well.

In this paper, we deal with the issue of complementary calculi in the context of description logics. More specifically, we consider the description logic $\mathcal{ALC}$ and present a sound and complete sequent-type rejection system for axiomatising *concept non-subsumption* for this logic. Note that, informally speaking, $\mathcal{ALC}$ is the least expressive of the so-called *expressive* description logics (for more details on the vast topic of description logics, we refer the reader to an overview article by Baader *et al.* [15]).

Concerning previous work on sequent-type calculi for description logics, we mention an axiomatisation of concept subsumption for different description logics, including $\mathcal{ALC}$, by Rademaker [16] and an earlier calculus for $\mathcal{ALC}$ by Borgida *et al.* [17].

As pointed out above, in our approach, we study an axiomatisation of concept *non-subsumption* for $\mathcal{ALC}$. We view this as a starting point for further investigations into complementary calculi for description logics as the more general case of dealing with reasoning from *knowledge bases*, which are usually the principal structures of description logics where reasoning operates on, would be a natural next step. In fact, our calculus is devised to axiomatise the invalidity of single *general concept inclusions* (GCIs) without reference to any knowledge base.

We also discuss the relation of our calculus to a well-known tableau procedure for $\mathcal{ALC}$ [18]. In general, as well known, sequent-type systems and tableau calculi are closely related—indeed, traditionally, tableau calculi are merely syntactic variants of standard sequent-type systems. However, popular tableau algorithms for description logics are formalised in order to axiomatise satisfiability rather than validity (cf. Baader and Sattler [18] for an overview). Hence, tableaux correspond in the case of description logics to *complementary* sequent systems. As a consequence, counter models for witnessing concept non-subsumption can easily be obtained from a rejection proof. We describe the relation of our calculus to the tableau algorithm for $\mathcal{ALC}$ as described by Baader and Sattler [18] in detail, and show how to construct a counter model from a proof in the complementary sequent system.

Finally, as also well-known, $\mathcal{ALC}$ can be translated into the multi-modal logic $\mathbf{K}_m$, which extends standard modal logic $\mathbf{K}$ by providing a countably infinite supply of modal operators of form $[\alpha]$, where α is a modality. In view of this correspondence, we obtain from our complementary calculus for $\mathcal{ALC}$ also a complementary sequent-type calculus for $\mathbf{K}_m$. This calculus generalises a similar one for modal logic $\mathbf{K}$ as introduced by Goranko [6]. In fact, Goranko's calculus served as a starting point for the development of our calculus for $\mathcal{ALC}$. We briefly discuss the complementary calculus for $\mathbf{K}_m$, thereby showing the relation of our calculus for $\mathcal{ALC}$ to Goranko's one for $\mathbf{K}$.

2 Notation and Basic Concepts

With respect to terminology and notation, we mainly follow Baader *et al.* [15].

Syntactically, $\mathcal{ALC}$ is formulated over countably infinite sets N_C, N_R, and N_O of *concept names*, *role names*, and *individual names*, respectively. The syntactic artefacts of $\mathcal{ALC}$ are *concepts*, which are inductively defined using the *concept constructors* $\sqcap$ ("concept intersection"), $\sqcup$ ("concept union"), $\neg$ ("concept negation"), $\forall$ ("value restriction"), $\exists$ ("existential restriction"), as well as the concepts $\top$ and $\bot$ as usual. For the sake of brevity, we agree upon omitting parentheses whenever possible and assign $\neg$, $\forall$, and $\exists$ the highest rank, and $\sqcap$ and $\sqcup$ the least binding priority. We use $C, D, \dots$ as metavariables for concepts and $p, q, r, \dots$ as metavariables for role names. When we consider concrete examples, we assume different metavariables to stand for distinct syntactic objects.

By an *interpretation* we understand an ordered pair $\mathcal{I} = \langle \Delta^{\mathcal{I}}, \cdot^{\mathcal{I}} \rangle$, where $\Delta^{\mathcal{I}}$ is a non-empty set called *domain* and $\cdot^{\mathcal{I}}$ is a function assigning each concept name $C \in \mathsf{N}_C$ a set $C^{\mathcal{I}} \subseteq \Delta^{\mathcal{I}}$, each role name $r \in \mathsf{N}_R$ a set $r^{\mathcal{I}} \subseteq \Delta^{\mathcal{I}} \times \Delta^{\mathcal{I}}$, and each individual name $a \in \mathsf{N}_O$ an element $a^{\mathcal{I}} \in \Delta^{\mathcal{I}}$. The function $\cdot^{\mathcal{I}}$ is furthermore required to obey the semantics of the concept constructors in the usual way. For a concept C and an interpretation $\mathcal{I}$, $C^{\mathcal{I}}$ is the *extension* of C under $\mathcal{I}$. A concept C is *satisfiable* if there exists an interpretation $\mathcal{I}$ such that $C^{\mathcal{I}} \neq \emptyset$, and *unsatisfiable* otherwise.

A *general concept inclusion* (GCI) is an expression of form $C \sqsubseteq D$, where C and D are arbitrary concepts. An interpretation $\mathcal{I}$ *satisfies* a GCI iff $C^{\mathcal{I}} \subseteq D^{\mathcal{I}}$, and *falsifies* it otherwise. In the former case, $\mathcal{I}$ is a *model* of $C \sqsubseteq D$, while in the latter case, $\mathcal{I}$ is a *counter model* of $C \sqsubseteq D$. The GCI $C \sqsubseteq D$ is *valid* if every interpretation satisfies it. In this case, we say that D *subsumes* C.

Example 1. Let hasChild be a role name, and Doctor and Professor concept names. Then,

$$\exists \mathsf{hasChild}.(\mathsf{Doctor} \sqcap \mathsf{Professor}) \sqsubseteq \exists \mathsf{hasChild}.\mathsf{Doctor} \sqcap \exists \mathsf{hasChild}.\mathsf{Professor}$$

is a valid GCI, while

$$\exists \mathsf{hasChild}.\mathsf{Doctor} \sqcap \exists \mathsf{hasChild}.\mathsf{Professor} \sqsubseteq \exists \mathsf{hasChild}.(\mathsf{Doctor} \sqcap \mathsf{Professor})$$

is not valid.   □

When considering proof systems, it is convenient and necessary to focus on interpretations of a special form when deciding semantic properties of language constructs. A *tree-shaped interpretation* is an interpretation $\mathcal{I}$ such that the set

$$\delta(\mathcal{I}) := \{(v, w) \mid (v, w) \in \bigcup_{r \in \mathsf{N}_R} r^{\mathcal{I}}\}$$

of ordered tuples forms a tree. If the tree is finite, then $\mathcal{I}$ is a *finite* tree-shaped interpretation. The *root* of a tree-shaped interpretation $\mathcal{I}$ is defined to be the root of $\delta(\mathcal{I})$ and the *length* of $\mathcal{I}$ is the length of the longest path in $\delta(\mathcal{I})$. The following property of $\mathcal{ALC}$ is well-known [18].

Proposition 2. *A concept C is satisfiable iff there exists a finite tree-shaped interpretation $\mathcal{T}$ such that $v_0 \in C^{\mathcal{T}}$, where v_0 is the root of $\mathcal{T}$.*

When we say that a tree-shaped interpretation $\mathcal{I}$ *satisfies* a concept, then we mean that the root of $\mathcal{I}$ is in the extension of the concept. Let $\mathcal{T}$ and $\mathcal{T}'$ be tree-shaped interpretations with roots v and v', respectively. Then, $\mathcal{T}'$ is an *r-subtree* of $\mathcal{T}$ (symbolically, $\mathcal{T}' \lhd_r \mathcal{T}$) if $(v, v') \in r^{\mathcal{T}}$ and $\delta(\mathcal{T}')$ is a subtree of $\delta(\mathcal{T})$ in the usual sense. $\mathcal{T}'$ is a *subtree* of $\mathcal{T}$ if there exists an $r \in \mathsf{N}_R$ such that $\mathcal{T}'$ is an *r-subtree* of $\mathcal{T}$.

Let $\mathcal{T}_1, \ldots, \mathcal{T}_n$ be tree-shaped interpretations such that for $i \neq j$ it holds that $\delta(\mathcal{T}_i) \cap \delta(\mathcal{T}_j) = \emptyset$ $(1 \leq i, j \leq n)$. Then, $\mathcal{T} = \langle v_0; r_1, \mathcal{T}_1; \ldots; r_n, \mathcal{T}_n \rangle$ expresses the fact that $\mathcal{T}$ is a tree-shaped interpretation with root v_0 and, for every $i = 1, \ldots, n$, the interpretation $\mathcal{T}_i$ is an r_i-subtree of $\mathcal{T}$. Furthermore, $\mathcal{T}_1, \ldots, \mathcal{T}_n$ are the only subtrees of $\mathcal{T}$.

Example 3. Reconsider the invalid GCI

$$\exists \mathsf{hasChild}.\mathsf{Doctor} \sqcap \exists \mathsf{hasChild}.\mathsf{Professor} \sqsubseteq \exists \mathsf{hasChild}.(\mathsf{Doctor} \sqcap \mathsf{Professor})$$

from Example 1. Since this GCI is invalid, the concept

$$C = \exists \mathsf{hasChild}.\mathsf{Doctor} \sqcap \exists \mathsf{hasChild}.\mathsf{Professor} \sqcap \neg \exists \mathsf{hasChild}.(\mathsf{Doctor} \sqcap \mathsf{Professor})$$

is satisfiable. A tree-shaped interpretation satisfying C is given by the interpretation $\mathcal{I} = \langle \{v_0, v_1, v_2\}, \cdot^{\mathcal{I}} \rangle$, with $\mathsf{Doctor}^{\mathcal{I}} = \{v_1\}$, $\mathsf{Professor}^{\mathcal{I}} = \{v_2\}$, and $\mathsf{hasChild}^{\mathcal{I}} = \{(v_0, v_1), (v_0, v_2)\}$. $\qquad\qquad\square$

3   A Rejection Calculus for $\mathcal{ALC}$

We now proceed defining our rejection calculus, which we denote by $\mathbf{SC}^c_{\mathcal{ALC}}$. The calculus will be devised to refute GCIs of form $C \sqsubseteq D$, where C and D are arbitrary concepts. Thereby, we define new syntactic artefacts, viz. *anti-sequents*.

Definition 4. *An* anti-sequent *is an ordered pair of form* $\Gamma \dashv \Delta$, *where* Γ *and* Δ *are finite multi-sets of concepts.* Γ *is the* antecedent *and* Δ *is the* succedent *of* $\Gamma \dashv \Delta$. *An anti-sequent* $\Gamma \dashv \Delta$ *is* propositional *if neither a concept of form* $\forall r.C$ *nor a concept of form* $\exists r.C$ *occurs as subconcept in any* $D \in \Gamma \cup \Delta$.

As usual, given a concept C or a set Σ of concepts, "$\Gamma, C \dashv \Delta$" denotes "$\Gamma \cup \{C\} \dashv \Delta$", and "$\Gamma, \Sigma \dashv \Delta$" denotes "$\Gamma \cup \Sigma \dashv \Delta$". Moreover, "$\dashv \Delta$" stands for "$\emptyset \dashv \Delta$" and "$\Gamma \dashv$" means "$\Gamma \dashv \emptyset$".

A *proof* in $\mathbf{SC}^c_{\mathcal{ALC}}$ is defined as usual in sequential systems. Furthermore, we will use terms which are common in sequent-type systems, like *end-sequent*, etc., without defining them explicitly (we refer the reader to Takeuti [19] for respective formal definitions of such concepts).

Definition 5. *An interpretation* $\mathcal{I}$ refutes *an anti-sequent* $\Gamma \dashv \Delta$ *if* $\mathcal{I}$ *is a counter model of the GCI* $\bigsqcap_{\gamma \in \Gamma} \gamma \sqsubseteq \bigsqcup_{\delta \in \Delta} \delta$, *where the empty concept intersection is defined to be* $\top$ *and the empty concept union is defined to be* $\bot$. *If there is an interpretation which refutes* $\Gamma \dashv \Delta$, *then we say that* $\Gamma \dashv \Delta$ *is* refutable. *Furthermore,* $\iota(\Gamma \dashv \Delta)$ *stands for* $\bigsqcap_{\gamma \in \Gamma} \gamma \sqcap \neg(\bigsqcup_{\delta \in \Delta} \delta)$.

In the following, we denote finite multi-sets of concepts by capital Greek letters $\Gamma, \Delta, \ldots$, while capital Latin letters $C, D, \ldots$ denote concepts.

It is easy to see that the problem of deciding whether an anti-sequent is refutable can be reduced to the problem of deciding whether a concept is satisfiable.

Theorem 6. *The anti-sequent* $s = \Gamma \dashv \Delta$ *is refutable iff* $\iota(s)$ *is satisfiable.*

An immediate consequence of this observation and Proposition 2 is that an anti-sequent is refutable iff it is refuted by some finite tree-shaped interpretation. Note also that a concept C is satisfiable iff the anti-sequent $C \dashv$ is refutable. Furthermore, a concept D does *not* subsume a concept C iff the anti-sequent $C \dashv D$ is refutable.

Now we turn to the postulates of $\mathbf{SC}^c_{\mathcal{ALC}}$. Roughly speaking, the axioms and rules of $\mathbf{SC}^{i0}_{\mathcal{ALC}}$ are generalisations of a sequential rejection system for modal logic $\mathbf{K}$ due to Goranko [6], by exploiting the well-known property that $\mathcal{ALC}$ is a syntactic variant of the multi-modal version of $\mathbf{K}$ [20] and by incorporating multiple modalities into Goranko's system. We discuss the relationship to

Goranko's system, in terms of a multi-modal generalisation of his system, in Sect. 5. Besides that, the rules for the propositional connectives $\sqcap$, $\sqcup$, and $\neg$ correspond directly to those of the rejection system for propositional logic [2, 3, 6]. Note that these rules exhibit non-determinism as opposed to exhaustive search in standard proof systems.

Let us fix some notation. For a set Σ of concepts and a role name r, we define $\neg\Sigma := \{\neg C \mid C \in \Sigma\}$, $\forall r.\Sigma := \{\forall r.C \mid C \in \Sigma\}$, and $\exists r.\Sigma := \{\exists r.C \mid C \in \Sigma\}$. For any role name r, Γ^r and Δ^r stand for multi-sets of concepts where every concept is either of form $\forall r.C$ or $\exists r.C$. Moreover, for such multi-sets, we define $\hat{\Gamma}^r := \{C \mid \forall r.C \in \Gamma\}$ and $\tilde{\Gamma}^r := \{C \mid \exists r.C \in \Gamma\}$.

Definition 7. *The axioms of* $\mathbf{SC}^c_{\mathcal{ALC}}$ *are anti-sequents of form*

$$\Gamma_0 \dashv \Delta_0 \ and \tag{1}$$

$$\forall r_1.\Gamma_1, \ldots, \forall r_n.\Gamma_n \dashv \exists r_1.\Delta_1, \ldots, \exists r_n.\Delta_n, \tag{2}$$

where Γ_0 and Δ_0 are disjoint multi-sets of concept names, $\Gamma_1, \ldots, \Gamma_n$ and $\Delta_1, \ldots, \Delta_n$ are multi-sets of concepts, and $r_1, \ldots, r_n$ are role names. Furthermore, the rules of $\mathbf{SC}^c_{\mathcal{ALC}}$ *are depicted in Fig. 1, where $r_1, \ldots, r_n$ are assumed to be distinct role names.*

Note that each axiom of form (1) is a propositional anti-sequent and, accordingly, we refer to an axiom of such a form as a *propositional axiom*.

Intuitively, in order to derive an anti-sequent s, our calculus tries to build a model which satisfies $\iota(s)$. Speaking in terms of modal logic, the mix rules guarantee that the resulting model contains "enough" worlds to satisfy $\iota(s)$. For example, a world which satisfies $\iota(s)$ for the anti-sequent $s = \exists r.C \dashv \exists r.(C \sqcap D)$ has to be connected to another world which is contained in the extension of C, but not in the extension of $C \sqcap D$. This is exactly what is achieved by the rules $(\textrm{MIX}, \forall)$ and $(\textrm{MIX}, \exists)$.

Example 8. A proof of $\exists r.C \sqcap \exists r.D \dashv \exists r.(C \sqcap D)$ in $\mathbf{SC}^c_{\mathcal{ALC}}$ is given by the following tree (with C and D being distinct concept names):

$$
\cfrac{
  \cfrac{D \dashv C}{D \dashv C \sqcap D}(\sqcap, r)_1 \quad
  \cfrac{
    \cfrac{\cfrac{C \dashv D}{C \dashv C \sqcap D}(\sqcap, r)_2 \quad \dashv \exists r.(C \sqcap D)}{\exists r.C \dashv \exists r.(C \sqcap D)}(\textrm{MIX}, \exists)
  }{}
}{
  \cfrac{\exists r.C, \exists r.D \dashv \exists r.(C \sqcap D)}{\exists r.C \sqcap \exists r.D \dashv \exists r.(C \sqcap D)}(\sqcap, l)
}(\textrm{MIX}, \exists)
$$

$\square$

We informally describe how a counter model can be obtained from a rejection proof. For simplicity, we consider the case where each rule application of $(\textrm{MIX}, \forall)$ and $(\textrm{MIX}, \exists)$ has $k = l = 1$. A tree-shaped counter model can be obtained from a proof by reading the proof from bottom to top and assigning each anti-sequent a node of the tree. Thereby, one starts by assigning the end-sequent of the proof the root node of the model, say v_0. In the proceeding steps, in

Logical Rules of $\mathbf{SC}^c_{\mathcal{ALC}}$

$$\frac{\Gamma, C, D \dashv \Delta}{\Gamma, C \sqcap D \dashv \Delta} \ (\sqcap, l) \qquad \frac{\Gamma \dashv C, \Delta}{\Gamma \dashv C \sqcap D, \Delta} \ (\sqcap, r)_1 \qquad \frac{\Gamma \dashv D, \Delta}{\Gamma \dashv C \sqcap D, \Delta} \ (\sqcap, r)_2$$

$$\frac{\Gamma \dashv C, D, \Delta}{\Gamma \dashv C \sqcup D, \Delta} \ (\sqcup, r) \qquad \frac{\Gamma, C \dashv \Delta}{\Gamma, C \sqcup D \dashv \Delta} \ (\sqcup, l)_1 \qquad \frac{\Gamma, D \dashv \Delta}{\Gamma, C \sqcup D \dashv \Delta} \ (\sqcup, l)_2$$

$$\frac{\Gamma \dashv C, \Delta}{\Gamma, \neg C \dashv \Delta} \ (\neg, l) \qquad \frac{\Gamma, C \dashv \Delta}{\Gamma \dashv \neg C, \Delta} \ (\neg, r) \qquad \frac{\Gamma \dashv \Delta}{\Gamma, \top \dashv \Delta} \ (\top) \qquad \frac{\Gamma \dashv \Delta}{\Gamma \dashv \bot, \Delta} \ (\bot)$$

$$\frac{\Gamma_0 \dashv \Delta_0 \qquad \Gamma^{r_1}, \dots, \Gamma^{r_n} \dashv \Delta^{r_1}, \dots, \Delta^{r_n}}{\Gamma_0, \Gamma^{r_1}, \dots, \Gamma^{r_n} \dashv \Delta_0, \Delta^{r_1}, \dots, \Delta^{r_n}} \ (\text{Mix}),$$

where $\Gamma_0 \dashv \Delta_0$ is a propositional axiom.

$$\frac{\hat{\Gamma}^{r_k} \dashv \tilde{\Delta}^{r_k}, C_k \ \cdots \ \hat{\Gamma}^{r_l} \dashv \tilde{\Delta}^{r_l}, C_l \qquad \Gamma^{r_1}, \dots, \Gamma^{r_n} \dashv \Delta^{r_1}, \dots, \Delta^{r_n}}{\Gamma^{r_1}, \dots, \Gamma^{r_n} \dashv \Delta^{r_1}, \dots, \Delta^{r_n}, \forall r_k.C_k, \dots, \forall r_l.C_l} \ (\text{Mix}, \forall),$$

$$\frac{\hat{\Gamma}^{r_k}, C_k \dashv \tilde{\Delta}^{r_k} \ \cdots \ \hat{\Gamma}^{r_l}, C_l \dashv \tilde{\Delta}^{r_l} \qquad \Gamma^{r_1}, \dots, \Gamma^{r_n} \dashv \Delta^{r_1}, \dots, \Delta^{r_n}}{\Gamma^{r_1}, \dots, \Gamma^{r_n}, \exists r_k.C_k, \dots, \exists r_l.C_l \dashv \Delta^{r_1}, \dots, \Delta^{r_n}} \ (\text{Mix}, \exists),$$

Structural Rules of $\mathbf{SC}^c_{\mathcal{ALC}}$

$$\frac{\Gamma, C \dashv \Delta}{\Gamma \dashv \Delta} \ (w^{-1}, l) \qquad \frac{\Gamma \dashv \Delta, C}{\Gamma \dashv \Delta} \ (w^{-1}, r) \qquad \frac{\Gamma, C \dashv \Delta}{\Gamma, C, C \dashv \Delta} \ (c^{-1}, l) \qquad \frac{\Gamma \dashv C, \Delta}{\Gamma \dashv C, C, \Delta} \ (c^{-1}, r)$$

Fig. 1. Rules of $\mathbf{SC}^c_{\mathcal{ALC}}$.

case of an application of a rule $(\text{Mix}, \forall)$ or $(\text{Mix}, \exists)$, a new child is created, where the parent of the new node is the node assigned to the conclusion of the rule application. The left premiss is then assigned the new node, while the right premiss is assigned the node of the conclusion of the rule application. The resulting arc is labelled with the role name r_1, as represented in the exposition of the rules $(\text{Mix}, \forall)$ and $(\text{Mix}, \exists)$ (cf. Fig. 1), indicating that the arc represents a tuple in the interpretation of r_1. In case of a rule application different from $(\text{Mix}, \forall)$ and $(\text{Mix}, \exists)$, the node assigned to the conclusion is also assigned to the premiss(es). In order to complete the specification of the counter model, it remains to define the extensions of the atomic concepts. If $\Gamma_0 \dashv \Delta_0$ is an axiom in our proof and v' its assigned node, then we simply ensure that (i) v' is in the extension of each $C \in \Gamma_0$ and (ii) v' does not occur in the extension of any $D \in \Delta_0$.

Example 9. A counter model for $\exists r.C \sqcap \exists r.D \sqsubseteq \exists r.(C \sqcap D)$ is given by the interpretation $\mathcal{I} = \langle \Delta^{\mathcal{I}}, \cdot^{\mathcal{I}} \rangle$, defined by $\Delta^{\mathcal{I}} = \{v_0, v_1, v_2\}$, $C^{\mathcal{I}} = \{v_2\}$, $D^{\mathcal{I}} = \{v_1\}$, and $r^{\mathcal{I}} = \{(v_0, v_1), (v_0, v_2)\}$. The reader may easily verify that this counter model can be read off the proof given in Example 8 by the informal method just described.   □

The next statements justify the soundness of particular rules and axioms. The knowledge about tree-shaped interpretations, i.e., the shape of models satisfying a concept, provide a semantic justification of the mix rules.

Lemma 10. *Let T be a tree-shaped interpretation with root v_0 which refutes an anti-sequent of form*

$$\Gamma^{r_1}, \ldots, \Gamma^{r_n} \dashv \Delta^{r_1}, \ldots, \Delta^{r_n}. \tag{3}$$

Then, the anti-sequent (3) is refuted by every tree-shaped interpretation T' such that

1. *for every role name $r \in N_R$, we have $r^T = r^{T'}$, and*
2. *for every concept name C, either $C^{T'} = C^T \cup \{v_0\}$, $C^{T'} = C^T \setminus \{v_0\}$, or $C^{T'} = C^T$ holds.*

Lemma 11. *Let $\Gamma_1, \ldots, \Gamma_n, \Delta_1, \ldots, \Delta_n$ be non-empty multi-sets of concepts and C a concept. Then,*

1. *every axiom of form*

$$\forall r_1.\Gamma_1, \ldots, \forall r_n.\Gamma_n \dashv \exists r_1.\Delta_1, \ldots, \exists r_n.\Delta_n \tag{4}$$

   is refuted by some tree-shaped interpretation, and
2. *for every $i = 1, \ldots, n$, if $\Gamma_i \dashv C$ is refuted by some tree-shaped interpretation T_0 and*

$$\forall r_1.\Gamma_1, \ldots, \forall r_n.\Gamma_n \dashv \forall r_1.\Delta_1, \ldots, \forall r_n.\Delta_n \tag{5}$$

   is refuted by some tree-shaped interpretation T such that there exist disjoint subtrees $T_1, \ldots, T_n$, where $T_j \lhd_{r_j} T$ $(j = 1, \ldots, n)$, then the tree-shaped interpretation $T' = \langle v'; r_1, T_1; \ldots; r_n, T_n; r_i, T_0 \rangle$ refutes the anti-sequent

$$\forall r_1.\Gamma_1, \ldots, \forall r_n.\Gamma_n \dashv \forall r_1.\Delta_1, \ldots, \forall r_n.\Delta_n, \forall r_i.C,$$

   where v' does not occur in the domain of any T_j $(j = 1, \ldots, n)$.

Proof. For Item 1, observe that $T_0 = \langle \{v_0\}, \cdot^T \rangle$, where $r^{T_0} = \emptyset$, for every $r \in N_R$, is a tree-shaped interpretation, where v_0 is arbitrary. By the definition of the semantics of $\mathcal{ALC}$, we have that for every concept of form $\forall r.C$, $(\forall r.C)^{T_0} = \{v_0\}$ holds, while for every concept of form $\exists r.C$ we clearly have $(\exists r.C)^{T_0} = \emptyset$. Hence, every axiom of the form (4) is refuted by T_0 as required.

For Item 2, assume, without loss of generality, that the roles r_i in an anti-sequent of form (5) are pairwise distinct $(i = 1, \ldots, n)$. To ease notation, for a given a set of concepts Γ, let us write $(\Gamma)_{\sqcap}$ for $\bigsqcap_{\gamma \in \Gamma} \gamma$. Let

$$s = \forall r_1.\Gamma_1, \ldots, \forall r_n.\Gamma_n \dashv \forall r_1.\Delta_1, \ldots, \forall r_n.\Delta_n$$

be refuted by T and $\Gamma_i \dashv C$ be refuted by the tree-shaped interpretation T_0 with root v_0. The tree T with root v' satisfies the concept $\iota(s)$, i.e., for every

$i = 1, \ldots, n$, we have $v' \in D_1^{\mathcal{T}}$ and $v' \in D_2^{\mathcal{T}}$ for $D_1 \in \forall r_i.\Gamma_i$ and $D_2 \in \exists r_i.\neg\Delta_i$. Hence, we have, for every non-empty Δ_i $(i = 1, \ldots, n)$, that there is some subtree $\mathcal{T}''$ with root v'' such that $(v', v'') \in r_i^{\mathcal{T}}$ and $v'' \notin D^{\mathcal{T}}$, for every $D \in \Delta_i$. By construction of $\mathcal{T}'$, this is also the case for $\mathcal{T}'$, hence it suffices to show that $v' \notin (\forall r_i.C)^{\mathcal{T}'}$ and $v' \in D^{\mathcal{T}'}$, for every $D \in \forall r_i.\Gamma_i$. Now let $i \in \{1, \ldots, n\}$ and consider the tree $\mathcal{T}'$. Clearly, $v' \in (\neg\forall r_i.C)^{\mathcal{T}'}$, i.e., $v' \in (\exists r_i.\neg C)^{\mathcal{T}'}$, since $(v', v_0) \in r_i^{\mathcal{T}}$ and $\Gamma_i \dashv C$ is refuted by $\mathcal{T}_0$. Furthermore, since (i) $v_0 \in D^{\mathcal{T}_0}$, for every $D \in \Gamma_i$, (ii) $v' \in D^{\mathcal{T}}$ for every $D \in \forall r_i.\Gamma_i$, and (iii) $(v', v_0) \in r_i^{\mathcal{T}'}$, we immediately obtain by the semantics of $\mathcal{ALC}$ that $v' \in D^{\mathcal{T}'}$, for every $D \in \forall r_i.\Gamma_i$. Since we chose the r_i to be distinct, $(\forall r_j.\Gamma_j)_{\sqcap}^{\mathcal{T}} = (\forall r_j.\Gamma_j)_{\sqcap}^{\mathcal{T}'}$ and $(\exists r_j.\neg\Delta_j)_{\sqcap}^{\mathcal{T}} = (\exists r_j.\neg\Delta_j)_{\sqcap}^{\mathcal{T}'}$, for all $j \neq i$. Hence, the anti-sequent

$$\forall r_1.\Gamma_1, \ldots, \forall r_n.\Gamma_n \dashv \forall r_1.\Delta_1, \ldots, \forall r_n.\Delta_n, \forall r_i.C$$

is refuted as desired. $\qquad\qquad\qquad\qquad\qquad\qquad\qquad\qquad\qquad\qquad\qquad\qquad\square$

Theorem 12. $\mathbf{SC}^c_{\mathcal{ALC}}$ *is sound, i.e., only the refutable anti-sequents are provable.*

Proof (Sketch). The proof proceeds by induction on proof length. This amounts to showing the refutability of the axioms and the soundness of each rule separately.

For the induction base, the refutability of a propositional axiom is obvious, while the refutability of an axiom of form $\forall r_1.\Gamma_1, \ldots, \forall r_n.\Gamma_n \dashv \exists r_1.\Delta_1, \ldots, \exists r_n.\Delta_n$ is exactly the statement of Item 1 of Lemma 11.

For the inductive step, we have to distinguish several cases depending on the last applied rule. It is a straightforward argument to show the soundness of the rules dealing with the propositional connectives $\sqcap$, $\sqcup$, and $\neg$. Hence, we just consider the mix rules briefly. For the rule (MIX), let $\Gamma_0 \dashv \Delta_0$ be a propositional axiom of $\mathbf{SC}^c_{\mathcal{ALC}}$ and $\Gamma^{r_1}, \ldots, \Gamma^{r_n} \dashv \Delta^{r_1}, \ldots, \Delta^{r_n}$ be refuted by a tree-shaped interpretation $\mathcal{T}$ with root v_0. By Lemma 10, we define an interpretation $\mathcal{T}'$ such that $C^{\mathcal{T}'} = C^{\mathcal{T}} \cup \{v_0\}$, for every $C \in \Gamma_0$, and $C^{\mathcal{T}'} = C^{\mathcal{T}} \setminus \{v_0\}$, for every $C \in \Delta_0$. Since $\Gamma_0 \cap \Delta_0 = \emptyset$, the interpretation $\mathcal{T}'$ is well-defined and refutes the anti-sequent $\Gamma_0, \Gamma^{r_1}, \ldots, \Gamma^{r_n} \dashv \Delta_0, \Delta^{r_1}, \ldots, \Delta^{r_n}$. For multi-sets Γ, Δ, and Π, it is easy to see that an anti-sequent of form $\Gamma \dashv \neg\Pi, \Delta$ is refutable iff $\Gamma, \Pi \dashv \Delta$ is. Considering this fact and that value restriction is dual to existential restriction (i.e., $(\forall r.C)^{\mathcal{I}} = (\neg\exists r.\neg C)^{\mathcal{I}}$, for every interpretation $\mathcal{I}$), the soundness of the rules (MIX, $\forall$) and (MIX, $\exists$) can easily be shown by repeated application of Item 2 of Lemma 11. $\qquad\qquad\qquad\qquad\qquad\qquad\qquad\qquad\qquad\qquad\qquad\qquad\square$

Following Goranko [6], the completeness argument for $\mathbf{SC}^c_{\mathcal{ALC}}$ is divided into two steps: first, one proves completeness of the propositional fragment of our calculus. This proceeds similar to the completeness proof of the rejection system for classical propositional logic as given by Bonatti [2] and is thus omitted for space reasons. The second step consists of showing completeness by induction on the length of the refuting tree-shaped interpretation.

In what follows, the *logical complexity*, $||C||$, of a concept C is defined to be the number of connectives occurring in C. The logical complexity $||\Gamma||$ of a multi-set Γ of concepts is the sum of the logical complexities of the concept occurring in Γ. For an anti-sequent $\Gamma \dashv \Delta$, we define the logical complexity $||\Gamma \dashv \Delta||$ to be $||\Gamma|| + ||\Delta||$.

Lemma 13. *Every refutable propositional anti-sequent is provable in* $\mathbf{SC}^c_{\mathcal{ALC}}$.

The next lemma states that refutable propositional anti-sequents possess a proof in which all concept names appear in a single axiom.

Lemma 14. *Let s be a refutable propositional anti-sequent which is refuted by an interpretation $\mathcal{I}$ such that $C_1, \ldots, C_n$ are exactly those distinct concept names among s whose extensions are non-empty under $\mathcal{I}$ and $D_1, \ldots, D_m$ are exactly those whose extensions are empty under $\mathcal{I}$, respectively. Then, there exists a proof of s in $\mathbf{SC}^c_{\mathcal{ALC}}$ which has as its only axiom $C_1, \ldots, C_n \dashv D_1, \ldots, D_m$.*

In what follows, an *atom* of an anti-sequent $\Gamma \dashv \Delta$ is either a concept name from $\Gamma \cup \Delta$ or some concept of form $Qr.C$ ($Q \in \{\forall, \exists\}$) which occurs in a concept from $\Gamma \cup \Delta$ and which is not in the scope of some $Q \in \{\forall, \exists\}$.

Theorem 15. $\mathbf{SC}^c_{\mathcal{ALC}}$ *is complete, i.e., every refutable anti-sequent is provable.*

Proof. Let $\Gamma \dashv \Delta$ be an arbitrary refutable anti-sequent and $\mathcal{T}$ a tree-shaped interpretation of length $\ell(\mathcal{T})$ with domain $\Delta^{\mathcal{T}}$ which refutes $\Gamma \dashv \Delta$. Since $\Gamma \dashv \Delta$ is refuted by $\mathcal{T}$, there must be some $c \in \Delta^{\mathcal{T}}$ such that (i) $c \in C^{\mathcal{T}}$, for all $C \in \Gamma$, and (ii) $c \notin D^{\mathcal{T}}$, for all $D \in \Delta$. In the following, we say that a concept C is *falsified by* $\mathcal{T}$ if $c \notin C^{\mathcal{T}}$ and *satisfied by* $\mathcal{T}$ otherwise. Now let $C_1, \ldots, C_k, \forall r_1.D_1, \ldots, \forall r_l.D_l, \exists s_1.F_1, \ldots, \exists s_\lambda.F_\lambda$ be the atoms of $\Gamma \cup \Delta$ which are satisfied by $\mathcal{T}$, and $E_1, \ldots, E_{k'}, \forall p_1.G_1, \ldots, \forall p_\mu.G_\mu, \exists q_1.H_1, \ldots, \exists q_m.H_m$ those which are falsified by $\mathcal{T}$, where $C_1, \ldots, C_k, E_1, \ldots, E_{k'}$ are atomic concepts and $r_1, \ldots, r_l, s_1, \ldots, s_\lambda, p_1, \ldots, p_\mu, q_1, \ldots, q_m$ are not necessarily distinct role names. We define the following sets:

$$\Gamma_0 = \{C_1, \ldots, C_k\}, \qquad\qquad \Delta_0 = \{E_1, \ldots, E_{k'}\},$$
$$\Sigma^\forall = \{\forall r_1.D_1, \ldots, \forall r_l.D_l\}, \qquad \Lambda^\exists = \{\exists s_1.F_1, \ldots, \exists s_\lambda.F_\lambda\},$$
$$\Pi^\forall = \{\forall p_1.G_1, \ldots, \forall p_\mu.G_\mu\}, \qquad \Phi^\exists = \{\exists q_1.H_1, \ldots, \exists q_m.H_m\}.$$

It suffices to infer

$$\Gamma_0, \Sigma^\forall, \Lambda^\exists \dashv \Delta_0, \Pi^\forall, \Phi^\exists \qquad\qquad (6)$$

since Lemma 13 and Lemma 14 allow us to infer $\Gamma \dashv \Delta$ from (6). This is accomplished by replacing all atoms of (6) by new distinct concept names which then becomes a propositional axiom. Then, we can infer a propositional anti-sequent $\Gamma' \dashv \Delta'$ by Lemma 13 and Lemma 14 which is obtained from $\Gamma \dashv \Delta$ by exactly the same replacement as mentioned before. Hence, there exists a proof of $\Gamma' \dashv \Delta'$—substituting the new concept names back we obtain a proof of $\Gamma \dashv \Delta$.

Obviously, we have that $\Gamma_0 \cap \Delta_0 = \emptyset$, hence, $\Gamma_0 \dashv \Delta_0$ is a propositional axiom. Furthermore, the anti-sequent

$$\Sigma^\forall \dashv \Phi^\exists \qquad (7)$$

constitutes an axiom of $\mathbf{SC}^c_{\mathcal{ALC}}$. We now proceed by induction on $\ell(\mathcal{T})$. For the base case, if $\ell(\mathcal{T}) = 0$, (6) must be of form $\Gamma_0, \Sigma^\forall \dashv \Delta_0, \Phi^\exists$, by the semantics of $\mathcal{ALC}$. This anti-sequent is inferred by the following rule application:

$$\frac{\Gamma_0 \dashv \Delta_0 \qquad \Sigma^\forall \dashv \Phi^\exists}{\Gamma_0, \Sigma^\forall \dashv \Delta_0, \Phi^\exists} \ (\text{MIX})$$

This completes the base case. Now assume that every anti-sequent which is refuted by some tree $\mathcal{T}$ with length $\ell(\mathcal{T}) \leq n$ is provable in $\mathbf{SC}^c_{\mathcal{ALC}}$. Furthermore, for every role name r, define $\Theta(r) = \{C \mid \forall r.C \in \Sigma^\forall\}$ and $\Xi(r) = \{C \mid \exists r.C \in \Phi^\exists\}$. If (6) is refuted by some tree $\mathcal{T}$ with length $\ell(\mathcal{T}) = n+1$, then the anti-sequents

$$\Theta(p_i) \dashv G_i, \Xi(p_i), \text{ for } i = 1, \ldots, \mu, \text{ and} \qquad (8)$$
$$\Theta(s_j), F_j \dashv \Xi(s_j), \text{ for } j = 1, \ldots, \lambda, \qquad (9)$$

are refuted by immediate subtrees of $\mathcal{T}$ with length $\ell(\mathcal{T}) = n$ and are therefore, by induction hypothesis, provable in $\mathbf{SC}^c_{\mathcal{ALC}}$. We first consider (8) and start for $i = 1$ applying $(\text{MIX}, \forall)$ to Axiom (7):

$$\frac{\Theta(p_1) \dashv G_1, \Xi(p_1) \qquad \Sigma^\forall \dashv \Phi^\exists}{\Sigma^\forall \dashv \forall p_1.G_1, \Phi^\exists} \ (\text{MIX}, \forall)$$

Now, for every $i = 2, \ldots, \mu$, we proceed constructing a proof of the anti-sequent $\Sigma^\forall \dashv \forall p_1.G_1, \ldots, \forall p_i.G_i, \Phi^\exists$ from $\Sigma^\forall \dashv \forall p_1.G_1, \ldots, \forall p_{i-1}.G_{i-1}, \Phi^\exists$ in the following way:

$$\frac{\Theta(p_i) \dashv G_i, \Xi(p_i) \qquad \Sigma^\forall \dashv \forall p_1.G_1, \ldots, \forall p_{i-1}.G_{i-1}, \Phi^\exists}{\Sigma^\forall \dashv \forall p_1.G_1, \ldots, \forall p_i.G_i, \Phi^\exists} \ (\text{MIX}, \forall)$$

For $i = \mu$, we obtain a proof of the anti-sequent $\Sigma^\forall \dashv \Pi^\forall, \Phi^\exists$. Building upon this anti-sequent, we proceed in a similar manner by performing the following inference:

$$\frac{\Theta(s_1), F_1 \dashv \Xi(s_1) \qquad \Sigma^\forall \dashv \Pi^\forall, \Phi^\exists}{\Sigma^\forall, \exists s_1.F_1 \dashv \Pi^\forall, \Phi^\exists} \ (\text{MIX}, \exists)$$

Again, for every $i = 2, \ldots, \lambda$, we proceed constructing a proof of the anti-sequent $\Sigma^\forall, \exists s_1.F_1, \ldots, \exists s_i.F_i \dashv \Pi^\forall, \Phi^\exists$ from $\Sigma^\forall, \exists s_1.F_1, \ldots, \exists s_{i-1}.F_{i-1} \dashv \Pi^\forall, \Phi^\exists$ as follows:

$$\frac{\Theta(s_i), F_i \dashv \Xi(s_i) \qquad \Sigma^\forall, \exists s_1.F_1, \ldots, \exists s_{i-1}.F_{i-1} \dashv \Pi^\forall, \Phi^\exists}{\Sigma^\forall, \exists s_1.F_1, \ldots, \exists s_i.F_i \dashv \Pi^\forall, \Phi^\exists} \ (\text{MIX}, \exists)$$

For $i = \lambda$, we obtain a proof of the anti-sequent $\Sigma^\forall, \Lambda^\exists \dashv \Pi^\forall, \Phi^\exists$. Finally, we apply the rule (MIX) in order to obtain a proof of the desired anti-sequent:

$$\frac{\Gamma_0 \dashv \Delta_0 \qquad \Sigma^\forall, \Lambda^\exists \dashv \Pi^\forall, \Phi^\exists}{\Gamma_0, \Sigma^\forall, \Lambda^\exists \dashv \Delta_0, \Pi^\forall, \Phi^\exists} \text{ (MIX)}$$

Hence, (6) is inferred and the induction step is completed. Since every refutable anti-sequent is refuted by some tree-shaped interpretation, every refutable anti-sequent is provable and $\mathbf{SC}^c_{\mathcal{ALC}}$ is complete as claimed.   □

4   Comparing $\mathbf{SC}^c_{\mathcal{ALC}}$ with an $\mathcal{ALC}$ Tableau Algorithm

The most common reasoning procedures which have been studied for description logics are tableau algorithms. They are well known for $\mathcal{ALC}$ and its extensions and have been implemented in state-of-the-art reasoners (like, e.g., in the FaCT system [21]). Tableau algorithms rely on the construction of a *canonical model* which witnesses the satisfiability of a concept or a knowledge base. We now briefly discuss the relationship of our calculus and the tableau procedure for concept satisfiability as discussed by Baader and Sattler [18].

The basic structure the algorithm works on is the so-called *completion graph*. A completion graph is an ordered triple $\langle V, E, \mathcal{L} \rangle$, where V is a set of *nodes*, $E \subseteq V \times V$ is a set of *edges*, and $\mathcal{L}$ is a *labelling function* which assigns a set of concepts to each node and a role name to each edge. Given a concept C in negation normal form (i.e., where negation occurs in C only in front of concept names), the *initial completion graph of* C is a completion graph $\langle V, E, \mathcal{L} \rangle$ where $V = \{v_0\}$, $E = \emptyset$, and $\mathcal{L}(v_0) = \{C\}$. A completion graph $G = \langle V, E, \mathcal{L} \rangle$ contains a *clash* if $\{D, \neg D\} \subseteq \mathcal{L}(v)$ for some node v and some concept D. G is *complete* if no rules are applicable any more. The algorithm operates at each instant on a set $\mathbf{G}$ of completion graphs. The *completion rules* specify the rules which may be applied to infer a new set of completion graphs $\mathbf{G}'$ from some set of completion graphs $\mathbf{G}$. Given a concept C, the algorithm starts with the set $\mathbf{G}_0 = \{G_0\}$, where G_0 is the initial completion graph of C, and successively computes a new set $\mathbf{G}_{i+1}$ of completion graphs from the set $\mathbf{G}_i$. Thereby, every completion graph which has a clash is immediately dropped. The algorithm halts if for some $j \geq 0$, $\mathbf{G}_j$ contains a complete completion graph or $\mathbf{G}_j = \emptyset$. In the former case, the algorithm answers that the concept C is satisfiable, in the latter case it answers that C is unsatisfiable. It is well-known that a model of the concept under consideration can be extracted from a complete completion graph and that (in the case of concept satisfiability) a complete completion graph represents a tree [18].

Example 16. Consider the anti-sequent from Example 8. This anti-sequent is refutable iff the concept $\hat{C} = \exists r.C \sqcap \exists r.D \sqcap \forall r.(\neg C \sqcup \neg D)$ is satisfiable. The following graph represents a complete completion graph for $\hat{C}$:

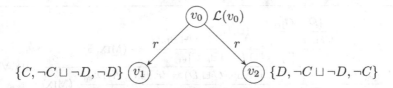

Here, $\mathcal{L}(v_0) = \{\hat{C}, \exists r.C, \exists r.D, \forall r.(\neg C \sqcap \neg D)\}$. Note that this graph can easily be turned into an interpretation satisfying $\hat{C}$. □

We now describe how to obtain a complete completion graph from a proof of $\mathbf{SC}^c_{\mathcal{ALC}}$ such that the root of the completion graph is labelled with the end-sequent of the proof (for the sake of readability, we consider without loss of generality the case where $k = l = 1$ in the rules (Mix, $\forall$) and (Mix, $\exists$)). Let $nnf(C)$ denote the negation normal form of a concept C. For a set of concepts Γ, define $nnf(\Gamma) = \{nnf(C) \mid C \in \Gamma\}$ and, for an anti-sequent $s = \Gamma \dashv \Delta$, define $nnf(s) = nnf(\Gamma \cup \neg\Delta)$. Furthermore, let $\tau[G]$ denote the root of a completion graph G. We define a mapping ξ which maps any proof of $\mathbf{SC}^c_{\mathcal{ALC}}$ to some complete completion graph. Let χ be a proof of $\mathbf{SC}^c_{\mathcal{ALC}}$ and s_χ be the end-sequent of χ. The mapping ξ is inductively defined as follows:

– If s_χ is a propositional axiom, then $\xi(\chi) = \langle V, E, \mathcal{L} \rangle$, where $V = \{v_0\}$, $E = \emptyset$, and $\mathcal{L}(v_0) = nnf(s_\chi)$.
– If s_χ results from an application of some binary rule ρ, and $\xi(\chi_1) = G_1 = \langle V_1, E_1, \mathcal{L}_1 \rangle$ and $\xi(\chi_2) = G_2 = \langle V_2, E_2, \mathcal{L}_2 \rangle$, where χ_1 is the proof of the left premiss, χ_2 is the proof of the right premiss, and G_1 and G_2 are disjoint, then we distinguish several cases:
  • If $\rho = $(Mix, $\forall$), then s_χ is of form $\Gamma^{r_1}, \ldots, \Gamma^{r_n} \dashv \Delta^{r_1}, \ldots, \Delta^{r_n}, \forall r.C$. Then, $\xi(\chi) = G = \langle V, E, \mathcal{L} \rangle$, where $\tau[G] = v'$ ($v' \notin V_1 \cup V_2$), $V = V_1 \cup V_2 \cup \{v'\}$, $E = E_1 \cup E_2 \cup \{e\}$, for $e = (v', \tau[G_1])$, and the labelling function preserves the labels from G_1 and G_2 but additionally satisfies $\mathcal{L}(e) = r$ and $\mathcal{L}(v') = nnf(s_\chi)$.
  • If $\rho = $(Mix, $\exists$), then s_χ is of form $\Gamma^{r_1}, \ldots, \Gamma^{r_n}, \exists r.C \dashv \Delta^{r_1}, \ldots, \Delta^{r_n}$, $V = V_1 \cup V_2 \cup \{v'\}$, $E = E_1 \cup E_2 \cup \{e\}$, for $e = (v', \tau[G_1])$, and the labelling function preserves the labels from G_1 and G_2 but additionally satisfies $\mathcal{L}(e) = r$ and $\mathcal{L}(v') = nnf(s_\chi)$.
  • If $\rho = $(Mix), then s_χ is of form $\Gamma_0, \Gamma^{r_1}, \ldots, \Gamma^{r_n} \dashv \Delta_0, \Delta^{r_1}, \ldots, \Delta^{r_n}$. Then, $\xi(\chi) = G = \langle V, E, \mathcal{L} \rangle$, where $\tau[G] = \tau[G_2]$, $V = V_1 \cup V_2$, $E = E_1 \cup E_2$, and $\mathcal{L}(\tau[G]) = nnf(s_\chi)$.
– If s_χ results from an application of a unary rule, and $\xi(\chi_1) = G_1 = \langle V_1, E_1, \mathcal{L}_1 \rangle$, where χ_1 is proof of the upper sequent, then $\xi(\chi) = G = \langle V, E, \mathcal{L} \rangle$, where $V = V_1$, $E = E_1$, and the labelling function preserves the labels from G_1 but additionally satisfies $\mathcal{L}(\tau[G]) = \mathcal{L}_1(\tau[G]) \cup nnf(s_\chi)$.

Theorem 17. *Let χ be a proof in $\mathbf{SC}^c_{\mathcal{ALC}}$ and s_χ the end-sequent of χ. Then, there exists a complete completion graph $G = \langle V, E, \mathcal{L} \rangle$ such that $\xi(\chi) = G$ and $nnf(s_\chi) \subseteq \mathcal{L}(\tau[G])$.*

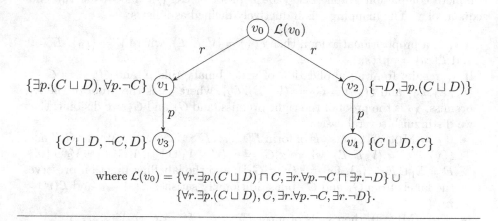

$$\dfrac{\dfrac{[D \dashv C]_{v_3}}{[C \sqcup D \dashv C]_{v_3}}\ (\sqcup, l)_2 \qquad [\dashv \exists p.C]_{v_1}}{\dfrac{[\exists p.(C \sqcup D) \dashv \exists p.C]_{v_1}}{[\forall r.\exists p.(C \sqcup D) \dashv \forall r.\exists p.C, \forall r.D]_{v_0}}\ (\text{MIX}, \exists)} \ (\text{MIX}, \forall)$$

$$\dfrac{[C \dashv]_{v_0} \qquad [\forall r.\exists p.(C \sqcup D) \dashv \forall r.\exists p.C, \forall r.D]_{v_0}}{\dfrac{[\forall r.\exists p.(C \sqcup D), C \dashv \forall r.\exists p.C, \forall r.D]_{v_0}}{\dfrac{[\forall r.\exists p.(C \sqcup D), C \dashv \forall r.\exists p.C \sqcup \forall r.D]_{v_0}}{[\forall r.\exists p.(C \sqcup D) \sqcap C \dashv \forall r.\exists p.C \sqcup \forall r.D]_{v_0}}\ (\sqcap, l)}\ (\sqcup, r)} \ (\text{MIX})$$

where α is the following proof:

$$\dfrac{[\dashv D]_{v_2} \qquad \dfrac{\dfrac{[C \dashv]_{v_4}}{[C \sqcup D \dashv]_{v_4}}\ (\sqcup, l)_1 \qquad [\dashv]_{v_2}}{\dfrac{[\exists p.(C \sqcup D) \dashv]_{v_2}}{[\exists p.(C \sqcup D) \dashv D]_{v_2}}\ (\text{MIX}, \exists)}}{\dfrac{[\exists p.(C \sqcup D) \dashv D]_{v_2}}{[\forall r.\exists p.(C \sqcup D) \dashv \forall r.D]_{v_0}}} \quad [\forall r.\exists p(C \sqcup D) \dashv]_{v_0} \ (\text{MIX}, \forall)$$

where $\mathcal{L}(v_0) = \{\forall r.\exists p.(C \sqcup D) \sqcap C, \exists r.\forall p.\neg C \sqcap \exists r.\neg D\} \cup$
$\{\forall r.\exists p.(C \sqcup D), C, \exists r.\forall p.\neg C, \exists r.\neg D\}.$

Fig. 2. A proof in $\mathbf{SC}^c_{\mathcal{ALC}}$ and a corresponding completion graph.

Example 18. In Fig. 2, we compare a proof of $\mathbf{SC}^c_{\mathcal{ALC}}$ with its corresponding complete completion graph $G = \langle V, E, \mathcal{L} \rangle$. For better readability, we labelled each anti-sequent in the proof with subscripts of form $[s]_v$ which means that $nnf(s) \subseteq \mathcal{L}(v)$. Note that the completion graph represents a model of the concept $\iota(s_\chi)$, where s_χ is the end-sequent of the depicted proof. In fact, a model is given by $\mathcal{I} = \langle \Delta^{\mathcal{I}}, \cdot^{\mathcal{I}} \rangle$, where $\Delta^{\mathcal{I}} = \{v_i \mid 0 \le i \le 4\}$, $D^{\mathcal{I}} = \{v_3\}$, $C^{\mathcal{I}} = \{v_0, v_4\}$, $r^{\mathcal{I}} = \{(v_0, v_1), (v_0, v_2)\}$, and $p^{\mathcal{I}} = \{(v_1, v_3), (v_2, v_4)\}$. □

5 A Multi-Modal Rejection Calculus

As mentioned above, the development of our calculus for $\mathcal{ALC}$ is based on a rejection calculus for modal logic $\mathbf{K}$, as introduced by Goranko [6], by taking into account that $\mathcal{ALC}$ can be translated into a multi-modal version of $\mathbf{K}$. In this section, we lay down the relation of our calculus to Goranko's system, thereby generalising his calculus to the multi-modal case.

We start with describing the multi-modal logic $\mathbf{K}_m$. In general, the signature of multi-modal logics usually provide a countably infinite supply of different *modalities* which we identify by lower case Greek letters $\alpha, \beta, \ldots$ as well as a countably infinite supply of *propositional variables* $p, q, \ldots$. Formulae in the language of a multi-modal logic are then built up using the propositional connectives $\wedge, \vee, \neg, \top$, and $\bot$, together with unary operators of form $[\alpha]$, where α is a modality. The latter kind of operators are referred to as *modal operators*, and we define $\langle \alpha \rangle := \neg[\alpha]\neg$, for every modality α.

Following Goranko and Otto [22], let τ be the set of all modalities. A *Kripke interpretation* is a triple $\mathcal{M} = \langle W, \{R_\alpha\}_{\alpha \in \tau}, V \rangle$, where W is a non-empty set of *worlds*, $R_\alpha \subseteq W \times W$ defines an *accessibility relation* for each $\alpha \in \tau$, and V maps any propositional variable to a subset of W, i.e., V defines which propositional variables are true at which worlds. The pair $\langle W, \{R_\alpha\}_{\alpha \in \tau} \rangle$ defines the *Kripke frame* on which $\mathcal{M}$ is *based*. Given any Kripke interpretation $\mathcal{M} = \langle W, \{R_\alpha\}_{\alpha \in \tau}, V \rangle$, we define the *truth of a formula* φ at a world $w \in W$, denoted by $\mathcal{M}, w \models \varphi$, inductively in the usual manner. Furthermore, the notions of *validity* in a frame and validity in a class of frames is defined as usual (cf. Goranko and Otto [22] for a detailed account). $\mathbf{K}_m$ is the multi-modal logic consisting of all formulae which are valid in all Kripke frames.

A concept of $\mathcal{ALC}$ can be translated into a formula of $\mathbf{K}_m$ by viewing concepts of form $\forall r.C$ as modal formulae of form $[\alpha]C'$, where C' is the corresponding translation of the concept C. Each role name corresponds to one and only one modality. Furthermore, the propositional connectives of $\mathcal{ALC}$ can easily seen to be translated into the usual connectives of classical propositional logic. From a semantic point of view, interpretations of $\mathcal{ALC}$ correspond to Kripke interpretations if we identify the domain of the interpretation with the corresponding set of worlds of the Kripke interpretation. Furthermore, the interpretation of each role name corresponds to some accessibility relation. The extension of an $\mathcal{ALC}$ concept contains then exactly those worlds of the corresponding Kripke interpretation where the translation of the concept is satisfied.

Let C be a concept, $\mathcal{I}$ an interpretation, and let the translations be given by the formula φ and the Kripke interpretation $\mathcal{M}$, respectively. It holds that $\mathcal{M}, w \models \varphi$ iff $w \in C^\mathcal{I}$. For a full treatment of the translation, we refer to Schild [20] and Baader *et al.* [15].

A rejection system for $\mathbf{K}_m$ can now be defined as follows. Let a *multi-modal anti-sequent* $\Gamma \dashv \Delta$ be defined as in the case of $\mathcal{ALC}$, but with Γ and Δ being finite multi-sets of multi-modal formulae. $\Gamma \dashv \Delta$ is *refutable* if there exists a Kripke interpretation $\mathcal{M} = \langle W, \{R_\alpha\}_{\alpha \in \tau}, V \rangle$ and some $w \in W$ such that $\mathcal{M}, w \models \varphi$, for every $\varphi \in \Gamma$, but $\mathcal{M}, w \not\models \psi$, for every $\psi \in \Delta$. Axioms of this system are

LOGICAL RULES

$$\frac{\Gamma,\varphi,\psi \dashv \Delta}{\Gamma,\varphi \wedge \psi \dashv \Delta} \ (\wedge,l) \qquad \frac{\Gamma \dashv \varphi, \Delta}{\Gamma \dashv \varphi \wedge \psi, \Delta} \ (\wedge,r)_1 \qquad \frac{\Gamma \dashv \psi, \Delta}{\Gamma \dashv \varphi \wedge \psi, \Delta} \ (\wedge,r)_2$$

$$\frac{\Gamma \dashv \varphi, \psi, \Delta}{\Gamma \dashv \varphi \vee \psi, \Delta} \ (\vee,r) \qquad \frac{\Gamma,\varphi \dashv \Delta}{\Gamma,\varphi \vee \psi \dashv \Delta} \ (\vee,l)_1 \qquad \frac{\Gamma,\psi \dashv \Delta}{\Gamma,\varphi \vee \psi \dashv \Delta} \ (\vee,l)_2$$

$$\frac{\Gamma \dashv \varphi, \Delta}{\Gamma,\neg\varphi \dashv \Delta} \ (\neg,l) \qquad \frac{\Gamma,\varphi \dashv \Delta}{\Gamma \dashv \neg\varphi, \Delta} \ (\neg,r) \qquad \frac{\Gamma \dashv \Delta}{\Gamma,\top \dashv \Delta} \ (\top) \qquad \frac{\Gamma \dashv \Delta}{\Gamma \dashv \bot, \Delta} \ (\bot)$$

$$\frac{\Gamma_0 \dashv \Delta_0 \qquad [\alpha_1]\Gamma_1,\ldots,[\alpha_n]\Gamma_n \dashv [\alpha_1]\Delta_1,\ldots,[\alpha_n]\Delta_n}{\Gamma_0,[\alpha_1]\Gamma_1,\ldots,[\alpha_n]\Gamma_n \dashv \Delta_0,[\alpha_1]\Delta_1,\ldots,[\alpha_n]\Delta_n} \ (\text{MIX})$$

where Γ_0, Δ_0 are disjoint sets of propositional variables.

$$\frac{\Gamma_k \dashv \varphi_k \ \cdots \ \Gamma_l \dashv \varphi_l \qquad [\alpha_1]\Gamma_1,\ldots,[\alpha_n]\Gamma_n \dashv [\alpha_1]\Delta_1,\ldots,[\alpha_n]\Delta_n}{[\alpha_1]\Gamma_1,\ldots,[\alpha_n]\Gamma_n \dashv [\alpha_1]\Delta_1,\ldots,[\alpha_n]\Delta_n,[\alpha_k]\varphi_k,\ldots,[\alpha_l]\varphi_l} \ (\text{MIX}^2)$$

where $1 \leq k \leq l \leq n$.

STRUCTURAL RULES

$$\frac{\Gamma,\varphi \dashv \Delta}{\Gamma \dashv \Delta} \ (w^{-1},l) \quad \frac{\Gamma \dashv \Delta, \varphi}{\Gamma \dashv \Delta} \ (w^{-1},r) \quad \frac{\Gamma,\varphi \dashv \Delta}{\Gamma,\varphi,\varphi \dashv \Delta} \ (c^{-1},l) \quad \frac{\Gamma \dashv \varphi, \Delta}{\Gamma \dashv \varphi,\varphi, \Delta} \ (c^{-1},r)$$

Fig. 3. Rules of a multi-modal variant of $\mathbf{SC}^c_{\mathcal{ALC}}$.

given by anti-sequents of form $\Gamma_0 \dashv \Delta_0$, with Γ_0 and Δ_0 being disjoint sets of propositional variables, and anti-sequents of form $[\alpha_1]\Gamma_1,\ldots,[\alpha_n]\Gamma_n \dashv$. The rules of the resulting calculus are depicted in Fig. 3 (for any multi-set Γ and modality α, we have $[\alpha]\Gamma := \{[\alpha]\varphi \mid \varphi \in \Gamma\}$; $\alpha_1,\ldots,\alpha_n$ are pairwise different modalities).

Note that, e.g., (MIX^2) corresponds to our rule $(\text{MIX},\forall)$,

$$\frac{\Gamma \dashv \varphi \qquad \Box\Gamma \dashv \Box\Delta}{\Box\Gamma \dashv \Box\Delta, \Box\varphi} \ MIX^2_{\mathbf{K}},$$

where $\Box\Sigma := \{\Box\varphi \mid \varphi \in \Sigma\}$ for any multi-set Σ. We did not explicitly include here a corresponding rule for $\langle\alpha\rangle$ since this can be derived using (MIX^2).

The following result can be shown:

Theorem 19. *A multi-modal anti-sequent is refutable iff it is provable.*

6   Conclusion

We presented a sequent-type calculus for deciding concept non-subsumption in $\mathcal{ALC}$. Sequent calculi are important means for proof-theoretic investigations. We pointed out that our calculus is in some sense equivalent to a well-known tableau procedure which is interesting from a conceptual point of view: as pointed out by Goranko [6], the reason why complementary calculi have been rarely studied may be found in the fact that often theories are recursively axiomatisable while being undecidable. Hence, for such logics, reasonable complementary calculi cannot be devised as the set of invalid propositions is not recursively enumerable there.

However, most of the description logics studied so far are decidable, hence they permit an axiomatisation of the satisfiable concepts. Indeed, the well-known tableau procedures for different description logics focus on checking satisfiability, while unsatisfiability is established by exhaustive search. For such logics, the sequent-style counterparts to tableau procedures are complementary calculi, since checking satisfiability can be reduced to checking invalidity. We note in passing that the tableau procedure for $\mathcal{ALC}$ has also been simulated by *hyper-resolution* by Fermüller *et al.* [23].

As for future work, a natural extension of our calculus is to take $\mathcal{ALC}$ TBox reasoning into account. However, this seems not to be straightforward in view of existing tableau algorithms for it. We conjecture that a feasible approach would need to employ a more complicated notion of anti-sequent, which is related to the problem of non-termination of the respective tableau algorithm without blocking rules.

References

1. Łukasiewicz, J.: Aristotle's Syllogistic from the Standpoint of Modern Formal Logic. Clarendon Press, Oxford (1957)
2. Bonatti, P.A.: A Gentzen system for non-theorems. Technical report CD-TR 93/52, Technische Universität Wien, Institut für Informationssysteme (1993)
3. Tiomkin, M.L.: Proving unprovability. In: Proceedings of the LICS '88, pp. 22–26. IEEE Computer Society (1988)
4. Dyckhoff, R.: Contraction-free sequent calculi for intuitionistic logic. J. Symbolic Logic **57**(3), 795–807 (1992)
5. Kreisel, G., Putnam, H.: Eine Unableitbarkeitsbeweismethode für den Intuitionistischen Aussagenkalkül. Archiv für Mathematische Logik und Grundlagenforschung **3**(1–2), 74–78 (1957)
6. Goranko, V.: Refutation systems in modal logic. Studia Logica **53**, 299–324 (1994)
7. Skura, T.: Refutations and proofs in S4. In: Wansing, H. (ed.) Proof Theory of Modal Logic, pp. 45–51. Kluwer, Dordrecht (1996)
8. Oetsch, J., Tompits, H.: Gentzen-type refutation systems for three-valued logics with an application to disproving strong equivalence. In: Delgrande, J.P., Faber, W. (eds.) LPNMR 2011. LNCS, vol. 6645, pp. 254–259. Springer, Heidelberg (2011)
9. Wybraniec-Skardowska, U.: On the notion and function of the rejection of propositions. Acta Universitatis Wratislaviensis Logika **23**(2754), 179–202 (2005)
10. Caferra, R., Peltier, N.: Accepting/rejecting propositions from accepted/rejected propositions: a unifying overview. Int. J. Intell. Syst. **23**(10), 999–1020 (2008)
11. Bonatti, P.A., Olivetti, N.: Sequent calculi for propositional nonmonotonic logics. ACM Trans. Comput. Logic **3**(2), 226–278 (2002)
12. Reiter, R.: A logic for default reasoning. Artif. Intell. **13**(1–2), 81–132 (1980)

13. Moore, R.C.: Semantical considerations on nonmonotonic logic. In: Proceedings of the IJCAI '83, pp. 272–279. William Kaufmann (1983)
14. McCarthy, J.: Circumscription - a form of non-monotonic reasoning. Artif. Intell. **13**(1–2), 27–39 (1980)
15. Baader, F., Calvanese, D., McGuinness, D.L., Nardi, D., Patel-Schneider, P.F.: The Description Logic Handbook: Theory, Implementation, and Applications. Cambridge University Press, New York (2003)
16. Rademaker, A.: A Proof Theory for Description Logics. Springer, New York (2012)
17. Borgida, A., Franconi, E., Horrocks, I., McGuinness, D.L., Patel-Schneider, P.F.: Explaining $\mathcal{ALC}$ subsumption. In: Proceedings of the DL '99. CEUR Workshop Proceedings, vol. 22 (1999)
18. Baader, F., Sattler, U.: An overview of tableau algorithms for description logics. Studia Logica **69**, 5–40 (2001)
19. Takeuti, G.: Proof Theory. Studies in Logic and the Foundations of Mathematics Series. North-Holland, Amsterdam (1975)
20. Schild, K.: A Correspondence theory for terminological logics: preliminary report. In: Proceedings of the IJCAI '91, pp. 466–471. Morgan Kaufmann Publishers Inc. (1991)
21. Horrocks, I.: The FaCT system. In: de Swart, H. (ed.) TABLEAUX 1998. LNCS (LNAI), vol. 1397, pp. 307–312. Springer, Heidelberg (1998)
22. Goranko, V., Otto, M.: Model theory of modal logic. In Blackburn, P., Wolter, F., van Benthem, J. (eds.) Handbook of Modal Logic, pp. 255–325. Elsevier (2006)
23. Fermüller, C., Leitsch, A., Hustadt, U., Tammet, T.: Resolution decision procedures. In: Robinson, A., Vorkonov, A. (eds.) Handbook of Automated Reasoning, pp. 1791–1849. Elsevier, Amsterdam (2001)

And... Action! – Monoid Actions
and (Pre)orders

Nikita Danilenko[✉]

Institut für Informatik, Christian-Albrechts-Universität Kiel,
Olshausenstraße 40, 24098 Kiel, Germany
nda@informatik.uni-kiel.de

Abstract. Orders occur naturally in many areas of computer science
and mathematics. In several cases it is very simple do describe an order
mathematically, but it may be cumbersome to implement in some programming language. On the other hand many order relations are defined
in terms of an existential quantification. We provide a simple abstraction
of such definitions using the well-known concept of monoid actions and
furthermore show that in fact every order relation can be obtained from
a specific monoid action.

1 Introduction

In beginners' courses on mathematics (for computer scientists) several ordering
relations are usually given as examples. Consider for instance the natural order
on the natural numbers $\leq_\mathbb{N}$ defined in terms of

$$x \leq_\mathbb{N} y :\Longleftrightarrow \exists z \in \mathbb{N} : x + z = y$$

for all $x, y \in \mathbb{N}$. It is a basic and straightforward task to verify the three laws
(reflexivity, transitivity and anti-symmetry)[1] of an order relation for $\leq_\mathbb{N}$. A short
time after this definition we might find ourselves confronted with the concept of
lists and prefix lists. Now suppose that "$+\!\!+$" denotes the list concatenation and
consider the definition of a prefix list: for any set M and any two lists $x, y \in M^*$
we define

$$x \trianglelefteq y :\Longleftrightarrow \exists z \in M^* : x +\!\!+ z = y$$

and call $\trianglelefteq$ the "is-prefix-of"-relation. Observe how the definition itself is basically
the one of $\leq_\mathbb{N}$ – we merely exchanged $\mathbb{N}$ and $+$ by M^* and $+\!\!+$ respectively. We
know that $(\mathbb{N}, +, 0)$ and $(M^*, +\!\!+, \varepsilon)$ are monoids and this facts seems to be
encoded in the definition of the orders somehow. Let us have a look at a final
example, namely the order on $\mathbb{Z}$, which is defined for all $x, y \in \mathbb{Z}$ by:

$$x \leq_\mathbb{Z} y :\Longleftrightarrow \exists z \in \mathbb{N} : x + z = y.$$

---

[1] A reflexive and transitive relation is called **preorder**.

M. Hanus and R. Rocha (Eds.): KDPD 2013, LNAI 8439, pp. 83–98, 2014.
DOI: 10.1007/978-3-319-08909-6_6, © Springer International Publishing Switzerland 2014

Again we notice the resemblance to the previous definitions, but in this case the addition is a "skew" one, since its type is $+ : \mathbb{Z} \times \mathbb{N} \to \mathbb{Z}$.

We observe that all of these orders are structurally defined as

$$x \sqsubseteq y :\Longleftrightarrow \exists z \in M : x \otimes z = y,$$

for all $x, y \in A$, where A is a set, $(M, *, e)$ is a monoid and $\otimes : A \times M \to A$ is a function that is associative in the sense that $(a \otimes m) \otimes m' = a \otimes (m * m')$ for all $a \in A, m, m' \in M$.

From these observations we can derive a simple concept for defining order relations and study order properties in terms of this concept. While the basic "ingredients" (as monoid actions) are well-known and have been much studied, to the best of our knowledge our approach to orders has not.

In the following we will use "functions"[2] called "swap", "curry" and "fps". For sets A, B, C and a function $f : A \times B \to C$ we have that

$$\mathrm{swap}(f) : B \times A \to C, \, (b, a) \mapsto f(a, b),$$

$$\mathrm{curry}(f) : A \to C^B, \, a \mapsto \left(b \mapsto f(a, b) \right),$$

$$\mathrm{fps} : A^A \to 2^A, \, g \mapsto \{ a \in A \mid g(a) = a \}.$$

Note that "curry" is essentially the homonymous function from functional (logic) programming and "swap" is the uncurried version of *flip*. However, the function "fps" (mnemonic: Fixed Point Set) is not to be confused with the least fixpoint operator *fix* :: $(a \to a) \to a$ which is also well established in from functional (logic) programming: for instance, we have $\mathrm{fps}(\mathrm{id}_A) = A$, while *fix id* diverges. We use $f(a, -) = (x \mapsto f(a, x))$ and $f(-, b) = (x \mapsto f(x, b))$ to denote partial applications and f^{-1} for the preimage of f.

The paper is structured as follows.

- We present the abstraction behind the orders we have just discussed.
- A characterisation of orders in terms of monoid actions is given.
- It is shown how to obtain an action that creates a given order.
- We provide an implementation of functions that can be used to obtain orders in the functional logic language Curry.

2   Monoid Actions

The concept of a structure (e.g. group, algebra) acting on some set (or other structure) is well-known in mathematics, particularly in algebraic contexts.

Let us begin with a simple abstraction of the observation concerning the function type of the addition in the last example of the introduction[3]. Recall

---

[2] All of these can be considered functional classes in the sense of set theory.

[3] Actually we should have written $+ : \mathbb{N} \times \mathbb{Z} \to \mathbb{Z}$ for congruence with the following definition. The reason we did not is that for the three examples the monoid is better placed in the second component, whereas from a mathematical point of view it is more convenient to place it in the first one.

that a monoid is an algebraic structure $(M, \cdot, e)$ where $\cdot$ is a binary, associative operation and e is a neutral element with respect to $\cdot$.

Definition 1 (Monoid action). *Let $(M, \cdot, e)$ be a monoid and A a non-empty set. A mapping $\varphi : M \times A \to A$ is called **monoid action of M on A**, if and only if the following conditions hold:*

(1) $\varphi(e, -) = id_A$,   *(preservation of unity)*
(2) $\forall x, y \in M : \forall a \in A : \varphi(x, \varphi(y, a)) = \varphi(x \cdot y, a)$ *(associativity).*

Thus a monoid action gives us an outer operation on A (cf. inner operations). If $\varphi : M \times A \to A$ is a monoid action, we will abbreviate $\varphi(m, a) =: m \cdot_\varphi a$ for all $m \in M$ and $a \in A$.

Considering a monoid action in its curried version $\text{curry}(\varphi) : M \to A^A$ gives us that $\text{curry}(\varphi)$ is a monoid homomorphism. Conversely every monoid homomorphism $f : M \to A^A$ can be converted into a monoid action by defining $\varphi_f : M \times A \to A, (m, a) \mapsto f(m)(a)$. In fact these two operations are mutually inverse. These properties are known so well that they constitute typical exercises for students.

We proceed to provide some examples of monoid actions.

Example (Monoid actions)

1. Let $(M, \cdot, e)$ be a monoid. Then $\cdot$ is a monoid action of M on M.
2. The mapping $+ : \mathbb{N} \times \mathbb{Z} \to \mathbb{Z}$ is a monoid action of $(\mathbb{N}, +, 0)$ on $\mathbb{Z}$.
3. Let (Q, Σ, δ) be a transition system. Then $\text{swap}(\delta^*) : \Sigma^* \times Q \to Q$ is a monoid action of $(\Sigma^*, \text{swap}(+\!\!+), \varepsilon)$ on Q.
4. Let A be a set and $\varphi : A^A \times A \to A, (f, x) \mapsto f(x)$. Then φ is a monoid action of $(A^A, \circ, id_A)$ on A.

These properties are easily checked: the first one is trivially true, the second one can be shown in a large variety of simple ways, the third one relies on the fact that δ^* is the homomorphic continuation of δ on Σ^* and the fourth one merely rephrases elementary properties of function composition and application.   □

It is little surprising that monoid actions have certain *permanence properties* e.g. direct products of monoid actions form monoid actions. Categorically speaking these properties state that the category of monoid acts is closed under certain operations. We will not deal with these properties since they are well known and not required in the remainder of this article. Instead we use the concept of monoid actions to define the (ordering) relations we have seen in the introduction.

Definition 2 (Action preorder). *Let $(M, \cdot, e)$ be a monoid, A a set and $\varphi : M \times A \to A$ a monoid act. We then define for all $u, b \in A$:*

$$a \sqsubseteq_\varphi b :\Longleftrightarrow \exists m \in M : m \cdot_\varphi a = b.$$

*The relation $\sqsubseteq_\varphi$ is called **action preorder**.*

Note that the definition captures the essence of all orders we have presented in the beginning of the paper. To justify the anticipatory name of the relation we need to show a simple lemma. The proof is very simple and we include it only for the purpose of demonstration.

Lemma 1 (Action preorder). *Let $(M, \cdot, e)$ be a monoid, A a set and $\varphi : M \times A \to A$ a monoid action. Then the following hold:*

(1) The action preorder $\sqsubseteq_\varphi$ is in fact a preorder on A.
(2) For all $m \in M$ the mapping $\varphi(m, -)$ is expanding, i.e. $\forall a \in A : a \sqsubseteq_\varphi m \cdot_\varphi a$.

Proof (1) Let $x \in A$. By the preservation of units we get $x = e \cdot_\varphi x$, thus $x \sqsubseteq_\varphi x$. Now let $a, b, c \in A$ such that $a \sqsubseteq_\varphi b$ and $b \sqsubseteq_\varphi c$. Then there are $m, n \in M$ such that $m \cdot_\varphi a = b$ and $n \cdot_\varphi b = c$. The associativity of the action gives us

$$c = n \cdot_\varphi b = n \cdot_\varphi (m \cdot_\varphi a) = (n \cdot m) \cdot_\varphi a.$$

Since $n \cdot m \in M$ we get $a \sqsubseteq_\varphi c$.
(2) Left as an exercise to the reader.   □

Let us make two observations concerning this lemma. First – showing the reflexivity and transitivity of actual relations (like $\leq_\mathbb{N}$ or $\trianglelefteq$) will always result in essentially the very proof of this lemma. Second – the two properties "preservation of units" and "associativity" of monoid acts supply the sufficient conditions for "reflexivity" and "transitivity" respectively.

So far we have seen some examples of action preorders (that incidentally were orders as well). In such a setting two questions suggest themselves:

1. When is an action preorder an order?
2. Is every order an action preorder?

Ideally the answer to the first question should be some kind of characterisation and the answer to the second should be a Boolean value followed by a construction in the positive case.

Before we turn to the use of these definitions and properties for implementation we would like to provide answers to both questions. The applications will follow in Sect. 4.

Finally, let us note that the *relation* defined by the action preorder is very well known in group theory in the context of group actions. In this context the relation above is always an equivalence relation (again this is well-known and often used as an exercise) and is commonly used to investigate action properties (cf. the orbit-stabiliser-theorem [1].). To the best of our knowledge little effort has been invested in the study of this relation in the presence of monoid actions.

3   Action Preorders

First of all let us deal with the question when an action preorder is an order. When we start to prove the antisymmetry of an ordering relation like $\leq_\mathbb{N}$ we take

$a, b \in \mathbb{N}$ s.t. $a \leq_\mathbb{N} b$ and $b \leq_\mathbb{N} a$. Then we find that there are $c, d \in \mathbb{N}$ satisfying $c + a = b$ and $d + b = a$. Thus we get

$$a = d + b = d + (c + a) = (d + c) + a.$$

So far we have used the associativity of $+$, but from the equation above we need to find that $a = b$. In case of the naturals we would probably proceed as follows: since $a = (d + c) + a$, we find that $0 = d + c$ and then $d = 0 = c$. We used injectivity of adding a number in the first step and some kind of "non-invertibility property" in the second one. Clearly, requiring these properties in an abstracted fashion immediately results in the proof of antisymmetry. Since we used a single proof layout for this abstraction it is not surprising that these two properties turn out to be sufficient, but not necessary conditions for the antisymmetry of the action preorder. Fortunately they can be abstracted into a single property that is applicable in the general case.

Proposition 1 (Characterisation of antisymmetry I). *Let $(M, \cdot, e)$ be a monoid, A a set and $\varphi : M \times A \to A$ a monoid action. Then the following statements are equivalent:*

(1) The action preorder $\sqsubseteq_\varphi$ is antisymmetric (i.e. an order).
(2) $\forall x, y \in M : \forall a \in A : (x \cdot y) \cdot_\varphi a = a \Rightarrow y \cdot_\varphi a = a$.

Proof $(1) \Longrightarrow (2)$: We invite the reader to verify this on his or her own.

$(2) \Longrightarrow (1)$: Assume that (2) holds. Let $a, b \in A$ such that $a \sqsubseteq_\varphi b$ and $b \sqsubseteq_\varphi a$. Then there are $x, y \in M$ such that $x \cdot_\varphi b = a$ and $y \cdot_\varphi a = b$, hence

$$a = x \cdot_\varphi b = x \cdot_\varphi (y \cdot_\varphi a) = (x \cdot y) \cdot_\varphi a.$$

By (2) this yields $y \cdot_\varphi a = a$, but on the other hand $b = y \cdot_\varphi a$, so $a = b$. □

Note how the proof of $(2) \Longrightarrow (1)$ resembles our exemplary proof from the beginning of this section. As we have mentioned before, $\mathrm{curry}(\varphi)$ is a monoid homomorphism and thus its image $S := \mathrm{curry}(\varphi)(M)$ is a submonoid of A^A. Observe that if there is a function $f \in S$ such that f is invertible in S and $f \neq \mathrm{id}_A$, there are $x, y \in M$ such that $f = \mathrm{curry}(\varphi)(y)$ and $\mathrm{curry}(\varphi)(x) \circ f = \mathrm{id}_A$. We then find that there is an $a \in A$ such that $y \cdot_\varphi a = f(a) \neq a$, but $(x \cdot y) \cdot_\varphi a = a$, so the above proposition states that $\sqsubseteq_\varphi$ is *not* antisymmetric. In other words: if S has non-trivial invertible elements, the corresponding preorder is not an order. In particular, if M is a group, so is S and if M additionally contains more than one element and $\mathrm{curry}(\varphi)$ is not the trivial homomorphism (i.e. $m \mapsto \mathrm{id}_A$) then S is a non-trivial group and thus $\sqsubseteq_\varphi$ is not an order, because it contains non-trivial invertible elements.

The property that is equivalent to the antisymmetry can be viewed as a kind of fixpoint property: for all $x, y \in M$ and $a \in A$ we have that if a is a fixpoint of $b \mapsto (x \cdot y) \cdot_\varphi b$ it is also a fixpoint of $b \mapsto y \cdot_\varphi b$ (which also implies that it is a fixpoint of $b \mapsto x \cdot_\varphi b$). This fact can be expressed as follows.

Proposition 2 (Characterisation of antisymmetry II). *Let $(M, \cdot, e)$ be a monoid, A a set and $\varphi : M \times A \to A$ a monoid action. Then the following statements are equivalent:*

(1) The action preorder $\sqsubseteq_\varphi$ is antisymmetric (i.e. an order).
(2) $\mathrm{fps} \circ \mathrm{curry}(\varphi) : (M, \cdot, e) \to (2^A, \cap, A)$ is a monoid homomorphism.

Proof Let $\psi := \mathrm{fps} \circ \mathrm{curry}(\varphi)$. First of all we find that

$$(\mathrm{fps} \circ \mathrm{curry}(\varphi))(e) = \mathrm{fps}(\mathrm{curry}(\varphi)(e)) = \mathrm{fps}(\mathrm{id}_A) = A \,.$$

Now let $x, y \in M$. Then we can reason as follows:

$\quad \forall\, a \in A : (x \cdot y) \cdot_\varphi a = a \Rightarrow y \cdot_\varphi a = a$
$\Longleftrightarrow \{$ note above this proposition $\}$
$\quad \forall\, a \in A : (x \cdot y) \cdot_\varphi a = a \Rightarrow y \cdot_\varphi a = a \wedge x \cdot_\varphi a = a$
$\Longleftrightarrow \{$ fixpoint rephrasing $\}$
$\quad \forall\, a \in A :$
$\qquad a \in \mathrm{fps}(\mathrm{curry}(\varphi)(x \cdot y)) \Rightarrow a \in \mathrm{fps}(\mathrm{curry}(\varphi)(x)) \cap \mathrm{fps}(\mathrm{curry}(\varphi)(y))$
$\Longleftrightarrow \{$ definition of $\subseteq$ $\}$
$\quad \mathrm{fps}(\mathrm{curry}(\varphi)(x \cdot y)) \subseteq \mathrm{fps}(\mathrm{curry}(\varphi)(x)) \cap \mathrm{fps}(\mathrm{curry}(\varphi)(y))$
$\Longleftrightarrow \{\ (*)\ \}$
$\quad \mathrm{fps}(\mathrm{curry}(\varphi)(x \cdot y)) = \mathrm{fps}(\mathrm{curry}(\varphi)(x)) \cap \mathrm{fps}(\mathrm{curry}(\varphi)(y))$
$\Longleftrightarrow \{$ definition of composition and application $\}$
$\quad \psi(x \cdot y) = \psi(x) \cap \psi(y).$

The equivalence denoted by $(*)$ is simple, since for any functions $f, g : A \to A$ we have that if $x \in \mathrm{fps}(f) \cap \mathrm{fps}(g)$ then $f(g(x)) = f(x) = x$ and thus $x \in \mathrm{fps}(f \circ g)$. Now we get

$\qquad \sqsubseteq_\varphi$ is antisymmetric
$\Longleftrightarrow \{$ by Lemma 1 $\}$
$\quad \forall\, m, n \in M : \forall\, a \in A : (m \cdot n) \cdot_\varphi a = a \Rightarrow n \cdot_\varphi a = a$
$\Longleftrightarrow \{$ equivalence above $\}$
$\quad \forall\, m, n \in M : \psi(m \cdot n) = \psi(m) \cap \psi(n)$
$\Longleftrightarrow \{$ see above $\}$
$\quad \psi$ is a monoid homomorphism.

$\qquad\qquad\qquad\qquad\qquad\qquad\qquad\qquad\qquad\qquad\qquad\qquad\qquad\qquad\qquad$ $\square$

Let us now show how to create a fitting monoid and a monoid action for a given preorder. The idea is quite simple – we want to define a transition system, such that its transition function is the action. To do that, we observe that orders and preorders are quite often drawn as their Hasse diagrams. These diagrams

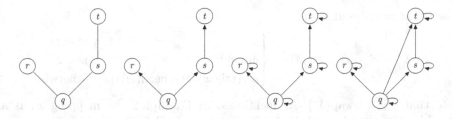

Fig. 1. A Hasse diagram, added directions, added reflexivity, added transitivity

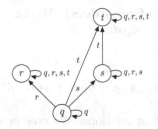

Fig. 2. An order transformed to a transition system

are designed specifically to omit reflexivity and transitivity since they can be restored in a trivial fashion as demonstrated in Fig. 1. The last image bears a striking resemblance with a transition system except that there is no alphabet that can be used to act upon the states. Still, we can introduce an alphabet as indicated in Fig. 2.

The sketched idea behind this alphabet can be formalised as follows. Let A be a non-empty set and $\preccurlyeq \subseteq A \times A$ a preorder, set $S := A$, $\Sigma := A$ and

$$\delta : S \times \Sigma \to S, \quad (s,\sigma) \mapsto \begin{cases} \sigma & : s \preccurlyeq \sigma \\ s & : \text{otherwise}. \end{cases}$$

Obviously δ is well-defined. Let $\mathfrak{A} := (S, \Sigma, \delta)$. Then $\mathfrak{A}$ is a transition system. Let $x, y \in A$ and consider x a state and y a letter. This consideration yields the equivalence:

$$\delta(x, y) = y \iff x \preccurlyeq y.$$

When we rewrite δ as an infix operation (i.e. $x \, \delta \, y$ instead of $\delta(x,y)$), the above equivalence is very similar to the following well-known property of sets:

$$A \cup B = B \iff A \subseteq B.$$

In any lattice $(L, \sqcup, \sqcap)$ (which generalises the powerset of a set) we have

$$a \sqcup b = b \iff a \sqsubseteq b$$

by definition and in every idempotent, commutative semigroup $(S, +)$ we have

$$a + b = b \iff a \leq b.$$

Now let $M := \Sigma^*$ and

$$\varphi_\delta : \Sigma^* \times S \to S, \quad (w,s) \mapsto \begin{cases} s & : w = \varepsilon \\ \delta(s,w) & : w \in \Sigma \\ \varphi_\delta\,(\mathsf{tail}(w), \delta(s, \mathsf{head}(w))) & : \text{otherwise}. \end{cases}$$

Note that $\varphi_\delta = \mathsf{swap}(\delta^*)$. As mentioned in Example 2 the mapping φ_δ is a monoid action, which yields that $\sqsubseteq_{\varphi_\delta}$ as defined in Definition 2 is a preorder on $S = A$. It turns out that this preorder is the same relation as the original one.

Theorem 1 (Generation of preorders). *We have $\sqsubseteq_{\varphi_\delta} = \preccurlyeq$, where φ_δ is the action and $\preccurlyeq$ is the preorder introduced above.*

Proof First of all we have:

$$\sqsubseteq_\varphi = \preccurlyeq \iff \forall a,b \in A : a \preccurlyeq b \Leftrightarrow \left(\exists w \in A^* : \varphi(w,a) = b\right).$$

by definition of the equality of relations. We now prove the second statement. "$\Rightarrow$": Let $a,b \in A$ such that $a \preccurlyeq b$ and $w := b$. Then $w \in \Sigma \subseteq \Sigma^*$ and we have:

$$\varphi(w,a) = \delta(a,w) = \delta(a,b) = b.$$

"$\Leftarrow$": To simplify the proof we observe that the following holds:

$$\forall a,b \in A : \left(\exists w \in A^* : \varphi(w,a) = b\right) \Rightarrow a \preccurlyeq b \qquad (*)$$

$$\iff \quad \forall a,b \in A : \forall w \in A^* : \varphi(w,a) = b \Rightarrow a \preccurlyeq b$$

$$\iff \quad \forall w \in A^* : \forall a,b \in A : \varphi(w,a) = b \Rightarrow a \preccurlyeq b.$$

The equivalence marked with $(*)$ holds since the conclusion is independent of w (if x is not free in Q, then $\forall x : (P(x) \Rightarrow Q)$ is equivalent to $(\exists x : P(x)) \Rightarrow Q$ by De Morgan's law and distributivity). The latter statement will be proved by induction on the word length.

Ind. beginning: Let $w \in A^*$ such that $|w| = 0$. Then we have that $w = \varepsilon$. Now let $a,b \in A$ such that $\varphi(w,a) = b$. This gives us $b = \varphi(w,a) = \varphi(\varepsilon,a) = a$ and the reflexivity of $\preccurlyeq$ yields $a \preccurlyeq b$.

Ind. hypothesis: Let $n \in \mathbb{N}$ and assume that the following holds:

$$\forall w \in A^* : |w| = n \Rightarrow \left(\forall a,b \in A : \varphi(w,a) = b \Rightarrow a \preccurlyeq b\right).$$

Ind. step: Let $v \in A^*$ such that $|v| = n + 1$. Then there are $x \in A$ and $w \in A^*$ such that $v = xw$ and $|w| = n$. Let $a,b \in A$ such that $\varphi(v,a) = b$. Then we have:

$$b = \varphi(v,a) = \varphi(xw,a) = \varphi(w, \delta(a,x)) = \begin{cases} \varphi(w,x) & : a \preccurlyeq x \\ \varphi(w,a) & : \text{otherwise}. \end{cases}$$

If $a \preccurlyeq x$, then by the induction hypothesis we have $x \preccurlyeq b$, which, by the transitivity of $\preccurlyeq$, results in $a \preccurlyeq b$. If on the other hand $a \not\preccurlyeq x$, then $b = \varphi(w, a)$ and by the induction hypothesis we immediately obtain $a \preccurlyeq b$, since $|w| = n$. This concludes the induction and the proof as well. $\square$

We now rephrase the result of this section for the purpose of legibility.

Corollary 1 (Preorders are monoidally generated). *Let A be a non-empty set and $\preccurlyeq \subseteq A \times A$ a preorder. Then there is a monoid M and a monoid action $\varphi : M \times A \to A$ such that $\preccurlyeq = \sqsubseteq_\varphi$.*

For any non-empty set A the set A^* is infinite (and countable iff A is countable). This monoid is somewhat large, since it is the free monoid generated by the set A. However we can use the well-known quotient construction (cf. [8]) to obtain an action in which different monoid elements act differently[4] (i.e. curry(φ) is injective; such an action is called *faithful*). To form a quotient one defines for all $m, n \in M$:

$$m \sim n :\Longleftrightarrow \text{curry}(\varphi)(m) = \text{curry}(\varphi)(n).$$

Clearly, $\sim$ is an equivalence relation, moreover it is a monoid congruence and thus $M/_\sim$ is a monoid as well. The new action

$$\varphi_{\text{quotient}} : M/_\sim \times A \to A, ([m]_\sim, a) \mapsto \varphi(m, a)$$

is then faithful. The quotient monoid is usually far smaller than the free monoid. Additionally, in many cases the quotient monoid can be described in a less generic (and more comprehensible) way than as the quotient modulo a congruence.

Let us revisit the antisymmetry of the action preorder once more in the context of transition systems. Consider the transition system in Fig. 3. The preorder induced by this action is not an order, because we have $\delta(1, q) = r$ and $\delta(1, r) = q$, which gives us $q \sqsubseteq_\delta r$ and $r \sqsubseteq_\delta q$ respectively, but $q \neq r$. What is the key ingredient to break antisymmetry in this example? It is the existence of a non-trivial cycle. With our previous results we can then prove the following lemma.

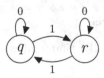

Fig. 3. An action that doesn't induce an order

---

[4] The cited source deals with finite state sets only, but the technique easily carries over to infinite sets as well.

Proposition 3 (Characterisation of antisymmetry III). *Let* $\mathfrak{A} = (S, \Sigma, \delta)$
be a transition system. Then the following statements are equivalent:

(1) $\sqsubseteq_{\text{swap}(\delta*)}$ *is antisymmetric.*
(2) $\mathfrak{A}$ *contains no non-trivial cycles, i.e. if a word does not change a state, so does every prefix of this word.*

This can be proved with Proposition 1, since the second condition is easily translated into the second statement of the cited proposition.

There is an interesting analogue of this proposition in graph theory. Consider a graph $G = (V, E)$, where V is a set and $E \subseteq V \times V$. The reachability relation $\leadsto_G$ of G is given by

$$x \leadsto_G y : \iff \text{there is a path from } x \text{ to } y$$

for all $x, y \in V$. It is easy to see that $\leadsto_G$ is reflexive and transitive[5]. Also, it is well-known that if G contains no non-trivial cycles (i.e. loops are allowed), then $\leadsto_G$ is an order relation. One can also show that the antisymmetry of $\leadsto_G$ results in no non-trivial cycles. These facts demonstrate that antisymmetry and cycle-freeness are closely related and can be considered in the very same light.

4   Implementation in Curry

In this section we use monoid actions to implement orders in the functional logical programming language Curry (cf. [7]). We provide a simple prototypical implementation of monoid actions and resulting preorders and discuss its shortcomings as well. To run our code we use the KiCS2 compiler (see [6]).

The components of a monoidally generated order (a monoid and a monoid action) can be expressed more generally for simplicity. We can use that to implement a very simple version of the general preorder.[6]

> **type** *MonoidAction* μ $\alpha = \mu \to \alpha \to \alpha$
> **type** *OrderS* $\alpha = \alpha \to \alpha \to Success$
> *preOrder* :: *MonoidAction* μ $\alpha \to OrderS$ α
> *preOrder* $(\otimes)$ x $y = z \otimes x =:= y$
> **where** z **free**

Let us consider an example. To that end we define the naturals as Peano numbers.

---

[5] In fact $\leadsto_G$ is the reflexive-transitive closure of E, cf. [9].
[6] Clearly this is a greatly simplified approach, since not every function $f :: \mu \to \alpha \to \alpha$ is a monoid action. A user has to verify that μ is a monoid and f is a monoid action to ensure that *preOrder f* is in fact a preorder. Alternatively, a proof assistant (e.g. Coq [4]) can be used to guarantee that *preOrder* is applicable only once the necessary conditions have been proved.

$$\textbf{data } \mathbb{N} = \mathbb{O} \mid S\ \mathbb{N}$$
$$(\oplus) :: \mathbb{N} \to \mathbb{N} \to \mathbb{N}$$
$$\mathbb{O} \ \oplus y = y$$
$$S\ x \oplus y = S\ (x \oplus y)$$

This addition yields a notion of comparison[7]:

$$(\sqsubseteq_\oplus) :: OrderS\ \mathbb{N}$$
$$(\sqsubseteq_\oplus) = preOrder\ (\oplus)$$

Loading these definitions in KiCS2 we get

kics2> $S\ (S\ \mathbb{O}) \sqsubseteq_\oplus \mathbb{O}$
No more values
kics2> $S\ (S\ \mathbb{O}) \sqsubseteq_\oplus S\ (S\ \mathbb{O})$
Success
No more values
kics2> $x \sqsubseteq_\oplus S\ \mathbb{O}$ **where** x **free**
$\{x = (S\ \mathbb{O})\}$ *Success*
$\{x = \mathbb{O}\}$ *Success*
No more values

Note that the very concept of an action preorder uses both concepts integrated in Curry – a functional component "$\otimes$" and a logical one "z **where** z **free**".

While this implementation is very close to its mathematical basis, the reader may have noticed that our order type does not have a relational look-and-feel, since orders are more likely to have the type $\alpha \to \alpha \to Bool$. Such a relational version can be obtained by using negation as failure which can be expressed elegantly in Curry using set functions [3]. Since our implementation is intended to be prototypical and the above type is rather natural in light of logic programming we omit the presentation of such a version.

Let us illustrate these "logical orders" with an example function. In total orders[8] each two elements are comparable, which allows the notion of a maximum.

$$maximumBy :: OrderS\ \alpha \to \alpha \to \alpha \to \alpha$$
$$maximumBy\ (\leqslant)\ x\ y \mid x \leqslant y = y$$
$$maximumBy\ (\leqslant)\ x\ y \mid y \leqslant x = x$$

The rules we gave are overlapping – in case of equality both rules are applicable, resulting in multiple results. Fortunately, when $\leq$ is an order all results are equal, since the constraints $x \leq y$ and $y \leq x$ imply that $x = y$.

Clearly, the previous example of an order generated by a monoid action was trivial, since the action was the monoid operation itself. Let us consider a nontrivial example next – the "has-suffix-order" $\unrhd$ on A^* for some given set A. The order is defined by

$$xs \unrhd ys :\Longleftrightarrow \exists zs \in A^* : xs = zs +\!\!+ ys\,.$$

---

[7] It is simple (but lengthy) to define the integers based on the naturals and to extend the definitions of $\oplus$ and $\sqsubseteq_\oplus$ to allow the implementation of $\leq_{\mathbb{Z}}$.

[8] An order $\leq\subseteq A \times A$ is called **total** iff for all $x, y \in A$ it is true that $x \leq y$ or $y \leq x$.

Note that the definition does not hint at a possible monoid action that generates this order, because such an action needs to apply a function to the *first* element of the comparison and not the second one. Still we can calculate the following:

$$xs \trianglerighteq ys \iff \exists zs \in A^* : xs = zs +\!\!+ ys$$
$$\iff \exists zs \in A^* : zs +\!\!+ ys = xs$$
$$\iff ys \sqsubseteq_{(+\!\!+)} xs,$$

which can be rephrased as $(\trianglerighteq) = flip\ (preOrder\ (+\!\!+))$. How is this relation more interesting? We were able to define this order as a flipped version of another order. Clearly this is interesting in its own right, but Corollary 1 states that there is a monoid and a monoid action yielding precisely the order we need. Let us define an auxiliary function.[9]

$$drop_N :: \mathbb{N} \to [\alpha] \to [\alpha]$$
$$drop_N\ \mathbb{O} \qquad x@(\_:\_) = x$$
$$drop_N\ \_ \qquad [] \qquad\quad = []$$
$$drop_N\ (S\ x)\ (\_ : ys) \quad = drop_N\ x\ ys$$

We observe that the following holds for all $xs, ys :: [\alpha]$:

$$xs \trianglerighteq ys \iff \exists n :: \mathbb{N} : drop_N\ n\ xs = ys.$$

This does look like an action preorder. But is $drop_N$ an action? It is indeed.

Lemma 2 (Properties of $drop_N$). *Let A be a non-empty set and*

$$\delta : \mathbb{N} \times A^* \to A^*,\ (n, l) \mapsto \begin{cases} l & : n = 0 \\ & : l = [] \\ \delta(n', l') & : \exists n' \in \mathbb{N} : n = 1 + n' \wedge \exists x \in A : l = x : l'. \end{cases}$$

Then δ is (well-defined and) a monoid action of $\mathbb{N}$ on A^* and $\trianglerighteq\ =\ \sqsubseteq_\delta$ is an order.

Using properties of natural numbers the proof is basically a straightforward induction. We omit it for two reasons – avoiding unnecessary clutter and the fact that δ from the above lemma is only a version of $drop_N$ that operates on finite and deterministic arguments, while $drop_N$ can be used on infinite or non-deterministic arguments as well.

We can implement the order $\trianglerighteq$ in terms of $drop_N$.

$$(\trianglerighteq) :: OrderS\ [\alpha]$$
$$(\trianglerighteq) = preOrder\ drop_N$$

For comparison we also define the version discussed above.

$$(\trianglerighteq_2) :: OrderS\ [\alpha]$$
$$(\trianglerighteq_2) = flip\ (preOrder\ (+\!\!+))$$

---

[9] The pattern matching in the first rule makes the rules non-overlapping.

The difference between these implementation is that $(\unrhd_2)$ searches in an upward fashion, while $(\unrhd)$ does the same in an downward direction. This is to say that for $(\unrhd_2)$ we perform the unification $z +\!\!+ y =\!:= x$ with a free variable z. This creates a list structure omitting unnecessary components, which essentially searches for a possible number of cons cells that can be ignored in x to obtain y. Using $(\unrhd)$ does precisely that explicitly, because we search for a natural number of elements to explicitly drop from x to obtain y.

Typically functionally similar parts of programs are abstracted as far as possible to be applicable in different situations. In functional languages such abstractions usually include higher-order functions that take necessary operations as additional arguments. Our implementation of *preOrder* is such a higher order function, that allows us to define *every* order in terms of a specific action.

A drawback of the general abstraction is its possible complexity or even indecidability. Concerning the complexity consider the comparison of $one = S\ \mathbb{O}$ to some value $(S\ x) :: \mathbb{N}$. Using $\sqsubseteq_\oplus$ this operation is quadratic in the size of $S\ x$. Clearly an implementation of the same order "by hand" requires precisely two comparisons – first match the outer S and then compare $\mathbb{O}$ to x, which is automatically true. As for the indecidability we consider integer lists. Now consider the list $ones = 1 : ones$. We can then define $x = 0 : 1 : ones$ and $y = 1 : 0 : ones$. When equipped with the lexicographical order $\leq_{\text{lex}}$, which we implement by hand, we can easily determine that $x \leq_{\text{lex}} y$. Now suppose we have found a monoid action φ that generates the order $\leq_{\text{lex}}$. Then there is a value z such that $\varphi\ z\ x = y$ (mathematically!), but this equality is not decidable.

In general the above implementation is applicable to orders on finite terms without any occurrences of $\bot$ (the latter is due to the semantics of $=\!:=$, see [7]).

The presented implementation is not exclusive to Curry (in particular because KiCS2 translates Curry to Haskell) and it is possible to translate it into a Haskell program by using a non-determinism monad (cf. [5]) after the replacement of logical variables by overlapping rules (e.g. [2]). It should be possible to translate the latter implementation to any language with higher-order functions.

5   An Alternative Abstraction

The reader may have noticed that until now we have presented orders on countable sets only. But what about, say, the order on $\mathbb{R}$? The usual definition states that for $x, y \in \mathbb{R}$ we have

$$x \leq_\mathbb{R} y :\Longleftrightarrow y - x \in \mathbb{R}_{\geq 0} \iff \exists z \in \mathbb{R}_{\geq 0} : z + x = y.$$

Structurally the latter definition looks exactly like the order on $\mathbb{Z}$. While it seems odd that we use $\mathbb{R}_{\geq 0}$ in the definition, we recall that the order on $\mathbb{R}$ requires the existence of a so-called *positive cone* that is named $\mathbb{R}_{>0}$ in this example and $\mathbb{R}_{\geq 0} = \mathbb{R}_{>0} \cup \{0\}$. A positive cone of a group $(G, +, 0)$ is a $P \subseteq G$ that satisfies

$$P + P \subseteq P \quad \wedge \quad 0 \notin P \quad \wedge \quad G = -P \cup \{0\} \cup P,$$

where $P + P$ and $-P$ are understood element-wise. Then $P_0 := P \cup \{0\}$ is a submonoid of G and $+|_{P_0 \times G}$ is a monoid action. This construction covers the action that creates the order on $\mathbb{Z}$ as well as the action that induces the order on $\mathbb{R}$.

Observe that the notion of a positive cone requires an inversion operation of the group operation. In case of monoids the operation of the monoid usually cannot be inverted. Replacing the monoid with a group transforms the monoid action into a group action (without any additional requirements). Then the action preorder becomes an equivalence relation and non-trivial orders are never equivalences. In fact the concept of cycle-freeness requires the monoids in our examples to be "anti-groups", which is to say that the image of the monoid under the curried action does not contain non-trivial invertible elements. What is more is that every order that is defined in terms of a positive cone is total, which is easily shown using the above definition. Since not every order is total, we cannot expect to find a "positive-cone representation" for every order.

As we have mentioned above the use of a positive cone to define an order is a special case of a monoid act. The latter requires less structure and is thus easily defined. The former automatically comes with more properties and is generally more natural in the context of ordering groups or fields.

6   Related Work and Discussion

Monoid actions and corresponding preorders appear in different contexts naturally e.g. transition systems [8] and algebra [1]. The examples from the introduction constitute well-known orders that are defined as action preorders, while the "has-suffix"-order is slightly less common. Orders are also ubiquitous in computer science and mathematics, but they are mostly treated in as tools, rather than objects of research.

From our results we know that every order can be defined in terms of a monoid action and we have given an exemplary implementation in Curry. In our implementation we merely checked whether there is a variable, that acts on the first argument in a way that results in the second one. Clearly, we could also ask for this variable as well and obtain the following function.

$$\textit{cofactor} :: \textit{MonoidAct } \mu \, \alpha \to \alpha \to \alpha \to \mu$$
$$\textit{cofactor} (\otimes) \; a \; b \mid z \otimes a =:= b = z$$
$$\textbf{where } z \textbf{ free}$$

The above function bears a striking similarity with the following definition of an "inverse function".

$$\textit{inverse} :: (a \to b) \to b \to a$$
$$\textit{inverse } f \; y \mid f \; x =:= y = x$$
$$\textbf{where } x \textbf{ free}$$

In fact we can redefine *cofactor* in terms of *inverse*, namely:

$$\textit{cofactor} (\otimes) \; a \; b = \textit{inverse} \; (\otimes a) \; b$$

When translated back into mathematical notation the above states that *cofactor* yields some $z \in \varphi(-, a)^{-1}(\{b\})$, where φ is $\otimes$ uncurried. Searching for preimage values (at least implicitly) is a common task in logic programming that appears naturally in a variety of definitions, e.g.:

$$predecessor :: \mathbb{N} \to \mathbb{N}$$
$$predecessor\ n \mid S\ m =\!:= n = m$$
$$\textbf{where } m \textbf{ free}$$
$$predecessorAsPreimage = inverse\ S$$

Both *cofactor* and *inverse* are not necessarily functions in the mathematical sense, but potentially non-deterministic functions that may return more or less than one value for a single input. Such functions constitute an important component of typical functional logic programs and are well supported by Curry.

As we stated earlier, the action preorder is a well-known relation and we have used very little of existing knowledge. If $\varphi : M \times A \to A$ is a monoid action, we can consider the orbit of some element $a \in A$ that is defined as

$$\mathrm{orbit}_M(a) := \{m \cdot_\varphi a \mid m \in M\} = \{b \in A \mid \exists m \in M : m \cdot_\varphi a = b\}$$
$$= \{b \in A \mid a \sqsubseteq_\varphi b\}.$$

The latter set is simply the majorant of a (sometimes denoted $\{a\}^{\uparrow}$ or (a)). This simple connection allows to study the notion of majorants in the context of monoid actions (and vice versa) thus combining two well-known and well-studied concepts. An additional similarity occurs when comparing our application of the free monoid with the construction of a free group of a set [1]. We omit the details here due to lack of space and immediate applicability.

In Curry every type is already ordered lexicographically w.r.t. the constructors (see definition of *compare* in [7]). Clearly this gives a total order on every type, but the actual comparison results depend on the order of constructors, which requires careful choice of this order. Defining integers as

$$\textbf{data } \mathbb{Z} = Pos\ \mathbb{N} \mid Neg\ \mathbb{N}$$

leads to positive numbers being smaller than negative ones. Actions provide a simple way to define (additional) orders in terms of functions that do not depend on the order of the constructors. In the above example a library that defines the data type $\mathbb{Z}$ is likely to define an addition on integers $add :: \mathbb{Z} \to \mathbb{Z} \to \mathbb{Z}$ and an embedding $toInteger :: \mathbb{N} \to \mathbb{Z}$. With these two functions one can easily define

$$act :: \mathbb{N} \to \mathbb{Z} \to \mathbb{Z}$$
$$act\ n\ z = add\ (toInteger\ n)\ z$$

which yields the usual order on $\mathbb{Z}$ as *preOrder act*. An additional gain in using actions is that one can define orders in a more declarative way – instead of thinking about bit representations or constructors one simply states what is necessary for one number to be smaller than another.

The prototypical stencil of the above implementation can be varied in different ways. One variation concerns the adjustment of the type to the relational version $\alpha \to \alpha \to Bool$ that we have mentioned before. Often an order on a set A is naturally exp ressed in terms of another order $(B, \leq_B)$ and an injective function $f : A \to B$ such that for all $a, a' \in A$ we have

$$a \leq_A a' :\Longleftrightarrow f(a) \leq_B f(a').$$

Additionally if M is a monoid and $\varphi : M \times B \to B$ is a monoid action that generates $\leq_B$ we can incorporate f into the existential quantification:

$$a \leq_A a' \Longleftrightarrow \exists z \in M : z \cdot_\varphi f(a) = f(a').$$

While the codomain of f has at least the cardinality of A (because of the injectivity of f), the comparison of values in B may be less complex.

On a more theoretical note it is interesting to study properties of orders in terms of properties of actions and vice versa. For instance one can show that a faithful action of a monoid that has no invertible elements except for its unit always yields an infinite order, which is not obvious at first glance. We suspect that there are a lot more connections between these seemingly different concepts.

Hats off to: Fabian Reck for the detailed explanation of Curry, Rudolf Berghammer for encouraging this work and additional examples, Insa Stucke for general discussions and the reviewers for their highly informative and helpful feedback.

References

1. Aluffi, P.: Algebra: Chapter 0. AMS, London (2009)
2. Antoy, S., Hanus, M.: Overlapping rules and logic variables in functional logic programs. In: Etalle, S., Truszczyński, M. (eds.) ICLP 2006. LNCS, vol. 4079, pp. 87–101. Springer, Heidelberg (2006)
3. Antoy, S., Hanus, M.: Set functions for functional logic programming. In: Proceedings of the 11th International ACM SIGPLAN Conference on Principle and Practice of Declarative Programming (PPDP'09), pp. 73–82. ACM Press (2009)
4. Bertot, Y., Castéran, P.: Interactive Theorem Proving and Program Development. Coq'Art: The Calculus of Inductive Constructions. Springer, Heidelberg (2004)
5. Braßel, B., Fischer, S., Hanus, M., Reck, F.: Transforming functional logic programs into monadic functional programs. In: Mariño, J. (ed.) WFLP 2010. LNCS, vol. 6559, pp. 30–47. Springer, Heidelberg (2011)
6. Braßel, B., Hanus, M., Peemöller, B., Reck, F.: KiCS2: a new compiler from Curry to Haskell. In: Kuchen, H. (ed.) WFLP 2011. LNCS, vol. 6816, pp. 1–18. Springer, Heidelberg (2011)
7. Hanus, M. (ed.).: Curry: An Integrated Functional Logic Language (Vers. 0.8.3) (2012). http://www.curry-language.org
8. Holcombe, W.M.L.: Algebraic Automata Theory. Cambridge University Press, Cambridge (1982)
9. Schmidt, G., Ströhlein, T.: Relations and Graphs. Springer, Heidelberg (1993)

HEX-Programs with Existential Quantification

Thomas Eiter, Michael Fink$^{(\boxtimes)}$, Thomas Krennwallner, and Christoph Redl

Institut für Informationssysteme, Technische Universität Wien,
Favoritenstraße 9-11, 1040 Vienna, Austria
{eiter,fink,tkren,redl}@kr.tuwien.ac.at

Abstract. HEX-programs extend ASP by external sources. In this paper, we present *domain-specific existential quantifiers* on top of HEX-programs, i.e., ASP programs with external access which may introduce new values that also show up in the answer sets. Pure logical existential quantification corresponds to a specific instance of our approach. Programs with existential quantifiers may have infinite groundings in general, but for specific reasoning tasks a finite subset of the grounding can suffice. We introduce a generalized grounding algorithm for such problems, which exploits domain-specific termination criteria in order to generate a finite grounding for bounded model generation. As an application we consider query answering over existential rules. In contrast to other approaches, several extensions can be naturally integrated into our approach. We further show how terms with *function symbols* can be handled by HEX-programs, which in fact can be seen as a specific form of existential quantification.

1 Introduction

Answer Set Programming (ASP) is a declarative programming approach which due to expressive and efficient systems like SMODELS, DLV and CLASP, has been gaining popularity for many applications [3]. Current trends in computing, such as context awareness or distributed systems, raised the need for access to external sources in a program, which, e.g., on the Web ranges from light-weight data access (e.g., XML, RDF, or data bases) to knowledge-intensive formalisms (e.g., description logics).

To cater for this need, HEX-programs [7] extend ASP with so-called external atoms, through which the user can couple any external data source with a logic program. Roughly, such atoms pass information from the program, given by predicate extensions, into an external source which returns output values of an (abstract) function that it computes. This convenient extension has been exploited for many different applications, including querying data and ontologies on the Web, multi-context reasoning, or e-government, to mention a few;

This research has been supported by the Austrian Science Fund (FWF) project P20840, P20841, P24090, and by the Vienna Science and Technology Fund (WWTF) project ICT08-020.

M. Hanus and R. Rocha (Eds.): KDPD 2013, LNAI 8439, pp. 99–117, 2014.
DOI: 10.1007/978-3-319-08909-6_7, © Springer International Publishing Switzerland 2014

however, it can also be used to realize built-in functions. The extension is highly expressive as also recursive data access is possible.

A particular feature of external atoms is *value invention*, i.e., that they introduce new values that do not occur in the program. Such values may also occur in an answer set of a HEX-program, e.g., if we have a rule like

$$lookup(X, Y) \leftarrow p(X), \&do\_hash[X](Y)$$

where intuitively, the external atom $\&do\_hash[X](Y)$ generates a hash key Y for the input X and records it in the fact $lookup(X, Y)$. Here, the variable Y can be seen under existential quantification, i.e., as $\exists Y$, where the quantifier is externally evaluated, by taking domain-specific information into account; in the example above, this would be a procedure to calculate the hashkey. Such domain-specific quantification occurs frequently in applications, be it e.g. for built-in functions (just think of arithmetic), the successor of a current situation in situation calculus, retrieving the social security number of a person etc. To handle such quantifiers in ordinary ASP is cumbersome; they amount to interpreted functions and require proper encoding and/or special solvers.

HEX-programs however provide a uniform approach to represent such domain-specific existentials. The external treatment allows to deal elegantly with datatypes (e.g., the social security number, or an IBAN of bank account, or strings and numbers like reals), to respect parameters, and to realize partial or domain-restricted quantification of the form $\exists Y.\phi(X) \supset p(X, Y)$ where $\phi(X)$ is a formula that specifies the domain of elements X for which an existential value needs to exist; clearly, also range-restricted quantification $\exists Y.\psi(Y) \supset p(X, Y)$ that limits the value of Y to elements that satisfy ψ can be conveniently realized.

In general, such value invention on an infinite domain (e.g., for strings) leads to infinite models, which cannot be finitely generated. Under suitable restrictions on a program Π, this can be excluded, in particular if a finite portion of the grounding of Π is equivalent to its full, infinite grounding. This is exploited by various notions of *safety* of HEX-programs that generalize safety of logic programs.

In particular, *liberal domain-expansion safety (de-safety)* [6] is a recent notion based on term-bounding functions, which makes it modular and flexible; various well-known notions of safety are subsumed by it. For example, consider the program

$$\Pi = \{\ s(a); \quad t(Y) \leftarrow s(X), \&concat[X, a](Y); \quad s(X) \leftarrow t(X), d(X)\ \}, \quad (1)$$

where $\&concat[X, a](Y)$ is true iff Y is the string concatenation of X and a. Program Π is safe (in the usual sense) but $\&concat[X, a](Y)$ could hold for infinitely many Y, if one disregards the semantics of *concat*; however, if this is done by a term bounding function in abstract form, then the program is found to be liberally de-safe and thus a finite part of Π's grounding is sufficient to evaluate it.

Building on a grounding algorithm for liberally de-safe programs [5], we can effectively evaluate HEX-programs with domain-specific existentials that fall in this class. Moreover, we in fact generalize this algorithm with *domain specific termination*, such that for non-safe programs, a finitely *bounded grounding* is generated. Roughly speaking, such a bounded grounding amounts to domain-restricted quantification $\exists Y.\phi(X) \supset p(X, Y)$ where the domain condition $\phi(X)$ is dynamically evaluated during the grounding, and information about the grounding process may be also considered. Thus, domain-specific termination leads to a partial (bounded) grounding of the program, Π', yielding *bounded models* of the program Π; the idea is that the grounding is faithful in the sense that every answer set of Π' can be extended to a (possibly infinite) answer set of Π, and considering bounded models is sufficient for an application. This may be fruitfully exploited for applications like query answering over existential rules, reasoning about actions, or to evaluate classes of logic programs with function symbols like FDNC programs [8]. Furthermore, even if bounded models are not faithful (i.e., may not be extendable to models of the full grounding), they might be convenient e.g. to provide strings, arithmetic, recursive data structures like lists, trees etc., or action sequences of bounded length resp. depth. The point is that the bound does not have to be "coded" in the program (like maxint in DLV to bound the integer range), but can be provided via termination criteria in the grounding, which gives greater flexibility. Considering domain specific termination criteria and potentially even non de-safe programs is beyond the previous work [5,6]. The resulting algorithm properly generalizes the previous work [5] and applies to a wider range of applications.

Organization. After necessary preliminaries we proceed as follows.

- We introduce domain-specific existential quantification in HEX-programs and consider its realization (Sect. 3). To this end, we introduce a generalized grounding algorithm with *hooks* for termination criteria, which enables bounded grounding. Notably, its output for de-safe programs (using trivial criteria) is equivalent to the original program, i.e., it has the same answer sets.

    We illustrate some advantages of our approach, which cannot easily be integrated into direct implementations of existential quantifiers.
- As an example, we consider the realization of *null values* (which are customary in databases) as a domain-specific existential quantifier, leading to HEX$^{\exists}$-programs (Sect. 4); they include existential rules of form $\forall X \forall Z \exists Y.\psi(Z, Y) \leftarrow \phi(X, Y, Z)$ (also known as tuple-generating dependencies), where $\psi(Z, Y)$ is an atom[1] and $\phi(X, Y, Z)$ is a conjunction of atoms. Our framework can be thus exploited for bounded grounding, and in combination with a HEX-solver for bounded model generation of such programs.
- As an immediate application, we consider query answering over existential rules (Sect. 5), which reduces for prominent settings to query answering over a universal model. Under de-safety, a finite such model can be generated using our framework; this allows to cover a range of acyclic existential rules, including the very general notion of model-faithful acyclicity [14]. For non-de

---

[1] In general, $\psi(Z, Y)$ might be a conjunction of atoms but this may be normalized.

safe programs, a bounded universal model may be generated under suitable conditions; we illustrate this for Shy-programs - a class of programs with existential rules for which query answering is decidable, cf. [17].[2]

- Furthermore, we show how terms with function symbols can be processed using an encoding as a HEX-program (Sect. 6). To this end, we use dedicated external atoms to construct and decompose functional terms; bounded grounding enables us here to elegantly restrict the term depth, which is useful for applications such as reasoning with actions in situation calculus under bounded horizon, or reasoning from FDNC programs.

We conclude with a discussion and an outlook on future work in Sect. 7. Our prototype system is available at http://www.kr.tuwien.ac.at/research/systems/dlvhex. For proofs of our formal results, while available, we refer to an extended version due to space reasons.

2   Preliminaries

HEX-**Program Syntax.** HEX-programs generalize (disjunctive) logic programs under the answer set semantics [13] with external source access; for details and background see [7]. They are built over mutually disjoint sets $\mathcal{P}$, $\mathcal{X}$, $\mathcal{C}$, and $\mathcal{V}$ of ordinary predicates, external predicates, constants, and variables, respectively. Every $p \in \mathcal{P}$ has an arity $ar(p) \geq 0$, and every external predicate $\&g \in \mathcal{X}$ has an input arity $ar_i(\&g) \geq 0$ of input parameters and an output arity $ar_o(\&g) \geq 0$ of output arguments.

An *external atom* is of the form $\&g[\mathbf{X}](\mathbf{Y})$, where $\&g \in \mathcal{X}$, $\mathbf{X} = X_1, \ldots, X_\ell$ ($\ell = ar_i(\&g)$) are input parameters with $X_i \in \mathcal{P} \cup \mathcal{C} \cup \mathcal{V}$ for all $1 \leq i \leq \ell$, and $\mathbf{Y} = Y_1, \ldots, Y_m$ ($m = ar_o(\&g)$) are output terms with $Y_i \in \mathcal{C} \cup \mathcal{V}$ for all $1 \leq i \leq m$; we use lower case $\mathbf{x} = x_1, \ldots, x_\ell$ resp. $\mathbf{y} = y_1, \ldots, y_m$ if $\mathbf{X}$ resp. $\mathbf{Y}$ is variable-free. We assume the input parameters of $\&g$ are typed by $type(\&g, i) \in \{\texttt{const}, \texttt{pred}\}$ for $1 \leq i \leq ar_i(\&g)$, and that $X_i \in \mathcal{P}$ if $type(\&g, i) = \texttt{pred}$ and $X_i \in \mathcal{C} \cup \mathcal{V}$ otherwise.

A HEX-program consists of rules

$$a_1 \vee \cdots \vee a_k \leftarrow b_1, \ldots, b_m, \text{not } b_{m+1}, \ldots, \text{not } b_n \,, \tag{2}$$

where each a_i is an (ordinary) atom $p(X_1, \ldots, X_\ell)$ with $X_i \in \mathcal{C} \cup \mathcal{V}$ for all $1 \leq i \leq \ell$, each b_j is either an ordinary atom or an external atom, and $k + n > 0$.

The *head* of a rule r is $H(r) = \{a_1, \ldots, a_n\}$ and the *body* is $B(r) = \{b_1, \ldots, b_m, \text{not } b_{m+1}, \ldots, \text{not } b_n\}$. We call b or not b in a rule body a *default literal*; $B^+(r) = \{b_1, \ldots, b_m\}$ is the *positive body*, $B^-(r) = \{b_{m+1}, \ldots, b_n\}$ is the *negative body*. For a program Π (rule r), let $A(\Pi)$ ($A(r)$) be the set of all ordinary atoms and $EA(\Pi)$ ($EA(r)$) be the set of all external atoms occurring in Π (in r).

HEX-**Program Semantics.** Following [11], a *(signed) ground literal* is a positive or a negative formula $\mathbf{T}a$ resp. $\mathbf{F}a$, where a is a ground ordinary atom.

---

[2] For space reasons we refer to [17] for the definition of Shy-programs.

For a ground literal $\sigma = \mathbf{T}a$ or $\sigma = \mathbf{F}a$, let $\overline{\sigma}$ denote its opposite, i.e., $\overline{\mathbf{T}a} = \mathbf{F}a$ and $\overline{\mathbf{F}a} = \mathbf{T}a$. An *assignment* $\mathbf{A}$ is a consistent set of literals $\mathbf{T}a$ or $\mathbf{F}a$, where $\mathbf{T}a$ expresses that a is true and $\mathbf{F}a$ that a is false. We also identify a complete assignment $\mathbf{A}$ with its true atoms, i.e., $\mathbf{T}(\mathbf{A}) = \{a \mid \mathbf{T}a \in \mathbf{A}\}$. The semantics of a ground external atom $\&g[\mathbf{x}](\mathbf{y})$ wrt. a complete assignment $\mathbf{A}$ is given by a $1+k+l$-ary Boolean-valued *oracle function*, $f_{\&g}(\mathbf{A}, \mathbf{x}, \mathbf{y})$. Parameter x_i with $type(\&g, i) = \mathtt{pred}$ is *monotonic* (*antimonotonic*), if $f_{\&g}(\mathbf{A}, \mathbf{x}, \mathbf{y}) \leq f_{\&g}(\mathbf{A}', \mathbf{x}, \mathbf{y})$ ($f_{\&g}(\mathbf{A}', \mathbf{x}, \mathbf{y}) \leq f_{\&g}(\mathbf{A}, \mathbf{x}, \mathbf{y})$) whenever $\mathbf{A}'$ increases $\mathbf{A}$ only by literals $\mathbf{T}a$, where a has predicate x_i; otherwise, x_i is called *nonmonotonic*.

Non-ground programs are handled by grounding as usual. The set of constants appearing in a program Π is denoted C_Π. The *grounding* $grnd_C(r)$ of a rule r wrt. $C \subseteq \mathcal{C}$ is the set of all rules $\{\sigma(r) \mid \sigma : \mathcal{V} \mapsto C\}$, where σ is a *grounding substitution*, and $\sigma(r)$ results if each variable X in r is replaced by $\sigma(X)$. The *grounding of a program* Π wrt. C is defined as $grnd_C(\Pi) = \bigcup_{r \in \Pi} grnd_C(r)$.

Satisfaction of rules and programs [13] is extended to HEX-rules r and programs Π in the obvious way. The *FLP-reduct* is defined as $fgrnd_C(\Pi)^{\mathbf{A}} = \{r \in grnd_C(\Pi) \mid \mathbf{A} \models B(r)\}$. An *answer set* of a program Π is a model of $fgrnd_C(\Pi)^{\mathbf{A}}$ that is subset-minimal in its positive part [9]. We denote by $\mathcal{AS}(\Pi)$ the set of all answer sets of Π.

Take as an example the program $\Pi = \{str(N) \leftarrow str(L), \&head[L](N); str(N) \leftarrow str(L), \&tail[L](N)\}$, where $\&head[L](N)$ ($\&tail[L](N)$) is true iff string N is string L without the last (first) character. For $str(x)$, Π computes all substrings of string x.

Safety. In general, $\mathcal{C}$ has constants that do not occur in Π and can even be infinite (e.g., the set of all strings). Safety criteria guarantee that a finite portion $\Pi' \subseteq grnd_C(\Pi)$ (also called *finite grounding* of Π; usually by restricting to a finite $C \subseteq \mathcal{C}$) has the same answer sets as Π. Ordinary safety requires that every variable in a rule r occurs either in an ordinary atom in $B^+(r)$, or in the output list $\mathbf{Y}$ of an external atom $\&g[\mathbf{X}](\mathbf{Y})$ in $B^+(r)$ where all variables in $\mathbf{X}$ are safe. However, this notion is not sufficient.

Example 1. Let $\Pi = \{s(a); t(Y) \leftarrow s(X), \&concat[X, a](Y); s(X) \leftarrow t(X), d(X)\}$, where $\&concat[X, a](Y)$ is true iff Y is the string concatenation of X and a. Then Π is safe but $\&concat[X, a](Y)$ can introduce infinitely many values.

The general notion of *(liberal) domain-expansion safety (de-safety)* subsumes a range of other well-known notions and can be easily extended in a modular fashion [6]. It is based on *term bounding functions* (TBFs), which intuitively declare terms in rules as *bounded*, if there are only finitely many substitutions for this term in a *canonical grounding* $CG(\Pi)$ of Π.[3] The latter is infinite in general but finite for de-safe programs.

More specifically we consider *attributes* and *ranges*. For an ordinary predicate $p \in \mathcal{P}$, let $p{\restriction}i$ be the *i-th attribute* of p for all $1 \leq i \leq ar(p)$. For an

---

[3] $CG(\Pi)$ is the least fixed point $G_\Pi^\infty(\emptyset)$ of a monotone operator $G_\Pi(\Pi') = \bigcup_{r \in \Pi} \{r\theta \mid r\theta \in grnd_C(r), \exists \mathbf{A} \subseteq \mathcal{A}(\Pi'), \mathbf{A} \not\models \bot, \mathbf{A} \models B^+(r\theta)\}$ on programs Π' [6].

external predicate $\&g \in \mathcal{X}$ with input list $\mathbf{X}$ in rule r, let $\&g[\mathbf{X}]_r \upharpoonright_T i$ with $T \in \{\mathrm{I}, \mathrm{O}\}$ be the *i-th input resp. output attribute of* $\&g[\mathbf{X}]$ *in* r for all $1 \leq i \leq ar_T(\&g)$. For a ground program Π, an attribute *range* is, intuitively, the set of ground terms which occur in the position of the attribute. Formally, for an attribute $p \upharpoonright i$ we have $range(p \upharpoonright i, \Pi) = \{t_i \mid p(t_1, \dots, t_{ar(p)}) \in A(\Pi)\}$; for $\&g[\mathbf{X}]_r \upharpoonright_T i$ it is $range(\&g[\mathbf{X}]_r \upharpoonright_T i, \Pi) = \{x_i^T \mid \&g[\mathbf{x}^{\mathrm{I}}](\mathbf{x}^{\mathrm{O}}) \in EA(\Pi)\}$, where $\mathbf{x}^s = x_1^s, \dots, x_{ar_s(\&g)}^s$. Now term bounding functions are introduced as follows:

Definition 1. (Term Bounding Function (TBF)). *A TBF* $b(\Pi, r, S, B)$ *maps a program* Π, *a rule* $r \in \Pi$, *a set* S *of already safe attributes, and a set* B *of already bounded terms in* r *to an enlarged set* $b(\Pi, r, S, B) \supseteq B$ *of bounded terms, s.t. every* $t \in b(\Pi, r, S, B)$ *has finitely many substitutions in* $CG(\Pi)$ *if (i) the attributes* S *have a finite range in* $CG(\Pi)$ *and (ii) each term in* $terms(r) \cap B$ *has finitely many substitutions in* $CG(\Pi)$.

Liberal domain-expansion safety of programs is then parameterized with a term bounding function, such that concrete syntactic and/or semantic properties can be plugged in; concrete term bounding functions are described in [6]. The concept is defined in terms of domain-expansion safe attributes $S_\infty(\Pi)$, which are stepwise identified as $S_n(\Pi)$ in mutual recursion with bounded terms $B_n(r, \Pi, b)$ of rules r in Π.

Definition 2. ((Liberal) Domain-expansion Safety). *Given a TBF* b, *the set of* bounded terms $B_n(r, \Pi, b)$ *in step* $n \geq 1$ *in a rule* $r \in \Pi$ *is* $B_n(r, \Pi, b) = \bigcup_{j \geq 0} B_{n,j}(r, \Pi, b)$ *where* $B_{n,0}(r, \Pi, b) = \emptyset$ *and for* $j \geq 0$, $B_{n,j+1}(r, \Pi, b) = b(\Pi, r, S_{n-1}(\Pi), B_{n,j})$.

The set of domain-expansion safe attributes $S_\infty(\Pi) = \bigcup_{i \geq 0} S_i(\Pi)$ *of a program* Π *is iteratively constructed with* $S_0(\Pi) = \emptyset$ *and for* $n \geq 0$:

- $p \upharpoonright i \in S_{n+1}(\Pi)$ *if for each* $r \in \Pi$ *and atom* $p(t_1, \dots, t_{ar(p)}) \in H(r)$, *it holds that* $t_i \in B_{n+1}(r, \Pi, b)$, *i.e.,* t_i *is bounded;*
- $\&g[\mathbf{X}]_r \upharpoonright_\mathrm{I} i \in S_{n+1}(\Pi)$ *if each* $\mathbf{X}_i$ *is a bounded variable, or* $\mathbf{X}_i$ *is a predicate input parameter* p *and* $p \upharpoonright 1, \dots, p \upharpoonright ar(p) \in S_n(\Pi)$;
- $\&g[\mathbf{X}]_r \upharpoonright_\mathrm{O} i \in S_{n+1}(\Pi)$ *if and only if* r *contains an external atom* $\&g[\mathbf{X}](\mathbf{Y})$ *such that* $\mathbf{Y}_i$ *is bounded, or* $\&g[\mathbf{X}]_r \upharpoonright_\mathrm{I} 1, \dots, \&g[\mathbf{X}]_r \upharpoonright_\mathrm{I} ar_\mathrm{I}(\&g) \in S_n(\Pi)$.

A program Π *is* (liberally) de-safe, *if it is safe and all its attributes are de-safe.*

Example 2. The program Π from Example 1 is liberally de-safe using the TBF b_{synsem} from [6] (see Appendix A) as the generation of infinitely many values is prevented by $d(X)$ in the last rule.

Every de-safe HEX-program has a finite grounding that preserves all answer sets [6].

3   HEX-Programs with Existential Quantification

In this section, we consider HEX-programs with *domain-specific existential quantifiers*. This term refers to the introduction of new values in rule bodies which are propagated to the head such that they may appear in the answer sets of a program. Logical existential quantification is a special case of our approach (used in Sect. 4 to illustrate a specific instance), where just the existence but not the structure of values is of interest. Instead, in our work also the structure of introduced values may be relevant and can be controlled by external atoms.

Instantiating, i.e., applying, our approach builds on an extension of the grounding algorithm for HEX-programs in [5] by additional *hooks*. They support the insertion of application-specific termination criteria, and thus can be exploited for computing a finite subset of the grounding in case of non-de-safe HEX-programs. The latter may be sufficient to consider a certain reasoning task, e.g., for bounded model building. For instance, we discuss *queries* over (positive) programs with (logical) existential quantifiers in Sect. 5, which can be answered by computing a finite part of a canonical model.

HEX-Program Grounding. For introducing our *bounded grounding algorithm* BGroundHEX, we make use of so-called input auxiliary rules. We say that an external atom $\&g[\mathbf{Y}](\mathbf{X})$ *joins* an atom b, if some variable from $\mathbf{Y}$ occurs in b, where in case b is an external atom the occurrence is in the output list of b.

Definition 3. (Input Auxiliary Rule). *Let Π be a HEX-program. Then for each external atom $\&g[\mathbf{Y}](\mathbf{X})$ occurring in rule $r \in \Pi$, a rule $r_{inp}^{\&g[\mathbf{Y}](\mathbf{X})}$ is composed as follows:*

- *The head is $H(r_{inp}^{\&g[\mathbf{Y}](\mathbf{X})}) = \{g_{inp}(\mathbf{Y})\}$, where g_{inp} is a fresh predicate; and*
- *The body $B(r_{inp}^{\&g[\mathbf{Y}](\mathbf{X})})$ contains all $b \in B^+(r) \setminus \{\&g[\mathbf{Y}](\mathbf{X})\}$ which join $\&g[\mathbf{Y}](\mathbf{X})$.*

Intuitively, input auxiliary rules are used to derive all ground input tuples $\mathbf{y}$, under which the external atom needs to be evaluated.

Our grounding approach is based on a grounder for ordinary ASP programs. Compared to the naive grounding $grnd_C(\Pi)$, we allow the ASP grounder GroundASP to eliminate rules if their body is always false, and ordinary body literals from the grounding that are always true, as long as this does not change the answer sets. More formally, a rule r' is an *o-strengthening* (ordinary-strengthening) of a rule r, if $H(r') = H(r)$, $B(r') \subseteq B(r)$ and $B(r) \setminus B(r')$ contains only ordinary literals, i.e., no external atom replacements.

Definition 4. *An algorithm GroundASP that takes as input a program Π and outputs a ground program Π' is a faithful ASP grounder for a safe program Π, if:*

- *$\mathcal{AS}(\Pi') = \mathcal{AS}(grnd_{C_\Pi}(\Pi))$;*
- *Π' consists of o-strengthenings of rules in $grnd_{C_\Pi}(\Pi)$;*

– if $r \in grnd_{C_\Pi}(\Pi)$ has no o-strengthening in Π', then every answer set of $grnd_{C_\Pi}(\Pi)$ falsifies some ordinary literal in $B(r)$; and
– if $r \in grnd_{C_\Pi}(\Pi)$ has some o-strengthening $r' \in \Pi'$, then every answer set of $grnd_{C_\Pi}(\Pi)$ satisfies $B(r) \setminus B(r')$.

Intuitively, the bounded grounding Algorithm BGroundHEX can be explained as follows. Program Π is the non-ground input program. Program Π_p is the non-ground ordinary ASP *prototype program*, which is an iteratively updated variant of Π enriched with additional rules. In each step, the *preliminary ground program* Π_{pg} is produced by grounding Π_p using a standard ASP grounding algorithm. Program Π_{pg} is intended to converge against a fixpoint, i.e., a final *ground* HEX-*program* Π_g. For this purpose, the loop at (b) and the abortion check at (f) introduce two *hooks* (Repeat and Evaluate) which allow for realizing application-specific termination criteria. They need to be substituted by concrete program fragments depending on the reasoning task; for now we assume that the loop at (f) runs exactly once and the check at (f) is always true (which is sound and complete for model computation of de-safe programs, cf. Proposition 1).

The algorithm first introduces input auxiliary rules $r_{inp}^{\&g[\mathbf{Y}](\mathbf{X})}$ for every external atom $\&g[\mathbf{Y}](\mathbf{X})$ in a rule r in Π in Part (a). Then, all external atoms $\&g[\mathbf{Y}](\mathbf{X})$ in all rules r in Π_p are replaced by ordinary *replacement atoms* $e_{r,\&g[\mathbf{Y}]}(\mathbf{X})$. This allows the algorithm to use an ordinary ASP grounder GroundASP in the main loop at (b). After the grounding step, it is checked whether the grounding constants. For this, the algorithm checks, for all external atoms (d) and all relevant input interpretations (e), potential output tuples at (f), if they contain any new value that was not yet respected in the grounding. (Note that, $\mathbf{Y}_m, \mathbf{Y}_a, \mathbf{Y}_n$ denote the sets of *monotonic*, *antimonotonic*, and *nonmonotonic* predicate input parameters in $\mathbf{Y}$, respectively.) It adds the relevant constants in form of guessing rules at (g) to Π_p (this may also be expressed by unstratified negation). Then the main loop starts over again. Eventually, the algorithm is intended to find a program respecting all relevant constants. Then at (h), auxiliary input rules are removed and replacement atoms are translated to external atoms.

Let us illustrate the grounding algorithm with the following example.

Example 3. Let Π be the following program:

$$f : d(a).\ d(b).\ d(c).\ r_1 : s(Y) \leftarrow d(X), \&diff[d,n](Y), d(Y).$$
$$r_2 : n(Y) \leftarrow d(X), \&diff[d,s](Y), d(Y).$$
$$r_3 : c(Z) \leftarrow \&count[s](Z).$$

Here, $\&diff[s_1, s_2](x)$ is true for all elements x, which are in the extension of s_1 but not in that of s_2, and $\&count[s](i)$ is true for the integer i corresponding to the number of elements in the extension of s. The program first partitions the domain (extension of d) into two sets (extensions of s and n) and then computes the size of s. Program Π_p at the beginning of the first iteration is as follows, where $e_1(Y)$, $e_2(Y)$ and $e_3(Z)$ are shorthands for $e_{r_1,\&diff[d,n]}(Y)$, $e_{r_2,\&diff[d,s]}(Y)$, and $e_{r_3,\&count[s]}(Z)$, respectively.

Algorithm. BGroundHEX

Input: A HEX-program Π
Output: A ground HEX-program Π_g

(a) $\Pi_p = \Pi \cup \{r_{inp}^{\&g[\mathbf{Y}](\mathbf{X})} \mid \&g[\mathbf{Y}](\mathbf{X}) \text{ in } r \in \Pi\}$
   Replace all external atoms $\&g[\mathbf{Y}](\mathbf{X})$ in all rules r in Π_p by $e_{r,\&g[\mathbf{Y}]}(\mathbf{X})$
   $i \leftarrow 0$

(b) **while** *Repeat()* **do**
   | $i \leftarrow i + 1$ // Remember already processed input tuples at iteration i
(c) | Set *NewInputTuples* $\leftarrow \emptyset$ and $PIT_i \leftarrow \emptyset$
   | **repeat**
   | | $\Pi_{pg} \leftarrow$ GroundASP(Π_p) // partial grounding
(d) | | **for** $\&g[\mathbf{Y}](\mathbf{X})$ *in a rule* $r \in \Pi$ **do** // evaluate all external atoms
(e) | | | // do this under all relevant assignments
   | | | $\mathbf{A}_{ma} = \{\mathbf{T}p(\mathbf{c}) \mid a(\mathbf{c}) \in A(\Pi_{pg}), p \in \mathbf{Y}_m\} \cup \{\mathbf{F}p(\mathbf{c}) \mid a(\mathbf{c}) \in A(\Pi_{pg}), p \in \mathbf{Y}_a\}$
   | | | **for** $\mathbf{A}_{nm} \subseteq \{\mathbf{T}p(\mathbf{c}), \mathbf{F}p(\mathbf{c}) \mid p(\mathbf{c}) \in A(\Pi_{pg}), p \in \mathbf{Y}_n\}$ *s.t.* $\nexists a : \mathbf{T}a, \mathbf{F}a \in \mathbf{A}_{nm}$ **do**
   | | | | $\mathbf{A} = (\mathbf{A}_{ma} \cup \mathbf{A}_{nm} \cup \{\mathbf{T}a \mid a \leftarrow \in \Pi_{pg}\}) \setminus \{\mathbf{F}a \mid a \leftarrow \in \Pi_{pg}\}$
(f) | | | | **for** $\mathbf{y} \in \{\mathbf{c} \mid r_{inp}^{\&g[\mathbf{Y}](\mathbf{X})}(\mathbf{c}) \in A(\Pi_{pg}) \text{ s.t. Evaluate}(r_{inp}^{\&g[\mathbf{Y}](\mathbf{X})}(\mathbf{c})) = true\}$ **do**
(g) | | | | | // add ground guessing rules and remember y-evaluation
   | | | | | $\Pi_p \leftarrow \Pi_p \cup \{e_{r,\&g[\mathbf{y}]}(\mathbf{x}) \vee ne_{r,\&g[\mathbf{y}]}(\mathbf{x}) \leftarrow \mid f_{\&g}(\mathbf{A}, \mathbf{y}, \mathbf{x}) = 1\}$
   | | | | | *NewInputTuples* $\leftarrow$ *NewInputTuples* $\cup \{r_{inp}^{\&g[\mathbf{Y}](\mathbf{X})}(\mathbf{y})\}$

   | | $PIT_i \leftarrow PIT_i \cup$ *NewInputTuples*
   | **until** Π_{pg} *did not change*

(h) Remove input auxiliary rules and external atom guessing rules from Π_{pg}
   Replace all $e_{\&g[\mathbf{y}]}(\mathbf{x})$ in Π_{pg} by $\&g[\mathbf{y}](\mathbf{x})$
   return Π_{pg}

$$f : d(a).\ d(b).\ d(c).\quad r_1 : s(Y) \leftarrow d(X), e_1(Y), d(Y).$$
$$r_2 : n(Y) \leftarrow d(X), e_2(Y), d(Y).$$
$$r_3 : c(Z)\ \leftarrow e_3(Z).$$

Program Π_{pg} contains no instances of r_1, r_2 and r_3 because the optimizer recognizes that $e_1(Y)$, $e_2(Y)$ and $e_3(Z)$ occur in no rule head and no ground instance can be true in any answer set. Then the algorithm moves to the checking phase. It evaluates the external atoms in r_1 and r_2 under $\mathbf{A} = \{d(a), d(b), d(c)\}$ (note that $\&diff[s_1, s_2](x)$ is monotonic in s_1 and antimonotonic in s_2) and adds the rules $\{e_i(Z) \vee ne_i(Z) \leftarrow \mid Z \in \{a, b, c\}, i \in \{1, 2\}\}$ to Π_p. Then it evaluates $\&count[s](Z)$ under all $\mathbf{A} \subseteq \{s(a), s(b), s(c)\}$ because it is nonmonotonic in s, and adds the rules $\{e_3(Z) \vee ne_3(Z) \leftarrow \mid Z \in \{0, 1, 2, 3\}\}$. It terminates after the second iteration. $\qquad\square$

The main difference to the algorithm from [5] is the addition of the two hooks at (c) (**Repeat**) and at (f) (**Evaluate**), that need to be defined for a concrete instance of the algorithm (which we do in the following). We assume that the hooks are substituted by code fragments with access to all local variables. Moreover, the set PIT_i contains the input atoms for which the corresponding external atoms have been evaluated in iteration i. **Evaluate** decides for a given input atom $r_{inp}^{\&g[\mathbf{Y}](\mathbf{X})}(\mathbf{c})$ if the corresponding external atom shall be evaluated under $\mathbf{c}$. This allows for abortion of the grounding even if it is incomplete, which can be exploited for reasoning tasks over programs with infinite groundings where a finite subset of the grounding is sufficient. The second hook Repeat allows for repeating the core algorithm multiple times such that Evaluate can distinguish between input tuples processed in different iterations.

Naturally, soundness and completeness of the algorithm cannot be shown in general, but depends on concrete instances for the hooks at (c) and (f) which in turn may vary for different reasoning tasks. Instantiating the hooks does not follow a general pattern but is strongly application dependent. However, we give some concrete examples in the remaining part of the paper.

Domain-specific Existential Quantification in HEX-Programs. We can realize domain-specific existential quantification naturally in HEX-programs by appropriate external atoms that introduce new values to the program. The realization exploits *value invention* as supported by HEX-programs, i.e., external atoms which return constants that do not show up in the input program. Realizing existentials by external atoms also allows to use constants different from Skolem terms, i.e., datatypes with a specific semantics. The values introduced may depend on input parameters passed to the external atom.

Example 4. Consider the following rule:

$$iban(B, I) \leftarrow country(B, C), bank(B, N), \&iban[C, B, N](I).$$

Suppose $bank(b, n)$ models financial institutions b with their associated national number n, and $country(b, c)$ holds for an institution b and its home country c. Then one can use $\&iban[C, B, N](I)$ to generate an IBAN (*International Bank Account Number*) I from the country C, the bank name B and account number N.

Here, the structure of the introduced value is relevant, but an algorithm which computes it can be hidden from the user. The introduction of new values may also be subject to additional conditions which cannot easily be expressed in the program.

Example 5. Consider the following rule:

$$lifetime(M, L) \leftarrow machine(M, C), \&lifetime[M, C](L).$$

It expresses that each purchased machine m with cost c ($machine(m, c)$) higher than a given limit has assigned an expected lifetime l ($lifetime(m, l)$) used for fiscal purposes, whereas purchases below that limit are fully tax deductible in the year of acquirement. Then testing for exceedance of the limit might involve real numbers and cannot easily be done in the logic program. However, the external atom can easily be extended in such a way that a value is only introduced if this side constraint holds.

Counting quantifiers may be realized in this way, i.e., expressing that there exist *exactly* k or *at least* k elements, which is used e.g. in description logics. While a direct implementation of existentials requires changes in the reasoner, a simulation using external atoms is easily extendable.

4 HEX$^\exists$-Programs

We now realize the logical existential quantifier as a specific instance of our approach, which can also be written in the usual syntax; a rewriting then simulates it by using external atoms which return dedicated *null values* to represent a representative for the unnamed values introduced by existential quantifiers. We start by introducing a language for HEX-programs with logical existential quantifiers, called HEX$^\exists$-*programs*.

A HEX$^\exists$-*programs* is a finite set of rules of form

$$\forall \mathbf{X} \exists \mathbf{Y} : p(\mathbf{X}', \mathbf{Y}) \leftarrow \mathbf{conj}[\mathbf{X}], \tag{3}$$

where $\mathbf{X}$ and $\mathbf{Y}$ are disjoint sets of variables, $\mathbf{X}' \subseteq \mathbf{X}$, $p(\mathbf{X}', \mathbf{Y})$ is an atom, and $\mathbf{conj}[\mathbf{X}]$ is a conjunction of default literals or default external literals containing all and only the variables $\mathbf{X}$; without confusion, we also omit $\forall \mathbf{X}$.

Intuitively speaking, whenever $\mathbf{conj}[\mathbf{X}]$ holds for some vector of constants $\mathbf{X}$, then there should exist a vector $\mathbf{Y}$ of (unnamed) individuals such that $p(\mathbf{X}', \mathbf{Y})$ holds. Existential quantifiers are simulated by using *new* null values which represent the introduced unnamed individuals. Formally, we assume that $\mathcal{N} \subseteq \mathcal{C}$ is a set of dedicated null values, denoted by ω_i with $i \in \mathbb{N}$, which do not appear in the program.

We transform HEX$^\exists$-programs to HEX-programs as follows. For a HEX$^\exists$-program Π, let $T_\exists(\Pi)$ be the HEX-program with each rule r of form (3) replaced by

$$p(\mathbf{X}', \mathbf{Y}) \leftarrow \mathbf{conj}[\mathbf{X}], \& exists^{|\mathbf{X}'|,|\mathbf{Y}|}[r](,)\mathbf{X}'\mathbf{Y},$$

where $f_{\& exists^{n,m}}(\mathbf{A}, r, \mathbf{x}, \mathbf{y}) = 1$ iff $\mathbf{y} = \omega_1, \ldots, \omega_m$ is a vector of *fresh and unique null values* for $r, \mathbf{x}$, and $f_{\& exists^{n,m}}(\mathbf{A}, r, \mathbf{x}, \mathbf{y}) = 0$ otherwise.

Each existential quantifier is replaced by an external atom $\& exists^{|\mathbf{X}'|,|\mathbf{Y}|}[r, \mathbf{X}'](\mathbf{Y})$ of appropriate input and output arity which exploits value invention for simulating the logical existential quantifier similar to the *chase* algorithm.

We call a HEX$^\exists$-program Π liberally de-safe iff $T_\exists(\Pi)$ is liberally de-safe. Various notions of cyclicity have been introduced, e.g., in [14]; here we use the one from [6].

Example 6. The following set of rules is a HEX$^\exists$-program Π:

$$employee(john). \qquad employee(joe).$$
$$r_1 : \exists Y : office(X, Y) \leftarrow employee(X). \quad r_2 : room(Y) \leftarrow office(X, Y)$$

Then $T_\exists(\Pi)$ is the following de-safe program:

$$employee(john). \qquad employee(joe).$$
$$r_1' : \qquad office(X, Y) \leftarrow employee(X), \& exists^{1,1}[r_1, X](Y).$$
$$r_2 : \qquad room(Y) \leftarrow office(X, Y)$$

Intuitively, each employee X has some unnamed office Y of X, which is a room. The unique answer set of $T_\exists(\Pi)$ is $\{employee(john), employee(joe), office(john, \omega_1), office(joe, \omega_2), room(\omega_1), room(\omega_2)\}$.

For grounding de-safe programs, we simply let Repeat test for $i < 1$ and Evaluate return *true*. Explicit model computation is in general infeasible for non-de-safe programs. However, the resulting algorithm GroundDESafeHEX always terminates for de-safe programs. For non-de-safe programs, we can support bounded model generation by other hook instantiations. This is exploited e.g. for query answering over cyclic programs (described next). One can show that the algorithm computes all models of the program.

Proposition 1. *For de-safe programs* Π, $\mathcal{AS}(\mathsf{GroundDESafeHEX}(\Pi)) \equiv^{pos}$ $\mathcal{AS}(\Pi)$, *where* $\equiv^{pos}$ *denotes equivalence of the answer sets on positive atoms.*

5   Query Answering over Positive HEX$^\exists$-Programs

The basic idea for query answering over programs with possibly infinite models is to compute a ground program with a single answer set that can be used for answering the query. Positive programs with existential variables are essentially grounded by simulating the *parsimonious chase procedure* from [17], which uses null values for each existential quantification. However, for termination of BGroundHEX we need to provide specific instances of the hooks in the grounding algorithm.

We start by restricting the discussion to a fragment of HEX$^\exists$-programs, called *Datalog$^\exists$*-programs [17]. A *Datalog$^\exists$*-program is a HEX$^\exists$-program where every rule body **conj[X]** consists of positive ordinary atoms. Thus compared to HEX$^\exists$-programs, default negation and external atoms are excluded.

As an example, the following set of rules is a *Datalog$^\exists$*-program:

$$person(john). \quad person(joe).$$
$$r_1 : \exists Y : father(X,Y) \leftarrow person(X). \quad r_2 : person(Y) \leftarrow father(X,Y). \tag{4}$$

Next, we recall *homomorphisms* as used for defining *Datalog$^\exists$*-semantics and query answering over *Datalog$^\exists$*-programs. A *homomorphism* is a mapping $h \colon \mathcal{N} \cup \mathcal{V} \to \mathcal{C} \cup \mathcal{V}$. For a homomorphism h, let $h|_S$ be its restriction to $S \subseteq \mathcal{N} \cup \mathcal{V}$, i.e., $h|_S(X) = h(X)$ if $X \in S$ and is undefined otherwise. For any atom a, let $h(a)$ be the atom where each variable and null value V in a is replaced by $h(V)$; this is likewise extended to $h(S)$ for sets S of atoms and/or vectors of terms. A homomorphism h is a *substitution*, if $h(N) = N$ for all $N \in \mathcal{N}$. An atom a is *homomorphic* (*substitutive*) to atom b, if some homomorphism (substitution) h exists such that $h(a) = b$. An isomorphism between two atoms a and b is a bijective homomorphism h s.t. $h(a) = b$ and $h^{-1}(b) = a$.

A set M of atoms is a model of a *Datalog$^\exists$*-program Π, denoted $M \models \Pi$, if $h(B(r)) \subseteq M$ for some substitution h and $r \in \Pi$ of form (3) implies that $h|_{\mathbf{X}}(H(r))$ is substitutive to some atom in M; the set of all models of Π is denoted by $mods(\Pi)$.

Next, we can introduce queries over *Datalog$^\exists$*-programs. A *conjunctive query* (CQ) q is an expression of form $\exists \mathbf{Y} : \ \leftarrow \mathbf{conj}[\mathbf{X} \cup \mathbf{Y}]$, where $\mathbf{Y}$ and $\mathbf{X}$ (the free variables) are disjoint sets of variables and $\mathbf{conj}[\mathbf{X} \cup \mathbf{Y}]$ is a conjunction of ordinary atoms containing all and only the variables $\mathbf{X} \cup \mathbf{Y}$.

The answer of a CQ q with free variables $\mathbf{X}$ wrt. a model M is defined as follows:

$$ans(q, M) = \{h|_{\mathbf{X}} \mid h \text{ is a substitution and } h(\mathbf{conj}[\mathbf{X} \cup \mathbf{Y}]) \subseteq M\}.$$

Intuitively, this is the set of assignments to the free variables such that the query holds wrt. the model. The answer of a CQ q wrt. a program Π is then defined as the set $ans(q, \Pi) = \bigcap_{M \in mods(\Pi)} ans(q, M)$.

Query answering can be carried out over some *universal model* U of the program that is embeddable into each of its models by applying a suitable homomorphism. Formally, a model U of a program Π is called *universal* if, for each $M \in mods(\Pi)$, there is a homomorphism h s.t. $h(U) \subseteq M$. Thus, a universal model may be obtained using null values for unnamed individuals introduced by existential quantifiers. Moreover, it can be used to answer any query according to the following proposition [10]:

Proposition 2. ([10]). *Let U be a universal model of $Datalog^{\exists}$-program Π. Then, for any CQ q, it holds that $h \in ans(q, \Pi)$ iff $h \in ans(q, U)$ and $h : \mathcal{V} \rightarrow \mathcal{C} \setminus \mathcal{N}$.*

Intuitively, the set of all answers to q wrt. U which map all variables to non-null constants is exactly the set of answers to q wrt. Π.

Example 7. Let Π be the program consisting of rules (4). The CQ $\exists Y :\leftarrow person(X), father(X, Y)$ asks for all persons who have a father. The model $U = \{person(john), person(joe), father(john, \omega_1), father(joe, \omega_2), person(\omega_1), person(\omega_2), \ldots\}$ is a universal model of Π. Hence, $ans(q, \Pi)$ contains answers $h_1(X) = john$ and $h_2(X) = joe$.

Thus, computing a universal model is a key issue for query answering. A common approach for this step is the chase procedure. Intuitively, it starts from an empty interpretation and iteratively adds the head atoms of all rules with satisfied bodies, where existentially quantified variables are substituted by fresh nulls. However, in general this procedure does not terminate. Thus, a restricted *parsimonious chase procedure* was introduced in [17], which derives less atoms, and which is guaranteed to terminate for the class of *Shy-programs*. The latter is a syntactic fragment of so-called parsimonious programs that can be easily recognized but still significantly extends Datalog programs and linear $Datalog^{\exists}$-programs. Moreover, the interpretation computed by the parsimonious chase procedure is, although not a model of the program in general, still sound and complete for query answering; and a *bounded model* in our view.

For query answering over $Datalog^{\exists}$-programs we reuse the translation in Sect. 4.

Example 8. Consider the $Datalog^\exists$-program Π and its HEX translation $T_\exists(\Pi)$:

$$
\begin{array}{ll}
\Pi: & T_\exists(\Pi): \\[4pt]
\quad person(john). \qquad person(joe). & \quad person(john). \qquad person(joe). \\
\exists Y: father(X,Y) \leftarrow \quad person(X). & \quad father(X,Y) \leftarrow person(X), \\
\qquad person(Y) \leftarrow father(X,Y). & \qquad\qquad\qquad \& exists^{1,1}[r_1,X](Y). \\
& \quad person(Y) \leftarrow father(X,Y).
\end{array}
$$

Intuitively, each person X has some unnamed father Y of X which is also a person.

Note that $T_\exists(\Pi)$ is *not* de-safe in general. However, with the hooks in Algorithm BGroundHEX one can still guarantee termination. Let GroundDatalog$^\exists(\Pi, k)$ = BGroundHEX($T_\exists(\Pi)$) where Repeat tests for $i < k + 1$ where k is the number of existentially quantified variables in the query, and Evaluate(PIT_i, x) = *true* iff atom x is *not* homomorphic to any $a \in PIT_i$. The produced program has a single answer set, which essentially coincides with the result of *pChase* [17] that can be used for query answering. The basic idea of *pChase* is to start with an empty assignment, and iteratively "repair" it by adding the head atoms of the rules which have a satisfied body. Thus, query answering over Shy-programs is reduced to grounding and solving of a HEX-program.

Proposition 3. *For a Shy-program Π, GroundDatalog$^\exists(\Pi, k)$ has a unique answer set which is sound and complete for answering CQs with up to k existential variables.*

The main difference to *pChase* in [17] is essentially due to the homomorphism check. Actually, *pChase* instantiates existential variables in rules with satisfied body to new null values only if the resulting head atom is not homomorphic to an already derived atom. In contrast, our algorithm performs the homomorphism check for the input to $\&exists^{n,m}$ atoms. Thus, homomorphisms are detected when constants are cyclically sent to the external atom. Consequently, our approach may need one iteration more than *pChase*, but allows for a more elegant integration into our algorithm.

Example 9. For the program and query from Example 8, the algorithm computes a program with answer set $\{person(john), person(joe), father(john, \omega_1), father(joe, \omega_2), person(\omega_1), person(\omega_2)\}$. In contrast, *pChase* would stop already earlier with the interpretation $\{person(john), person(joe), father(john, \omega_1), father(joe, \omega_2)\}$ because $person(\omega_1), person(\omega_2)$ are homomorphic to $person(john)$, $person(joe)$.

More formally, one can show that GroundDatalog$^\exists(\Pi, k)$ yields, for a Shy-program Π, a program with a single answer set that is equivalent to *pChase*(Π, $k + 1$) in [17]. Lemma 4.9 in [17] implies that the resulting answer set can be used for answering queries with k different existentially quantified variables, which proves Proposition 3.

While *pChase* intermingles grounding and computing a universal model, our algorithm cleanly separates the two stages; modularized program evaluation by the solver will however also effect such intermingling. We nevertheless expect the more clean separation to be advantageous for extending Shy-programs to programs that involve existential quantifiers and other external atoms, which we leave for future work.

6   HEX-**Programs with Function Symbols**

In this section we show how to process terms with function symbols by a rewriting to de-safe HEX-programs. We will briefly discuss advantages of our approach compared to a direct implementation of function symbols.

We consider HEX-programs, where the arguments X_i for $1 \leq i \leq \ell$ of ordinary atoms $p(X_1, \ldots, X_\ell)$, and the constant input arguments in $\mathbf{X}$ and the output $\mathbf{Y}$ of an external atom $\&g[\mathbf{X}](\mathbf{Y})$ are from a set of *terms* $\mathcal{T}$, that is the least set $\mathcal{T} \supseteq \mathcal{V} \cup \mathcal{C}$ such that $f \in \mathcal{C}$ (constant symbols are also used as function symbols) and $t_1, \ldots, t_n \in \mathcal{T}$ imply $f(t_1, \ldots, t_n) \in \mathcal{T}$.

Following [4], we introduce for every $k \geq 0$ two external predicates $\&comp_k$ and $\&decomp_k$ with $ar_i(\&comp_k) = 1 + k$, $ar_o(\&comp_k) = 1$, $ar_i(\&decomp_k) = 1$, and $ar_o(\&decomp_k) = 1 + k$. We define

$$f_{\&comp_k}(\mathbf{A}, f, X_1, \ldots, X_k, T) = f_{\&decomp_k}(\mathbf{A}, T, f, X_1, \ldots, X_k) = 1,$$

iff $T = f(X_1, \ldots, X_k)$.

Composition and decomposition of function terms can be simulated using these external predicates. Function terms are replaced by new variables and appropriate additional external atoms with predicate $\&comp_k$ or $\&decomp_k$ in rule bodies to compute their values. More formally, we introduce the following rewriting.

For any HEX-program Π with function symbols, let $T_f(\Pi)$ be the HEX-program where each occurrence of a term $t = f(t_1, \ldots, t_n)$ in a rule r such that $B(r) \neq \emptyset$ is recursively replaced by a new variable V, and if V occurs afterwards in $H(r)$ or the input list of an external atom in $B(r)$, we add $\&comp_n[f, t_1, \ldots, t_n]$ (V) to $B(r)$; otherwise (i.e., V occurs afterwards in some ordinary body atom or the output list of an external atom), we add $\&decomp_n[V](f, t_1, \ldots, t_n)$ to $B(r)$.

Intuitively, $\&comp_n$ is used to construct a nested term from a function symbol and arguments, which might be nested terms themselves, and $\&decomp_n$ is used to extract the function symbol and the arguments from a nested term. The translation can be optimized wrt. evaluation efficiency, but we disregard this here for space reasons.

Example 10. Consider the HEX-program Π with function symbols and its translation:

$$\Pi : \quad q(z).\, q(y).$$
$$p(f(f(X))) \leftarrow q(X).$$
$$r(X) \leftarrow p(X).$$
$$r(X) \leftarrow r(f(X)).$$

$$T_f(\Pi) : q(z)\, q(y)$$
$$p(V) \leftarrow q(X), \&comp_1[f, X](U),$$
$$\&comp_1[f, U](V).$$
$$r(X) \leftarrow p(X).$$
$$r(X) \leftarrow r(V), \&decomp_1[V](f, X).$$

Intuitively, $T_f(\Pi)$ builds $f(f(X))$ for any X on which q holds using two atoms over $\&comp_1$, and it extracts terms X from derived $r(f(X))$ facts using a $\&decomp_1$-atom.

Note that $\&decomp_n$ supports a well-ordering on term depth such that its output has always a strictly smaller depth than its inputs. This is an important property for proving finite groundability of a program by exploiting the TBFs introduced in [6].

Example 11. The program $\Pi = \{q(f(f(a)));\ q(X) \leftarrow q(f(X))\}$ is translated to $T_f(\Pi) = \{q(f(f(a)));\ q(X) \leftarrow q(V), \&decomp_1[V](f, X)\}$. Since $\&decomp_1$ supports a well-ordering, the cycle is *benign* [6], i.e., it cannot introduce infinitely many values because the nesting depth of terms is strictly decreasing with each iteration.

The realization of function symbols via external atoms (which can in fact also be seen as domain-specific existential quantifiers) has the advantage that their processing can be controlled. For instance, the introduction of new nested terms may be restricted by additional conditions which can be integrated in the semantics of the external predicates $\&comp_k$ and $\&decomp_k$. A concrete example is *data type checking*, i.e., testing whether the arguments of a function term are from a certain domain. In particular, values might also be rejected, e.g., bounded generation up to a maximal term depth is possible. Another example is to compute some of the term arguments automatically from others, e.g., constructing the functional term $num(7, vii)$ from 7, where the second argument is the Roman representation of the first one.

Another advantage is that the use of external atoms for functional term processing allows for exploiting de-safety of HEX-programs to guarantee finiteness of the grounding. An expressive framework for handling domain-expansion safe programs [6] can be reused without the need to enforce safety criteria specific for function terms.

7   Discussion and Conclusion

We presented model computation and query answering over HEX-programs with domain-specific existential quantifiers, based on external atoms and a new grounding algorithm. In contrast to usual handling of existential quantifiers, ours especially allows for an easy integration of extensions such as additional constraints (even of non-logical nature) or data types. This is useful e.g. for model building applications where particular data is needed for existential values, and gives one the possibility to implement domain-restricted quantifiers and introduce null values, as in databases. The new grounding algorithm allows for controlled bounded grounding; this can be exploited for *bounded model generation*, which might be sufficient (or convenient) for applications. Natural candidates are configuration or, at an abstract level, generating finite models of general first-order formulas as in [12], where an incremental computation of finite models is provided by a translation into incremental ASP. There, grounding and solving is interleaved by continuously increasing the bound on the number of elements in the domain. (Note that, although not designed for interleaved evaluation, our approach is flexible enough to also mimic exactly this technique with suitable external atoms.) The work in [1] aims at grounding first-order sentences with complex terms such as functions and aggregates for model expansion tasks. Similar to ours, it is based on bottom-up computation, but we do not restrict to finite structures and allow for potentially infinite domains. As a show case, we considered purely logical existentials (null values), for which our grounding algorithm amounts to a simulation of the one in [17] for $Datalog^{\exists}$-programs. However, while [17] combine grounding and model building, our approach clearly separates the two steps; this may ease possible extensions.

We then realized function symbol processing as in [4], by using external atoms to manipulate nested terms. In contrast to other approaches, no extension of the reasoner is needed for this. Furthermore, using external atoms has the advantage that nested terms can be subject to (even non-logical) constraints given by the semantics of the external atoms, and that finiteness of the grounding follows from de-safety of HEX-programs.

In model-building over $HEX^{\exists}$-programs, we can combine existentials with function symbols, as $HEX^{\exists}$-programs can have external atoms in rule bodies. To allow this for query answering over $Datalog^{\exists}$-programs remains to be considered. More generally, also combining existentials with arbitrary external atoms and the use of default-negation in presence of existentials is an interesting issue for future research. This leads to *nonmonotonic existential rules*, which most recently are considered in [18] and in [15], equipping the $Datalog^{\pm}$formalism, which is tailored to ontological knowledge representation and tractable query answering, with well-founded negation. Another line for future research is to allow disjunctive rules and existential quantification as in $Datalog^{\exists,\vee}$ [2], leading to a generalization of the class of Shy-programs. Continuing on the work on guardedness conditions as in open answer set programming [16], $Datalog^{\exists}$, and $Datalog^{\pm}$ should prove useful to find important techniques for constructing more expressive variants of HEX-programs with domain-specific existential quantifiers. The separation of grounding and solving in our approach should be an advantage for such enhancements.

A   Appendix: Term Bounding Function b_{synsem}

The TBF b_{synsem} [6] builds on the *positive attribute dependency graph* $G_A(\Pi)$, whose nodes are the attributes of Π and whose edges model the information flow between them. E.g., if for rule r we have $p(\mathbf{X}) \in H(r)$ and $q(\mathbf{Y}) \in B^+(r)$ such that $X_i = Y_j$ for some $X_i \in \mathbf{X}$ and $Y_j \in \mathbf{Y}$, then we have a flow from $q\lceil j$ to $p\lceil i$. A cycle K in $G_A(\Pi)$ is *benign* wrt. a set of safe attributes S, if there exists a well-ordering $\leq_C$ of C, such that for every $\&g[\mathbf{X}]_r\lceil_o j \notin S$ in the cycle, $f_{\&g}(\mathbf{A}, x_1, \ldots, x_m, t_1, \ldots t_n) = 0$ whenever

– some x_i for $1 \leq i \leq m$ is a predicate parameter, $\&g[\mathbf{X}]_r\!\restriction_I i \notin S$ is in K, and we
  have $(s_1, \ldots, s_{ar(x_i)}) \in ext(\mathbf{A}, x_i)$, and $t_j \not\leq_C s_k$ for some $1 \leq k \leq ar(x_i)$; or
– for some $1 \leq i \leq m$, $type(\&g, i) = \mathrm{const}$, $\&g[\mathbf{X}]_r\!\restriction_I i \notin S$ is in K, and $t_j \not\leq_C x_i$.

A cycle in $G_A(\varPi)$ is called *malign* wrt. S if it is not benign. Then b_{synsem} is as
follows.

Definition 5 (Syntactic and Semantic Term Bounding Function).
We define the TBF $b_{synsem}(\varPi, r, S, B)$ such that $t \in b_{synsem}(\varPi, r, S, B)$ iff

(i) *t is a constant in r; or*
(ii) *there is an ordinary atom $q(s_1, \ldots, s_{ar(q)}) \in B^+(r)$ such that $t = s_j$, for
     some $1 \leq j \leq ar(q)$ and $q\!\restriction j \in S$; or*
(iii) *for some external atom $\&g[\mathbf{X}](\mathbf{Y}) \in B^+(r)$, we have that $t = Y_i$ for some
     $Y_i \in \mathbf{Y}$, and for each $X_i \in \mathbf{X}$, $X_i \in B$, if $\tau(\&g, i) = \mathrm{const}$, and $X_i\!\restriction 1, \ldots,$
     $X_i\!\restriction ar(X_i) \in S$ if $\tau(\&g, i) = \mathbf{pred}$; or*
(iv) *t is captured by some attribute α in $B^+(r)$ that is not reachable from malign
     cycles in $G_A(\varPi)$ wrt. S, i.e., if $\alpha = p\!\restriction i$ then $t = t_i$ for some $p(t_1, \ldots, t_\ell) \in$
     $B^+(r)$, and if $\alpha = \&g[\mathbf{X}]_r\!\restriction_T i$ then $t = X_i^T$ for some $\&g[\mathbf{X}^I](\mathbf{X}^O) \in B^+(r)$
     where the input and output vectors are $\mathbf{X}^T = X_1^T, \ldots, X_{ar(\&g)}^T$; or*
(v) *$t = Y_i$ for some $\&g[\mathbf{X}](\mathbf{Y}) \in B^+(r)$, where $\{y_i \mid \mathbf{x} \in (\mathcal{P} \cup \mathcal{C})^{ar_i(\&g)}, \mathbf{y} \in$
     $\mathcal{C}^{ar_o(\&g)}, f_{\&g}(\mathbf{A}, \mathbf{x}, \mathbf{y}) = 1\}$ is finite for all assignments $\mathbf{A}$.*
(vi) *$t \in \mathbf{X}$ for some $\&g[\mathbf{X}](\mathbf{Y}) \in B^+(r)$, where $U \in B$ for every $U \in \mathbf{Y}$ and $\{\mathbf{x} \mid$
     $\mathbf{x} \in (\mathcal{P} \cup \mathcal{C})^{ar_i(\&g)}, f_{\&g}(\mathbf{A}, \mathbf{x}, \mathbf{y}) = 1\}$ is finite for every $\mathbf{A}$ and $\mathbf{y} \in \mathcal{C}^{ar_o(\&g)}$.*

References

1. Aavani, A., Wu, X.N., Ternovska, E., Mitchell, D.: Grounding formulas with complex terms. In: Butz, C., Lingras, P. (eds.) Canadian AI 2011. LNCS, vol. 6657, pp. 13–25. Springer, Heidelberg (2011)
2. Alviano, M., Faber, W., Leone, N., Manna, M.: Disjunctive datalog with existential quantifiers: semantics, decidability, and complexity issues. TPLP **12**(4–5), 701–718 (2012)
3. Brewka, G., Eiter, T., Truszczyński, M.: Answer set programming at a glance. Commun. ACM **54**(12), 92–103 (2011)
4. Calimeri, F., Cozza, S., Ianni, G.: External sources of knowledge and value invention in logic programming. Ann. Math. Artif. Intell. **50**(3–4), 333–361 (2007)
5. Eiter, T., Fink, M., Krennwallner, T., Redl, C.: Grounding HEX-programs with expanding domains. In: Workshop on Grounding and Transformations for Theories with Variables (GTTV'13), pp. 3–15 (2013)
6. Eiter, T., Fink, M., Krennwallner, T., Redl, C.: Liberal safety for answer set programs with external sources. In: AAAI'13, pp. 267–275. AAAI Press (2013)
7. Eiter, T., Ianni, G., Schindlauer, R., Tompits, H.: A uniform integration of higher-order reasoning and external evaluations in answer-set programming. In: IJCAI, pp. 90–96 (2005)
8. Eiter, T., Simkus, M.: FDNC: decidable nonmonotonic disjunctive logic programs with function symbols. ACM Trans. Comput. Log. **11**(2), 14:1–14:50 (2010)

9. Faber, W., Leone, N., Pfeifer, G.: Semantics and complexity of recursive aggregates in answer set programming. Artif. Intell. **175**(1), 278–298 (2011)

10. Fagin, R., Kolaitis, P., Miller, R., Popa, L.: Data exchange: semantics and query answering. Theor. Comput. Sci. **336**(1), 89–124 (2005)

11. Gebser, M., Kaufmann, B., Schaub, T.: Conflict-driven answer set solving: from theory to practice. Artif. Intell. **187**, 52–89 (2012)

12. Gebser, M., Sabuncu, O., Schaub, T.: An incremental answer set programming based system for finite model computation. AI Commun. **24**(2), 195–212 (2011)

13. Gelfond, M., Lifschitz, V.: Classical negation in logic programs and disjunctive databases. New Gener. Comput. **9**(3–4), 365–386 (1991)

14. Grau, B.C., Horrocks, I., Krötzsch, M., Kupke, C., Magka, D., Motik, B., Wang, Z.: Acyclicity conditions and their application to query answering in description logics. In: KR'12, pp. 243–253. AAAI Press (2012)

15. Hernich, A., Kupke, C., Lukasiewicz, T., Gottlob, G.: Well-founded semantics for extended datalog and ontological reasoning. In: PODS'13, pp. 225–236. ACM (2013)

16. Heymans, S., Nieuwenborgh, D.V., Vermeir, D.: Open answer set programming with guarded programs. ACM Trans. Comput. Logic **9**(4), 26:1–26:53 (2008)

17. Leone, N., Manna, M., Terracina, G., Veltri, P.: Efficiently computable datalog$^{\exists}$ programs. In: KR'12, pp. 13–23. AAAI Press (2012)

18. Magka, D., Krötzsch, M., Horrocks, I.: Computing stable models for nonmonotonic existential rules. In: IJCAI'13, pp. 1031–1038. AAAI Press (2013)

Introducing Real Variables and Integer Objective Functions to Answer Set Programming

Guohua Liu, Tomi Janhunen$^{(\boxtimes)}$, and Ilkka Niemelä

Helsinki Institute for Information Technology HIIT, Department of Information and Computer Science, Aalto University, FI-00076 Aalto, Finland
{Guohua.Liu,Tomi.Janhunen,Ilkka.Niemela}@aalto.fi

Abstract. Answer set programming languages have been extended to support linear constraints and objective functions. However, the variables allowed in the constraints and functions are restricted to integer and Boolean domains, respectively. In this paper, we generalize the domain of linear constraints to real numbers and that of objective functions to integers. Since these extensions are based on a translation from logic programs to mixed integer programs, we compare the translation-based answer set programming approach with the native mixed integer programming approach using a number of benchmark problems.

1 Introduction

Answer set programming (ASP) [16], also known as logic programming under *stable model* semantics [8], is a declarative programming paradigm where a given problem is solved by devising a logic program whose *answer sets* capture the solutions of the problem and then by computing the answer sets using *answer set solvers*. The paradigm has been exploited in a rich variety of applications [2].

Linear constraints have been introduced to ASP [1,7,13,14] in order to combine the high-level modeling capabilities of ASP languages with the efficient constraint solving techniques developed in the area of constraint programming. In particular, a language ASP(LC) is devised in [13] which allows linear constraints to be used within the original ASP language structures. The answer set computation for ASP(LC) programs is based on mixed integer programming (MIP) where an ASP(LC) program is first translated into a MIP program and then the solutions of the MIP program are computed using a MIP solver. Finally, answer sets can be recovered from the solutions found (if any).

In this paper, we extend and evaluate the ASP(LC) language in the following aspects. First, we generalize the domain of variables allowed in linear constraints from integers to reals. Real variables are ubiquitous in applications, e.g., timing variables in scheduling problems. However, the MIP-based answer set

The support from the Finnish Centre of Excellence in Computational Inference Research (COIN) funded by the Academy of Finland (under grant #251170) is gratefully acknowledged.

M. Hanus and R. Rocha (Eds.): KDPD 2013, LNAI 8439, pp. 118–135, 2014.
DOI: 10.1007/978-3-319-08909-6_8, © Springer International Publishing Switzerland 2014

computation confines the variables in linear constraints to the integer domain. We overcome this limitation by developing a translation of ASP(LC) programs to MIP programs so that constraints over real variables are enabled in the language. Second, we introduce MIP objective functions, i.e., linear functions of integer variables, to the ASP(LC) language. The original ASP(LC) language allows objective functions of Boolean variables only, but integer variables are more convenient than Booleans in many applications [13]. To model optimization problems in these areas, we enable MIP objective functions in ASP by giving semantics for ASP programs with these functions. Third, we compare ASP(LC) to MIP. This is interesting as ASP(LC) provides a richer language than MIP where ASP language structures are extended with linear constraints but the implementation technique is based on translating an ASP(LC) program to a MIP program to solve. We choose some representative problems, study the ASP(LC) and MIP encodings of the problems, and evaluate their computational performance by experiments.

The rest of the paper is organized as follows. Some basic definitions related with linear constraints and ASP(LC) are recalled in Sect. 2. Then we extend ASP(LC) language with real variables in Sect. 3, introduce MIP objective functions in Sect. 4, and compare ASP(LC) and native MIP formulations of some benchmark problems in Sect. 5. The experiments are reported in Sect. 6 followed by a discussion on related works in Sect. 7. The paper is concluded by Sect. 8.

2 Preliminaries

In this section, we review the basic concepts of linear constraints, mixed integer programming, and the ASP(LC) language. We study *linear constraints* of the form

$$\sum_{i=1}^{n} u_i x_i \sim k \tag{1}$$

where the u_i's and k are real numbers and the x_i's are variables ranging over real numbers (including integers). We distinguish the variables to be *real* and *integer* variables when necessary. The operator $\sim$ is in $\{<, \leq, \geq, >\}$[1]. Constraints involving "$<$" and "$>$" are called *strict* constraints. A valuation ν from variables to numbers is a *solution* of (or *satisfies*) a constraint C of the form (1), denoted $\nu \models C$, iff $\sum_{i=1}^{n} u_i \nu(x_i) \sim k$ holds. A valuation ν is a solution of a set of constraints $\Pi = \{C_1, ..., C_m\}$, denoted $\nu \models \Pi$, iff $\nu \models C_i$ for each $C_i \in \Pi$. A set of linear constraints is *satisfiable* iff it has a solution.

A *mixed integer program* (or a *MIP program*), takes the form

$$\text{optimize} \quad \sum_{i=1}^{n} u_i x_i \tag{2}$$

$$\text{subject to} \quad C_1, ..., C_m. \tag{3}$$

---

[1] The operator "$=$" can be represented by "$\leq$" and "$\geq$".

where the keyword `optimize` is `minimize` or `maximize`, the u_i's are numbers, the x_i's are variables, and the C_i's are linear constraints. The operators in the constraints are in $\{\leq, =, \geq\}$. The function $\sum_{i=1}^{n} u_i x_i$ is called an *objective function*. The constraints $C_1, ..., C_m$ may be written as a set $\{C_1, ..., C_m\}$. A valuation ν is a solution of a MIP program iff $\nu \models \{C_1, ..., C_m\}$. A solution is *optimal* iff it minimizes (or maximizes) the value of the objective function. The objective function could be empty (missing from a MIP program), in which case the function is trivially optimized by any solution. The keywords `optimize` and `subject to` may be omitted if the objective function is empty. The goal of MIPs is to find the optimal solutions of a MIP program.

An *ASP(LC) program* is a set of rules of the form

$$a \leftarrow b_1, \ldots, b_n, \text{not } c_1, \ldots, \text{not } c_m, t_1, \ldots, t_l \tag{4}$$

where each a, b_i, and c_i is a propositional atom and each t_i, called a *theory atom*, is a linear constraint of the form (1). Propositional atoms and theory atoms may be uniformly called *atoms*. Atoms and atoms preceded by "not" are also referred to as *positive* and *negative literals*, respectively. Given a program P, the set of propositional and theory atoms appearing in P are denoted by $\mathcal{A}(P)$ and $\mathcal{T}(P)$, respectively. For a rule r of the form (4), the *head* and the *body* of r are defined by $\mathrm{H}(r) = \{a\}$ and $\mathrm{B}(r) = \{b_1, \ldots, b_n, \text{not } c_1, \ldots, \text{not } c_m, t_1, \ldots, t_l\}$. Furthermore, the *positive*, *negative*, and *theory* parts of the body are defined as $\mathrm{B}^+(r) = \{b_1, \ldots, b_n\}$, $\mathrm{B}^-(r) = \{c_1, \ldots, c_m\}$, and $\mathrm{B}^t(r) = \{t_1, \ldots, t_l\}$, respectively. The body and the head of a rule could be empty: a rule without body is a *fact* whose head is true unconditionally and a rule without head is an *integrity constraint* enforcing the body to be false.

A set of atoms M satisfies an atom a, denoted $M \models a$, iff $a \in M$, and it satisfies a negative literal "not a", denoted $M \models \text{not } a$, iff $a \notin M$. The set M satisfies a set of literals $L = \{l_1, \ldots, l_n\}$, denoted $M \models L$, iff $M \models l_i$ for each $l_i \in L$. An *interpretation* of ASP(LC) program P is a pair $\langle M, T \rangle$, where $M \subseteq \mathcal{A}(P)$ and $T \subseteq \mathcal{T}(P)$, such that $T \cup \bar{T}$ is satisfiable in linear arithmetics where $\bar{T} = \{\neg t \mid t \in \mathcal{T}(P) \text{ and } t \notin T\}$ and $\neg t$ denotes the constraint obtained by changing the operator of t to the complementary one. Two interpretations $I_1 = \langle M_1, T_1 \rangle$ and $I_2 = \langle M_2, T_2 \rangle$ are *equal*, denoted $I_1 = I_2$, iff $M_1 = M_2$ and $T_1 = T_2$. An interpretation $I = \langle M, T \rangle$ satisfies a literal l iff $M \cup T \models l$. An interpretation I satisfies a rule r, denoted $I \models r$, iff $I \models \mathrm{H}(r)$ or $I \nvDash \mathrm{B}(r)$. An integrity constraint is satisfied by I iff $I \nvDash \mathrm{B}(r)$. An interpretation I is a *model* of a program P, denoted $I \models P$, iff $I \models r$ for each $r \in P$.

Similar to and as an extension of the semantics of *normal* logic programs [8,16], answer sets of ASP(LC) programs are defined using the concept of *reduct* as follows.

Definition 1 (Liu et al. [13]). Let P be an ASP(LC) program and $\langle M, T \rangle$ an interpretation of P. The *reduct* of P with respect to $\langle M, T \rangle$, denoted $P^{\langle M,T \rangle}$, is defined as $P^{\langle M,T \rangle} = \{\mathrm{H}(r) \leftarrow \mathrm{B}^+(r) \mid r \in P, \mathrm{H}(r) \neq \emptyset, \mathrm{B}^-(r) \cap M = \emptyset, \text{ and } \mathrm{B}^t(r) \subseteq T\}$.

Definition 2 (Liu et al.[13]). Let P be an ASP(LC) program. An interpretation $\langle M, T \rangle$ is an *answer set* of P iff $\langle M, T \rangle \models P$ and M is the subset minimal model of $P^{\langle M, T \rangle}$. The set of answer sets of P is denoted by $AS(P)$.

Example 1. Let P be an ASP(LC) program consisting of the rules

$$a \leftarrow x - y \leq 2. \quad b \leftarrow x - y \geq 5. \quad \leftarrow x - y \geq 0.$$

The interpretation $I_1 = \langle \{a\}, \{x - y \leq 2\} \rangle$ is an answer set of P since $\{(x - y \leq 2), \neg(x - y \geq 5), \neg(x - y \geq 0)\}$ is satisfiable in linear arithmetics, $I_1 \models P$, and $\{a\}$ is the minimal model of $P^{I_1} = \{a \leftarrow .\}$. The interpretation $I_2 = \langle \{b\}, \{x - y \geq 5\} \rangle$ is not an answer set since $\{(x - y \geq 5), \neg(x - y \leq 2), \neg(x - y \geq 0)\}$ is unsatisfiable. Finally, $I_3 = \langle \emptyset, \{x - y \geq 0\} \rangle$ is not an answer set, since $I_3 \nvDash P$. □

Syntactically, theory atoms are allowed as heads of rules in the implementation of [13]. We use such rules in this paper for more intuitive reading. As regards their semantics, a rule with a theory atom as the head is equivalent to an integrity constraint, i.e., a rule $t \leftarrow a_1, \ldots, a_m, \text{not } b_1, \ldots, \text{not } b_n, t_1, \ldots, t_l$ where t is a theory atom is treated as the rule $\leftarrow a_1, \ldots, a_m, \text{not } b_1, \ldots, \text{not } b_n, t_1, \ldots, t_l, \neg t$ in answer set computation.

Note that, according to (1), the body of a rule (4) does not contain theory atoms with the operator "=", which will always be replaced by atoms with "$\geq$" and "$\leq$". Such a design relieves us from dealing with the operator "$\neq$". Similarly, a rule with an equality in the head (e.g. (36)), is a shorthand for a pair of rules with the operators "$\leq$" and "$\geq$" in their heads respectively.

Answer set computation for ASP(LC) programs is based on a translation to MIP programs [13]. We will refer to the translation as *MIP-translation* and denote the translation of a program P by $\tau(P)$. Due to space limitations, we skip a thorough review of $\tau(P)$ and focus on the fragment most relevant for this paper, i.e., the rules of the form

$$a \leftarrow t \tag{5}$$

where a is an propositional atom or not present at all and t is a theory atom. Recall that a rule without head is an integrity constraint.

In the translation, special linear constraints called *indicator constraints* are used. An indicator constraint is of the form $d = v \rightarrow C$ where d is a *binary variable* (integer variable with the domain $\{0, 1\}$), v is either 0 or 1, and C is a linear constraint. An indicator constraint is *strict* if C is strict and *non-strict* otherwise. An indicator constraint can be written as a constraint of the form (1) using the so-called *big-M* formulation.

For a program P consisting of simple rules (5) only, $\tau(P)$ is formed as follows:

1. For each theory atom t, we include a pair of indicator constraints

$$d = 1 \rightarrow t \qquad d = 0 \rightarrow \neg t \tag{6}$$

where d is a new binary variable introduced for t. The idea is to use the variable d to represent the constraint t in the sense that, for any solution ν of the constraints in (6), $\nu(d) = 1$ iff $\nu \models t$. Thus d can be viewed as a kind of a *name* for t.

2. Assuming that $a \leftarrow t_1, \ldots, a \leftarrow t_k$ are all the rules in P that have a as head, we include

$$a - d_1 \geq 0, \quad \ldots, \quad a - d_k \geq 0, \tag{7}$$
$$d_1 + \ldots + d_k - a \geq 0 \tag{8}$$

where $d_1, \ldots, d_k$ are the binary variables corresponding to $t_1, \ldots, t_k$ in (6). The constraints in (7) and (8) enforce that the joint head a holds iff some of the bodies $t_1, \ldots, t_k$ holds which is compatible with Clark's completion [3]. If $k = 1$, i.e., the atom a has a unique defining rule, the constraints of (7) and (8) reduce to $a - d_1 = 0$ which makes d_1 synonymous with a. Moreover, if the rule (5) is an integrity constraint, then $d_1 = 0$ is sufficient, as intuitively implied by $k = 1$ and $a = 0$.

In the implementation of $\tau(P)$, more variables and constraints are used to cover the rules of the general form (4). We refer the reader to [13] for details.

The solutions of the MIP-translation of a program capture its answer sets as follows. Let P be an ASP(LC) program and ν a mapping from variables to numbers. We define the ν-*induced interpretation* of P, denoted I_P^ν, by setting $I_P^\nu = \langle M, T \rangle$ where

$$M = \{a \mid a \in \mathcal{A}(P), \nu(a) = 1\} \text{ and} \tag{9}$$
$$T = \{t \mid t \in \mathcal{T}(P), \nu \models t\}. \tag{10}$$

Theorem 1 (Liu et al. [13]). Let P be an ASP(LC) program.

1. If ν is a solution of $\tau(P)$, then $I_P^\nu \in AS(P)$.
2. If $I \in AS(P)$, then there is a solution ν of $\tau(P)$ such that $I = I_P^\nu$.

Example 2. For the program P from Example 1, the translation $\tau(P)$ consists of:

$$
\begin{array}{lll}
d_1 = 1 \rightarrow x - y \leq 2, & d_1 = 0 \rightarrow x - y > 2, & a - d_1 = 0, \\
d_2 = 1 \rightarrow x - y \geq 5, & d_2 = 0 \rightarrow x - y < 5, & b - d_2 = 0, \\
d_3 = 1 \rightarrow x - y \geq 0, & d_3 = 0 \rightarrow x - y < 0, & d_3 = 0.
\end{array}
$$

For any solution ν of $\tau(P)$, we have $\nu(a) = 1$, $\nu(b) = 0$, and $\nu(x) - \nu(y) \leq 2$ which capture the unique answer set $\langle \{a\}, \{x - y \leq 2\} \rangle$ of P. □

3 Extension with Real Variables

In this section, we first illustrate the reasons why the strict constraints involved in the MIP-translation prevent real variables from ASP(LC) programs. Then, we develop a translation from strict constraints to non-strict ones. Finally, we apply the translation to remove strict constraints from the MIP-translation so that the use of real variables is enabled in ASP(LC) programs.

3.1 Problems Caused by Real Variables

Real variables are widely used in knowledge representation and reasoning. However, the computation of answer sets based on the MIP-translation becomes problematic in their presence. The reason is that typical MIP systems do not fully support strict constraints involving real variables, e.g., by treating strict constraints as non-strict ones. Consequently, the correspondence between solutions and answer sets may be lost.

Example 3. Consider the condition that Tom gets a bonus if he works at least 8.25 h and the fact that he works for that long. By formalizing these constraints we obtain an ASP(LC) program P consisting of the following rules:

$$bonus(tom) \leftarrow h(tom) \geq 8.25. \tag{11}$$
$$\leftarrow h(tom) < 8.25. \tag{12}$$

In the above, the ground term $h(tom)$ is treated as a real variable recording the working hours of Tom and $bonus(tom)$[2] is a ground (propositional) atom meaning that Tom will be paid a bonus. The MIP-translation $\tau(P)$ of P has the following constraints:

$$d_1 = 1 \rightarrow h(tom) \geq 8.25 \tag{13}$$
$$d_1 = 0 \rightarrow h(tom) < 8.25 \tag{14}$$
$$bonus(tom) - d_1 = 0 \tag{15}$$
$$d_2 = 1 \rightarrow h(tom) < 8.25 \tag{16}$$
$$d_2 = 0 \rightarrow h(tom) \geq 8.25 \tag{17}$$
$$d_2 = 0 \tag{18}$$

Given $\tau(P)$ as input, CPLEX provides a solution ν where $\nu(bonus(tom)) = \nu(d_1) = \nu(d_2) = 0$ and $\nu(h(tom) = 8.25)$. However $I_P^\nu = \langle \emptyset, \{h(tom) = 8.25\}\rangle$ is not an answer set of P since it does not satisfy the rule (11). This discrepancy is due to the fact that ν actually does not satisfy the strict constraint (14), but CPLEX treats it as the non-strict one $d_1 = 0 \rightarrow h(tom) \leq 8.25$ and so that gives ν as a solution unexpectedly.   □

The current implementation of the MIP-translation [13] addresses only integer-valued constraints where the coefficients and variables range over integers. Given this restriction, strict constraints of the form $\sum_{i=1}^{n} u_i x_i < k$ (resp. $> k$) can be implemented as non-strict ones $\sum_{i=1}^{n} u_i x_i \leq k - 1$ (resp. $\geq k + 1$).

It might be tempting to convert the domain of a problem from reals to integers by multiplying each term in a constraint with a (constant) coefficient. For instance, by multiplying the constraints with 100 and replacing the variable

---

[2] We use different fonts for function and predicate symbols, such as "h" and "bonus" in this example, for clarity.

$h(tom)$ by another variable $h'(tom)$ holding a hundredfold value, the program P in Example 3 turn to:

$$\text{bonus}(tom) \leftarrow h'(tom) \geq 825. \tag{19}$$
$$\leftarrow h'(tom) < 825. \tag{20}$$

Thereafter constraints (13) and (14) could be rewritten as non-strict constraints:

$$d_1 = 1 \rightarrow h'(tom) \geq 825. \tag{21}$$
$$d_1 = 0 \rightarrow h'(tom) \leq 824. \tag{22}$$

This approach, however, does not work in general. First, the translated program cannot cover the domain of the original problem due to the continuity of real numbers. For example, the rules (21) and (22) do not give any information about the working hours 8.245 which is covered by (13) and (14). Second, determining the required coefficients is infeasible in general since the real numbers occurring in constraints can be specified up to arbitrary precision which could vary from problem instance to another.

Because CPLEX treats strict constraints as non-strict ones, the MIP-translation becomes inapplicable for answer set computation in the presence of real-valued variables. To enable such computations, a revised translation which consists of non-strict constraints only is needed. Such a translation is devised in sections to come.

3.2 Non-strict Translation of Strict Constraints

We focus on strict constraints of the form $y > 0$. This goes without loss of generality because any constraint $\sum_{i=1}^{n} u_i x_i > k$ can be rewritten as a conjunction of a non-strict constraint $\sum_{i=1}^{n} u_i x_i - y = k$ and a strict one $y > 0$ where y is fresh. Also, a constraint of the form $\sum_{i=1}^{n} u_i x_i < k$ is equivalent to $-\sum_{i=1}^{n} u_i x_i > -k$.

Lemma 1. *Let Γ be a set of non-strict constraints, $S = \{x_1 > 0, \ldots, x_n > 0\}$, and δ a new variable. Then, the set $\Gamma \cup S$ is satisfiable iff for any bound $b > 0$, the set $\Gamma \cup S_\delta \cup \{0 < \delta \leq b\}$ where $S_\delta = \{x_1 \geq \delta, \ldots, x_n \geq \delta\}$ is satisfiable.*

Proof. We prove the direction "$\Rightarrow$" since the other direction is obvious. Since $\Gamma \cup S$ is satisfiable, there is a valuation ν such that $\nu \models \Gamma$ and $\nu(x_i) > 0$ for each $1 \leq i \leq n$. Let $b > 0$ be any number and $m = \min\{\nu(x_1), \ldots, \nu(x_n)\}$. Then $\nu(x_i) \geq m$ holds for any $1 \leq i \leq n$ and $m > 0$. Two cases arise to analyze:

1. $m \leq b$. Then $\Gamma \cup S_\delta \cup \{0 < \delta \leq b\}$ has a solution ν' which extends ν by the assignment $\nu'(\delta) = m$.
2. $m > b$. We define ν' as an extension of ν such that $\nu'(\delta) = b$. Thus $b > 0$ implies $\nu' \models 0 < \delta \leq b$. Moreover, for any $1 \leq i \leq n$, $\nu'(x_i) = \nu(x_i) \geq \nu'(\delta) = b$ since $m > b$. Therefore $\nu' \models x_i \geq \delta$ for any $1 \leq i \leq n$.

It follows that $\nu' \models \Gamma \cup S_\delta \cup \{0 < \delta \leq b\}$. $\square$

The result of Lemma 1 can be lifted to the case of indicator constraints since indicator constraints are essentially linear constraints.

Lemma 2. *Let Γ be a set of non-strict constraints,*

$$S = \{d_i = v_i \to x_i > 0 \mid 1 \leq i \leq n\} \tag{23}$$

a set of strict indicator constraints, and δ a new variable. Then, $\Gamma \cup S$ is satisfiable iff for any bound $b > 0$, $\Gamma \cup S_\delta \cup \{0 < \delta \leq b\}$ is satisfiable where

$$S_\delta = \{d_i = v_i \to x_i \geq \delta \mid 1 \leq i \leq n\}. \tag{24}$$

Lemma 2 shows that a set of strict indicator constraints can be transformed to a set of non-strict ones by introducing a new bounded variable $0 < \delta \leq b$. Below, we relax the last remaining strict constraint $\delta > 0$ to $\delta \geq 0$ using a MIP objective function.

Definition 3. *Let $\Pi = \Gamma \cup S$ be a set of constraints where Γ is a set of non-strict ones and S is the set of strict indicator constraints (23), δ a new variable, and $b > 0$ a bound. The non-strict translation of Π with respect to δ and b, denoted Π_δ^b, is:*

$$\begin{aligned}
&\textit{maximize} &&\delta \\
&\textit{subject to} &&\Gamma \cup S_\delta \cup \{0 \leq \delta \leq b\}
\end{aligned} \tag{25}$$

where S_δ is defined by (24).

Given Lemma 2 and Definition 3, the satisfiability of a set of constraints can be captured by its non-strict translation as formalized by the following theorem.

Theorem 2. *Let Π, S, and Π_δ^b be defined as in Definition 3. Then, Π is satisfiable iff Π_δ^b has a solution ν such that $\nu(\delta) > 0$.*

Theorem 2 enables the use of current MIP systems for checking the satisfiability of a set of strict constraints, i.e., by computing an optimal solution for the non-strict translation of the set and by checking if the objective function has a positive value.

3.3   Non-strict Translation of Programs

Next, we develop the non-strict translation of ASP(LC) programs using Definition 3.

Definition 4. *Let P be an ASP(LC) program, δ a new variable, and $b > 0$ a bound. The non-strict translation of P with respect to δ and b, is $\tau(P)_\delta^b$ where $\tau(P)$ is the MIP-translation of P.*

We show that the solutions of $\tau(P)_\delta^b$ and $\tau(P)$ are in a tight correspondence.

Lemma 3. *Let P be an ASP(LC) program that may involve real variables, δ a new variable, and $b > 0$ a bound.*

1. *For any solution $\nu \models \tau(P)$, there is a solution $\nu' \models \tau(P)_\delta^b$ such that $\nu(a) = \nu'(a)$ for each $a \in \mathcal{A}(P)$, $\nu \models t$ iff $\nu' \models t$ for each $t \in \mathcal{T}(P)$, and $\nu'(\delta) > 0$.*
2. *For any solution $\nu \models \tau(P)_\delta^b$ where $\nu(\delta) > 0$, there is a solution ν' of $\tau(P)$ such that $\nu(a) = \nu'(a)$ for each $a \in \mathcal{A}(P)$ and $\nu \models t$ iff $\nu' \models t$ for each $t \in \mathcal{T}(P)$.*

Proof. We prove Item 1 and omit the proof of Item 2 which is similar. Let ν be a solution of $\tau(P)$. Given ν, we extend $\tau(P)$ to $\tau'(P)$ by adding for each atom $a \in \mathcal{A}(P)$, a constraint $a = \nu(a)$, and for each theory atom $t \in \mathcal{T}(P)$ and the variable d introduced for t in (6), $d = \nu(d)$. It is clear that $\nu' = \nu$ is a solution of $\tau'(P)$. Let $\tau''(P)$ be the analogous extension of $\tau(P)_\delta^b$. Applying Theorem 2 to $\tau'(P)$, there is a solution ν'' of $\tau''(P)$ such that $\nu''(\delta) > 0$ and for each a, $\nu''(a) = \nu'(a) = \nu(a)$, and for each d, $\nu''(d) = \nu'(d) = \nu(d)$. The valuation ν'' is also a solution of $\tau(P)_\delta^b$, as $\tau(P)_\delta^b \subset \tau''(P)$. Note that for any $t \in \mathcal{T}(P)$ and the respective atom d, $\nu(d) = 1$ iff $\nu \models t$, and $\nu''(d) = 1$ iff $\nu'' \models t$. Then $\nu \models t$ iff $\nu'' \models t$ due to $\nu(d) = \nu''(d)$. □

Now, we relate the solutions of $\tau(P)_\delta^b$ and the answer sets of P. As a consequence of Lemma 3 and the generalization of Theorem 1 for real variables, we obtain:

Theorem 3. *Let P be an ASP(LC) program that may involve real variables, δ a new variable, and $b > 0$ a bound.*

1. *If ν is a solution of $\tau(P)_\delta^b$ such that $\nu(\delta) > 0$, then $I_P^\nu \in AS(P)$.*
2. *If $I \in AS(P)$, then there is a solution ν of $\tau(P)_\delta^b$ such that $I = I_P^\nu$ and $\nu(\delta) > 0$.*

Example 4. Let us revisit Example 3. By setting $b = 1$ as the bound, we obtain the non-strict translation $\tau(P)_\delta^1$ as follows:

```
maximize    δ
subject to  0 ≤ δ ≤ 1
```
$$
\begin{aligned}
d_1 = 1 &\to h(tom) \geq 8.25, & d_1 = 0 &\to h(tom) + \delta \leq 8.25, \\
d_2 = 1 &\to h(tom) + \delta \leq 8.25, & d_2 = 0 &\to h(tom) \geq 8.25, \\
\mathsf{bonus}(tom) &- d_1 = 0, & d_2 &= 0.
\end{aligned}
$$

For any optimal solution ν of $\tau(P)_\delta^1$, we have $\nu(\mathsf{bonus}(tom)) = \nu(d) = \nu(\delta) = 1$ and $\nu(h(tom)) \geq 8.25$ that corresponds to the intended answer set $\langle \{\mathsf{bonus}(tom)\}, \{h(tom) \geq 8.25\} \rangle$. We note that CPLEX provides exactly this solution for $\tau(P)_\delta^1$. □

It can be verified that the non-strict translation $\tau(P)_\delta^b$ reduces to the MIP-translation if the variables in P and the new variable δ are integers and the bound b is set to 1.

4   Extension with Objective Functions

In this section, we define optimal answer sets for ASP(LC) programs enhanced by *objective functions* of the form (2) and illustrate the resulting concept by examples.

Definition 5. *Let P be an ASP(LC) program with an objective function f and $\langle M, T \rangle \in AS(P)$. The answer set $\langle M, T \rangle$ is optimal iff there is a solution of $T \cup \bar{T}$ that gives the optimal value to f among the set of valuations*

$$\{\nu \mid \nu \models T \cup \bar{T} \text{ for some } \langle M, T \rangle \in AS(P)\}.$$

Example 5. Let P be an ASP(LC) program

$$\texttt{minimize } x. \quad a \leftarrow x \geq 5. \quad b \leftarrow x \geq 7. \quad \leftarrow x < 5.$$

The answer sets of P are $I_1 = \langle \{a\}, \{x \geq 5\} \rangle$ and $I_2 = \langle \{a, b\}, \{x \geq 5, x \geq 7\} \rangle$. Let $T_1 = \{x \geq 5\}$ and $T_2 = \{x \geq 5, x \geq 7\}$. The solutions of $T_1 \cup \bar{T}_1 = \{x \geq 5, x < 7\}$ admit a smaller value of the objective function $f(x) = x$, i.e., 5 than any solution of $T_2 \cup \bar{T}_2 = \{x \geq 5, x \geq 7\}$. Therefore the answer set I_1 is optimal. □

According to Definition 5, each optimal answer set identifies an optimal objective value. In other words, if the objective function is unbounded with respect to an answer set, then the answer set is not optimal. This is illustrated by our next example.

Example 6. Let P be an ASP(LC) program

$$\texttt{minimize } x. \quad a \leftarrow x \leq 4. \quad b \leftarrow x > 4.$$

The program P has two answer sets $I_1 = \langle \{a\}, \{x \leq 4\} \rangle$ and $I_2 = \langle \{b\}, \{x > 4\} \rangle$. Let $T_1 = \{x \leq 4\}$ and $T_2 = \{x > 4\}$. We have $T_1 \cup \bar{T}_1 = T_1$ and $T_2 \cup \bar{T}_2 = T_2$. Although $T_1 \cup \bar{T}_1$ admits smaller values of x than $T_2 \cup \bar{T}_2$, I_1 is not optimal, since x may become infinitely small subject to $T_1 \cup \bar{T}_1$. Therefore, P has no optimal answer set. □

For a program that does not involve real variables, we can establish an approach to computing the optimal answer sets as stated in the theorem below. This result essentially follows from Definition 5 and Theorem 1.

Theorem 4. *Let P be an ASP(LC) program involving integer variables only and f the objective function of P. Then $\langle M, T \rangle$ is an optimal answer set of P iff there is a solution $\nu \models \tau(P)$ such that $I_P^\nu = \langle M, T \rangle$ and ν gives the optimal value to f.*

However, when real variables are involved, the non-strict translation from ASP to MIP programs cannot be employed to compute the optimal answer sets, since the variable δ introduced in the translation may affect the optimal objective function value.

The ASP languages implemented in [6,18] support objective functions of the form

$$\texttt{\#optimize } [a_1 = w_{a_1}, ..., a_m = w_{a_m}, \texttt{not } b_1 = w_{b_1}, ..., \texttt{not } b_n = w_{b_n}] \qquad (26)$$

where the keyword $\texttt{optimize}$ is $\texttt{minimize}$ or $\texttt{maximize}$, a_i and "not b_i" are literals, and w_{a_i} and w_{b_i} are integer weights associated with the respective literals. The difference between the functions (2) and (26) is that x_i's in the former are integer variables whereas a_i's and b_i's in the latter are Boolean variables, i.e., propositional atoms from the ASP viewpoint. Integer variables are more convenient than Booleans for modeling some optimization problems as demonstrated by the following example.

Example 7. Let x be an integer variable taking a value from 1 to n and we want to minimize x. Using an objective function, this can be concisely encoded by:

$$\texttt{minimize } x. \quad x \geq 1. \quad x \leq n.$$

Following [16,19], to encode the same using the function (26), we need:

$$\texttt{\#minimize } [x(1) = 1, \ldots, x(n) = n]. \qquad (27)$$
$$x(i) \leftarrow \texttt{not } x(1), \ldots, \texttt{not } x(i-1), \texttt{not } x(i+1), \ldots \texttt{not } x(n). \quad 1 \leq i \leq n \qquad (28)$$

where the Boolean variable $x(i)$ represents that x takes value i and the rules in (28) encode that x takes exactly one value from 1 to n. The size of the ASP(LC) encoding is constant, while that of the original ASP encoding is quadratic[3] in the size of the domain of x, i.e., the objective function of length n plus n rules of length n.   □

In fact, a restriction to Boolean variables affects the performance of ASP systems when dealing with problems that involve large numerical domains [13].

5   A Comparison of ASP(LC) with MIP

Using the ASP(LC) paradigm, one solves a problem rather indirectly, i.e., by first modeling it as an ASP program, then by translating the program into a MIP program, and finally by solving the problem instance using a MIP solver. A question is how this approach compares with the native MIP approach where a problem is modeled and solved using just a mixed integer program. In this section, we study the modeling capabilities of ASP(LC) and MIP languages.

---

[3] Linear and logarithmic encodings can be achieved using *cardinality constraints* [6,18] and *bit vectors* [15], respectively. But both are more complex than the given ASP(LC) encoding.

We focus on two widely used primitives in modeling: *reachability* and *disjunctivity*. For the former, we study a Hamiltonian Routing Problem (HRP) and for the latter, a Job Shop Problem (JSP). The main observation is that ASP(LC) can provide more intuitive and compact encodings in debt to its capability to model non-trivial logical relations. But, the compactness does not always offer computational efficiency as perceived in the sequel.

In the HRP, we have a network and a set of *critical* vertices. The goal is to route a package along a Hamiltonian cycle in the network so that the package reaches each critical vertex within a given vertex-specific time limit. The network is represented by a set of vertices $V = \{1, \ldots, n\}$ and a set of weighted edges E consisting of elements (i, j, d) where $1 \leq i, j \leq n$ and d is a real number representing that the delay of the edge from i to j is d. The set of critical vertices CV consists of pairs (i, t) where $1 < i \leq n$ and t is a real number representing that the time limit of vertex i is t.

The ASP(LC) encoding of HRP in Fig. 1 is obtained by extending the encoding of Hamiltonian cycle problem [16] with timing constraints[4]. The rules (29)–(35) specify a Hamiltonian cycle. To model the timing constraints, we use a real variable $t(X)$ to denote the time when a vertex X is reached. Rules (36) and (37) determine the respective times of reaching vertices. The rule (38) ensures each critical vertex to be reached in time. The MIP program of HRP in Fig. 2 is from [17] with extensions of timing constraints[5]. The binary variable x_{ij} represents whether an edge (i, j, d) is on the Hamiltonian cycle ($x_{ij} = 1$) or not ($x_{ij} = 0$) and the integer variable p_i denotes the position of the vertex i on the cycle. The real variable r_i denotes the time of reaching vertex i. The constraints (39) and (40) say that each node has exactly one incoming and outgoing edge on a Hamiltonian cycle; (41) and (42) guarantee that each node has a unique position in the cycle; (43) and (44) ensure all the nodes in V form a single cycle (avoid separate cycles), where the condition $\nexists d : (i, j, d) \in E$ captures a pair of nodes (i, j) that are not connected by an edge with any delay d. The above constraints encode a Hamiltonian cycle. The remaining ones (45)–(47) specify the timing property, which are the counterparts of (36)–(38), respectively.

In comparison, reachability is modeled by the recursive rules (33) and (34) in the ASP(LC) program. Since MIP language cannot express such recursion directly, the reachability condition is captured otherwise—by constraining the positions of the nodes in (41)–(44). Note that the node positions are actually irrelevant for the existence of a cycle. In fact, modeling Hamiltonian cycles in non-complete graphs is challenging in MIP and the above encoding is the most compact one to the best of our knowledge.

In the JSP, we have a set of tasks $T = \{1, \ldots, n\}$ to be executed by a machine. Each task is associated with an earliest starting time and a processing duration. The goal is to schedule the tasks so that each task starts at its earliest starting

---

[4] More compact encoding can be obtained using *choice rules*, but for our purposes the current one is sufficient.

[5] Disequalities and implications can be represented using the operators "$\leq$" and "$\geq$" [17].

$$\text{hc}(X, Y) \leftarrow \text{e}(X, Y, D), \text{not } \text{nhc}(X, Y). \tag{29}$$

$$\text{nhc}(X, Y) \leftarrow \text{e}(X, Y, D_1), \text{e}(X, Z, D_2), \text{hc}(X, Z), Y \neq Z. \tag{30}$$

$$\text{nhc}(X, Y) \leftarrow \text{e}(X, Y, D_1), \text{e}(Z, Y, D_2), \text{hc}(Z, Y), X \neq Z. \tag{31}$$

$$\text{initial}(1). \tag{32}$$

$$\text{reach}(X) \leftarrow \text{reach}(Y), \text{hc}(Y, X), \text{not } \text{initial}(Y), \text{e}(Y, X, D). \tag{33}$$

$$\text{reach}(X) \leftarrow \text{hc}(Y, X), \text{initial}(Y), \text{e}(Y, X, D). \tag{34}$$

$$\leftarrow \text{v}(X), \text{not } \text{reach}(X). \tag{35}$$

$$t(1) = 0. \tag{36}$$

$$t(X) - t(Y) = D \leftarrow \text{hc}(Y, X), \text{e}(Y, X, D), X \neq 1. \tag{37}$$

$$t(X) \leq T \leftarrow \text{critical}(X, T). \tag{38}$$

Fig. 1. An ASP(LC) encoding of HRP

$$\sum_{(i,j,d)\in E} x_{ij} = 1 \qquad\qquad\qquad\qquad\qquad\qquad i \in V \tag{39}$$

$$\sum_{(j,i,d)\in E} x_{ji} = 1 \qquad\qquad\qquad\qquad\qquad\qquad i \in V \tag{40}$$

$$1 \leq p_i \leq n \qquad\qquad\qquad\qquad\qquad\qquad i \in V \tag{41}$$

$$p_i \neq p_j \qquad\qquad\qquad\qquad i \in V, j \in V, i \neq j \tag{42}$$

$$p_j \neq p_i + 1 \qquad\qquad i \in V, j \in V, i \neq j, \ \nexists d : (i,j,d) \in E \tag{43}$$

$$(p_i = n) \rightarrow (p_j \geq 2) \qquad\qquad i \in V, j \in V, i \neq j, \ \nexists d : (i,j,d) \in E \tag{44}$$

$$r_1 = 0 \tag{45}$$

$$x_{ij} = 1 \rightarrow r_j - r_i = d \qquad\qquad\qquad\qquad (i,j,d) \in E \tag{46}$$

$$r_i \leq t \qquad\qquad\qquad\qquad\qquad\qquad (i,t) \in CV \tag{47}$$

Fig. 2. A MIP encoding of HRP

time or later, the processing of the tasks do not overlap, and all tasks are finished by a given deadline. Using ASP(LC) we model the problem in Fig. 3 where the predicate $\text{task}(I, E, D)$ denotes that a task I has an earliest starting time E and a duration D and the real variables $s(I)$ and $e(I)$ denote the starting and ending times of task I, respectively. The rule (48) says that a task starts at its earliest starting time or later. The rule (49) ensures that each task is processed long enough. The rule (50) encodes the mutual exclusion of the tasks, i.e., for any two tasks, one must be finished before the starting of the other. The rule (51) enforces each task being finished by the deadline. Figure 4 adopts a recent MIP encoding of JSP [10] in CPLEX language, where the variables s_i and e_i represent the starting and ending times of task i, respectively; est_i and d_i are the earliest starting time and processing duration of task i; the binary variable x_{ij} denotes that task i ends before task j starts. The constraints (52), (53), and (57) are the counterparts of (48), (49), and (51) respectively. The constraints (54)–(56) exclude overlapping tasks.

$$s(I) \geq E \leftarrow \mathsf{task}(I, E, D). \tag{48}$$

$$e(I) - s(I) \geq D \leftarrow \mathsf{task}(I, E, D). \tag{49}$$

$$\leftarrow \mathsf{task}(I, E_1, D_1), \mathsf{task}(J, E_2, D_2), I \neq J, s(I) - s(J) \leq 0, s(J) - e(I) \leq 0. \tag{50}$$

$$e(I) \leq deadline \leftarrow \mathsf{task}(I, E, D). \tag{51}$$

Fig. 3. An ASP(LC) encoding of JSP

$$s_i \geq est_i \qquad\qquad i \in T \tag{52}$$

$$e_i - s_i \geq d_i \qquad\qquad i \in T \tag{53}$$

$$x_{ij} = 1 \rightarrow e_i - s_j < 0 \qquad\qquad i \in T, j \in T, i \neq j \tag{54}$$

$$x_{ij} = 0 \rightarrow e_i - s_j \geq 0 \qquad\qquad i \in T, j \in T, i \neq j \tag{55}$$

$$x_{ij} + x_{ji} = 1 \qquad\qquad i \in T, j \in T, i \neq j \tag{56}$$

$$e_i \leq deadline \qquad\qquad i \in T \tag{57}$$

Fig. 4. A MIP encoding of JSP

In the ASP(LC) program of JSP, the mutual exclusion of $s(I) - s(J) \leq 0$ and $s(J) - e(I) \leq 0$ is expressed by one rule (50). In contrast, MIP language lacks direct encoding of relations between constraints and therefore, to encode the exclusion of $e_i < s_j$ and $e_j < s_i$, one has to first represent them by new variables x_{ij} and x_{ji} in (54) and (55) respectively and then encode the relations of the variables in (56). Note that, for computation, the ASP(LC) and the native MIP encodings are essentially the same, since the translation of the rule (50) includes two indicator constraints to represent the constraints $s(I) - s(J) \leq 0$ and $s(J) - e(I) \leq 0$, respectively, and other additional constraints to encode the rule in the style of [13].

The presented encodings for HRP and JSP illustrate how ASP(LC) programs can provide a higher level of abstraction for modeling in contrast with MIP. The effect is clear for problem domains which involve recursively defined concepts and for which ASP-style rules subject to minimal model semantics provide a natural representation. Encoding such features in MIP require additional variables making the resulting representation more complex and difficult to maintain. A formulation with less variables is typically simpler and easier to understand which also favors easier maintainability and elaboration tolerance in face of additional constrains to be incorporated into the model. Furthermore, in contrast with traditional ASP, the possibility of using integer and/or real variables brings about another dimension of compaction if the problem domain of interest involves quantities having infinitely many or continuous values.

6 Experiments

We implemented the non-strict translation of ASP(LC) programs and the MIP objective functions by modifying the MINGO system [13]. The new system is called MINGO^r. We tested MINGO^r with a number of benchmarks[6]: the HRP and JSP detailed in Sect. 5, the Newspaper, Routing Max, and Routing Min problems from [13], and the Disjunctive Scheduling problem from [4]. These problems were selected as they involve either reachability or disjunctivity. The original instances are revised to include real numbers. The objective functions of the optimization versions of HRP and JSP are to minimize the time of reaching some critical node and the ending time of some task, respectively. The optimization problems involve integers only. The experiments were run on a Linux cluster having 112 AMD Opteron and 120 Intel Xeon nodes with Linux 6.1. In each run, the memory is limited to 4 GB and the cutoff time is set to 600 s.

In Table 1, we evaluate MINGO^r with different values of the parameter b, the bound used in the non-strict translation. The goal is to find a default setting for b. Table 1 suggests that $b = 10^{-6}$ is the best, considering the number of solved instances and the running time. We tested 100 random instances for each problem except Disjunctive Scheduling where we used the 10 instances given in [4]. Each instance was run 5 times and the average number of solved instances and the running time are reported. We also include the average sizes of the *ground* programs in kilobytes to give an idea on the space complexity of the instances. It is also worth pointing out that none of the problems reported in Table 1 is solvable by existing ASP systems since they involve real variables and thus comparisons of MINGO^r with other ASP systems are infeasible.

Tables 2 and 3 provide comparisons of the translation-based ASP approach with native MIP approach, where both MINGO^r and CPLEX are run with default settings. For the HRP problem in Table 2, we tested 50 randomly generated graphs of 30 nodes for each *density* (the ratio of the number of edges to the number of edges in the complete graph). The results show that the instances with medium densities are unsolvable to CPLEX before the cutoff but MINGO^r can solve them in reasonable time. We also note that CPLEX performs better for the graphs of high densities. This is because the MIP program encodes the positions of nonadjacent nodes. For the JSP problem in Table 3, we tested 50 random instances for each number of tasks and MINGO^r is slower than CPLEX by roughly a order of magnitude but, in spite of this, scaling is similar.

In summary, some observations are in order. On one hand, the ASP(LC) language enables compact encodings in debt to its capability of expressing non-trivial logical relations. Thus some redundant information, such as the order of nodes in a cycle in HRP, can be left out in favor of computational performance. On the other hand, the translation of ASP(LC) programs into MIP is fully general and thus some unnecessary extra variables could be introduced, e.g., in the case of JSP, the structure of the translation is more complex than the native MIP encoding. As a consequence, the translation-based approach is likely to be slower due to the extra time spent on propagation.

---

[6] A prototype implementation of the MINGO^r system and benchmarks can be found under http://research.ics.aalto.fi/software/asp/mingoR.

Table 1. The effect of the bound $b > 0$

Benchmark	$b = 10^{-9}$		$b = 10^{-6}$		$b = 10^{-3}$		$b = 1$		$b = 10^3$		Size
	Solved	Time	Solved	Time	Solved	Time	Solved	Time	Solved	Time	
Disj. Scheduling	10	0.60	10	0.78	10	0.99	10	0.89	10	12.60	206
Ham. Routing	100	30.79	100	24.52	100	23.16	100	41.63	0	NA	155
Job Shop	100	9.76	100	9.56	100	25.98	100	14.61	10	66.08	387
Newspaper	100	22.64	100	21.61	93	77.90	100	40.76	0	NA	846
Routing Max.	100	0.11	100	0.14	100	0.25	100	0.55	100	0.69	7
Routing Min.	76	109.58	77	102.98	80	127.12	46	95.07	20	79.07	368

7 Related Work

Dutertre and de Moura [5] translate strict linear constraints into non-strict ones using a new variable δ in analogy to Lemma 1. However, the variable remains unbounded from above in their proposal. In contrast to this, an explicit upper bound is introduced by our translation. The bound facilitates computation: if the translated set of constraints has no solution ν with $\nu(\delta) > 0$ under a particular bound b, this result is conclusive and no further computations are needed. Unbounded variables are problematic for typical MIP systems and thus having an upper bound for δ is important.

Given the extension of ASP(LC) programs with objective functions, MIP programs can be seen as a special case of ASP(LC) programs, i.e., for any MIP program P with an objective function (2) and constraints in (3), there is an ASP(LC) program P' whose objective function is (2) and whose rules simply list the constraints in (3) as theory atoms (facts in ASP terminology). Note that in this setting the non-strict translation of P' is identical to P, since P' involves non-strict constraints only.

In theory, all similar paradigms proposed in [1,7,14] cover real-valued constraints. Moreover, a recent ASP system CLINGCON [7] is implemented where constraints over integers are allowed in logic programs. But, to the best of our

Table 2. Hamiltonian Routing Problem

Density	Decision		Optimization	
	MINGO^r	CPLEX	MINGO^r	CPLEX
10	0.03	0.01	0.07	0.01
20	0.05	0.01	0.12	0.01
30	0.92	NA	50.81	NA
40	41.62	NA	NA	NA
50	13.94	NA	NA	NA
60	64.91	NA	NA	NA
70	35.78	NA	NA	NA
80	8.02	95.40	NA	NA
90	181.33	24.74	NA	NA
100	146.18	13.88	NA	NA

Table 3. Job Shop Problem

Tasks	Decision		Optimization	
	MINGO^r	CPLEX	MINGO^r	CPLEX
10	0.42	0.14	0.35	0.08
20	4.04	0.18	1.56	0.14
30	6.78	0.40	4.69	0.49
40	13.74	0.72	12.18	1.62
50	27.37	1.36	16.15	1.16
60	45.44	1.72	30.82	2.01
70	51.56	1.57	47.65	1.00
80	88.72	2.34	68.99	2.83
90	114.32	2.97	79.28	6.43
100	192.09	4.19	112.09	8.05

knowledge, there has not been any system that supports real-valued constraints nor integer-based objective functions. With the non-strict translation of ASP programs into mixed integer programs, we have implemented these primitives in the context of ASP.

Remark. Real numbers are implemented as floating point numbers in CPLEX, which uses numerically stable methods to perform its linear algebra so that round-off errors usually do not cause problems [11]. Note that—since computers are finite precision machines—the imprecision of floating point computations referred to as *numerical difficulties* is common to any computer systems and/or languages [9]. We do not address this issue in the paper and rather rely on the solutions implemented in CPLEX.

8 Conclusion and Future Work

In this paper, we generalize a translation from ASP(LC) programs to MIP programs so that linear constraints over real variables are enabled in answer set programming. Moreover, we introduce integer objective functions to ASP(LC) language. These results extend the applicability of answer set programming. We also compare the ASP approach with the native MIP approach and the results show that ASP extensions in question facilitate modeling and offer computational advantage at least for some problems. Our results suggest that MIP and ASP paradigms can benefit mutually: on one hand, efficient MIP formulations can be obtained by translating a compact ASP(LC) program; on the other hand, the ASP language can be extended with MIP constraints and objective functions to deal with problems that are not directly solvable using standard ASP languages.

The future work will be focused on system development. The possible directions include reducing the number of extra variables introduced by translations; stopping the solving phase early (as soon as δ is positive); and combining ASP and MIP solver technology to obtain a native solver for ASP(LC). Moreover, a comparative study of ASP(LC) encodings and MINGO^r with other constraint logic programming encodings and systems [12] can also provide insights into improving MINGO^r.

References

1. Balduccini, M.: Industrial-size scheduling with ASP+CP. In: Delgrande, J.P., Faber, W. (eds.) LPNMR 2011. LNCS, vol. 6645, pp. 284–296. Springer, Heidelberg (2011)
2. Brewka, G., Eiter, T., Truszczynski, M.: Answer set programming at a glance. Commun. ACM **54**(12), 92–103 (2011)
3. Clark, K.L.: Negation as failure. In: Logics and Databases, pp. 293–322 (1978)
4. Denecker, M., Vennekens, J., Bond, S., Gebser, M., Truszczyński, M.: The second answer set programming competition. In: Erdem, E., Lin, F., Schaub, T. (eds.) LPNMR 2009. LNCS, vol. 5753, pp. 637–654. Springer, Heidelberg (2009)

5. Dutertre, B., de Moura, L.: A fast linear-arithmetic solver for DPLL(T). In: Ball, T., Jones, R.B. (eds.) CAV 2006. LNCS, vol. 4144, pp. 81–94. Springer, Heidelberg (2006)
6. Gebser, M., Kaminski, R., König, A., Schaub, T.: Advances in *gringo* series 3. In: Delgrande, J.P., Faber, W. (eds.) LPNMR 2011. LNCS, vol. 6645, pp. 345–351. Springer, Heidelberg (2011)
7. Gebser, M., Ostrowski, M., Schaub, T.: Constraint answer set solving. In: Hill, P.M., Warren, D.S. (eds.) ICLP 2009. LNCS, vol. 5649, pp. 235–249. Springer, Heidelberg (2009)
8. Gelfond, M., Lifschitz, V.: The stable model semantics for logic programming. In: Kowalski, R.A., Bowen, K.A. (eds.) Logic Programming, Proceedings of the Fifth International Conference and Symposium, Seattle, Washington, August 15–19, 1988, vol. 2, pp. 1070–1080. MIT Press (1988). ISBN: 0-262-61056-6
9. Goldberg, D.: What every computer scientist should know about floating-point arithmetic. ACM Comput. Surv. **23**(1), 5–48 (1991)
10. Grimes, D., Hebrard, E., Malapert, A.: Closing the open shop: contradicting conventional wisdom. In: Gent, I.P. (ed.) CP 2009. LNCS, vol. 5732, pp. 400–408. Springer, Heidelberg (2009)
11. IBM: CPLEX performance tuning for linear programs. http://www-01.ibm.com/support/docview.wss?uid=swg21400034
12. Jaffar, J., Maher, M.J.: Constraint logic programming: a survey. J. Logic Program. **19/20**, 503–581 (1994)
13. Liu, G., Janhunen, T., Niemelä, I.: Answer set programming via mixed integer programming. In: Brewka, G., Eiter, T., McIlraith, S.A. (eds.) Principles of Knowledge Representation and Reasoning: Proceedings of the Thirteenth International Conference, KR 2012, Rome, Italy, June 10–14, 2012, pp. 32–42. AAAI Press (2012). ISBN: 978-1-57735-560-1
14. Mellarkod, V.S., Gelfond, M., Zhang, Y.: Integrating answer set programming and constraint logic programming. AMAI **53**(1–4), 251–287 (2008)
15. Nguyen, M., Janhunen, T., Niemelä, I.: Translating answer-set programs into bit-vector logic. In: Tompits, H., Abreu, S., Oetsch, J., Pührer, J., Seipel, D., Umeda, M., Wolf, A. (eds.) INAP/WLP 2011. LNCS, vol. 7773, pp. 91–109. Springer, Heidelberg (2013)
16. Niemelä, I.: Logic programs with stable model semantics as a constraint programming paradigm. AMAI **25**(3–4), 241–273 (1999)
17. Niemelä, I.: Linear and integer programming modelling and tools. Lecture Notes for the course on Search Problems and Algorithms, Helsinki University of Technology (2008)
18. Simons, P., Niemelä, I., Soininen, T.: Extending and implementing the stable model semantics. Artif. Intell. **138**(1–2), 181–234 (2002)
19. You, J.-H., Hou, G.: Arc-consistency + unit propagation = lookahead. In: Demoen, B., Lifschitz, V. (eds.) ICLP 2004. LNCS, vol. 3132, pp. 314–328. Springer, Heidelberg (2004)

Coverage Driven Test Generation and Consistency Algorithm

Jomu George Mani Paret[⊠] and Otmane Ait Mohamed

Department of Electrical and Computer Engineering, Concordia University,
1455 de Maisonneuve West, Montreal, QC H3G 1M8, Canada
{jo_pare,ait}@ece.concordia.ca

Abstract. Coverage driven test generation (CDTG) is an essential part
of functional verification where the objective is to generate input stimuli
that maximize the functional coverage of a design. CDTG techniques
analyze coverage results and adapt the stimulus generation process to
improve the coverage. One of the important components of CDTG based
tools is the constraint solver. The efficiency of the verification process
depends on the performance of the solver. The speed of the solver can
be increased if inconsistent values can be removed from the domain of
input variables. In this paper, we propose a new efficient consistency
algorithm called GACCC-op (generalized arc consistency on conjunction
of constraints-optimized) which can be used along with the constraint
solver of CDTG tools. The experimental results show that the proposed
technique helps to reduce the time required for solution generation of
CSPs by 19 %.

1 Introduction

As semiconductor technology improves, electronic designs are becoming more
complex. In order to ensure the functional correctness of a design, finding and
fixing design errors is important. Functional verification is the task of verifying
whether the hardware design confirms the required specification. This is a complex task which consumes the majority of the time and effort in most of the
electronic system design projects. Many studies show that up to 70 % of design
development time and resources are spent on functional verification [11]. There
are several methods to tackle the functional verification problem and one among
them is Constraint Random Test (CRT) generation.

Coverage is a measure used to determine the completeness of the input stimulus, generated by using CRT. Each measurable action is called a coverage task.
However, as the complexity of design increases, achieving a required coverage
goal even with CRT can still be very challenging. Limitations of the constrained
random approach led to the development of coverage based test generation techniques.

In coverage based test generation techniques, coverage tools are used side by
side with a stimulus generator (constraint solver) in order to assess the progress

M. Hanus and R. Rocha (Eds.): KDPD 2013, LNAI 8439, pp. 136–151, 2014.
DOI: 10.1007/978-3-319-08909-6_9, © Springer International Publishing Switzerland 2014

of the verification plan during the verification cycle. Coverage analysis allows for the modification of the directives (constraints) for the stimulus generators and to target areas of the design that are not covered well. This process of adapting the directives of stimulus generator, according to the feedback based on coverage reports, is called Coverage Driven Test Generation (Fig. 1). Though CDTG is essential for the completion of the verification cycle, it is a time consuming and an exhaustive process.

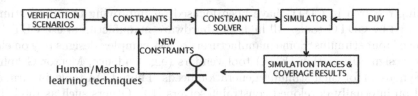

Fig. 1. Coverage driven test generation technique

Let us look at one example involving verification of the floating-point unit present in microprocessors. Stimulus generation for floating-point unit verification involves targeting corner cases, which can often be solved only through complex constraint solving. Hence the main task of the constraint solver is to generate a set of input stimulus that comprises a representative sample of the entire space, taking into account the many corner cases. A floating point unit with two input operands and 20 major FP instruction types (e.g. $\pm$zero, $\pm$min-denorm) yields 400 (20^2) cases that must be covered. With four floating point instructions (addition, subtraction, division and multiplication) there are about 1600 cases to be covered. The probability that a CDTG tool will generate a sequence that covers a particular combination is very low ($1:2^{17}$) [12]. Hence a CDTG tool will take many hours to generate the input stimuli required to attain the needed coverage.

The most important component in CDTG is the constraint solver. The efficiency of a CDTG is heavily dependent on the constraint solver. The stimulus generation methods of CDTGs are similar to a constraint satisfaction problem (CSP). But the CSPs arising from stimulus generation are different from typical CSPs [9]. One striking difference is the existence of variables with huge domains. Another difference is the requirement to produce multiple different solutions, distributed uniformly, for the same CSP. Hence available general purpose constraint solvers cannot be used along with CDTG tools to improve the stimuli generation time.

Certain values, in the domain of input variables of the CSP used for stimulus generation, cannot be part of the solution. The inconsistent values can be found out by consistency search and can be removed from the domain of input variables. If the reduced domain is given to the constraint solver of CDTG, then the solutions for CSP can be generated in less time and with reduced memory consumption. In this paper, we propose a consistency search algorithm which can be used along with CDTG tools, to reduce the domain of input variables. The remainder of this paper is organized as follows. We will explain some of the related work in Sect. 2. Section 3 describes the proposed consistency algorithm. Finally, we present our experimental results in Sect. 4, and give some concluding remarks and future work in Sect. 5.

2 Related Work

Existing research in CDTG have been focused on improving the input stimuli generated by CDTG tools. All high-end hardware manufacturers use CDTG to produce input stimulus. Some manufacturers of less complex designs rely on electronic design automation (EDA) tool vendors (e.g. Cadence, Mentor Graphics and Synopsys) for their stimulus generation needs. Those EDA tools, in turn, are based on internally developed constraint solvers [13]. Others such as Intel [15], adapt external off-the-shelf solvers to the stimulus generation problem. Some manufacturers such as IBM rely on proprietary constraint solvers developed in-house to solve this problem [12].

One approach for solving CSP is based on removing inconsistent values from the domain of variables till the solution is obtained. These methods are called consistency techniques. The most widely used consistency technique is called arc consistency (AC). In AC, for each constraint (C_i) in the CSP, for each variable (v_i) in the constraint C_i, for each domain value of the variable v_i, the algorithm will try to find a list of variable values which satisfies the constraint.

The arc consistency algorithms are divided into two categories: coarse-grained algorithms and fine-grained algorithms. Coarse grained algorithms are algorithms in which the removal of a value from the domain of a variable will be propagated to other variables in the problem. The first consistency algorithms AC-1 [14] and AC-3 [14] belong to this category. These two consistency algorithms are succeeded by AC2000 [8], AC2001-OP [4], AC3-OP [3] and AC3d [10].

Fine grained consistency algorithms are algorithms in which removal of a value from the domain of a variable X will be propagated to only other variables which are related to the variable X. Since only the variables that are affected by the change in domain value are revisited, this algorithm is faster than coarse grained algorithms. Algorithms AC-4 [16], AC4-OP [2], AC-5 [17] and AC-6 [5] belong to this category. AC-7 [6] is an algorithm developed based on AC-6. It uses the knowledge about the constraint properties to reduce the cost of consistency check.

All the above algorithms are developed for binary constraints. GAC-scheme [7] is a consistency algorithm developed for n-arity (n variables are there in the constraint) constraints. It is the extension of AC-7 for n-arity constraints.

Conjunctive Consistency [7] enforces GAC-scheme on conjunctions of constraints. We chose GAC scheme on conjunction of constraints for our purpose because:

1. We need to eliminate as much invalid domain values as possible. This can be done by conjunction of constraints.
2. During constraint propagation, the algorithm has to keep track of variables and domains which are affected by the removal of an inconsistent value. This requires complex data structure. GAC scheme do not require any specific data structure.
3. The constraints used in CDTG can have more than two variables and GAC-scheme can handle constraint of n-arity.
4. The constraints used in CDTG are not of a fixed type and GAC-scheme can be used with any type of constraints.

3 Consistency Algorithm

3.1 Preliminaries

Tuple: A tuple τ on an ordered set of variables is an ordered list which contains values for all the variables. $X(\tau)$ represents the set of variables in the tuple τ.

Constraint: A constraint C_i on an ordered set of variables gives the list of allowed tuples for the set of variables. $X(C_i)$ represents the set of variables in the constraint C_i.

Constraint Network: A constraint network is defined as a tuple $N = \langle X, D, C \rangle$ where:

X is a set of n variables, $X = \{x_1, \ldots, x_n\}$, $n > 0$
D is a finite set of domains for the n variables $= \{D(x_1), \ldots, D(x_n)\}$, $n > 0$
C is a set of constraints between variables $= \{C_1, \ldots, C_k\}$, $k > 0$

Valid Tuple: The value of variable x in a tuple τ is denoted by $\tau[x]$. A tuple τ on C_i is valid iff $\forall x \in X(C_i)$, $\tau[x] \in D(x)$ and τ satisfies the constraint C_i.

Support: If a $\in D(x_i)$ and τ be a valid tuple on C_j, then τ is called a support for (x_i, a) on C_j.

Arc Consistency: A value a $\in D(x_j)$ is consistent with C_i iff $x_j \in X(C_i)$ and $\exists \tau$ such that τ is a support for (x_j, a) on C_i. C_i is arc consistent iff $\forall x_j \in X(C_i)$, $D(x_j) \neq 0$ and $\forall a \in D(x_j)$, a is consistent with C_i.

Generalized Arc Consistency of a Network: A CSP is generalized arc consistent iff $\forall C_i \in C$ is arc consistent.

Conjunctive Consistency: If S_j is conjunction of a subset of constraints in C, then S_j is conjunctively consistent iff $\forall a \in D(x_k)$, $x_k \in X(S_j)$ and there exists a tuple τ such that $a = \tau[x_k]$ and τ is a support $\forall x_k$.

Conjunctive Consistency of a Network: Let P = <X, D, S> be a constraint network. P is conjunctive consistent network iff $\forall S_j \in S$ is conjunctive consistent.

3.2   GACCC

In GACCC [7], first a variable in a conjunction of constraint is selected and the selected variable will be assigned a value from its domain. The algorithm will generate tuples in lexicographical order (the selected variable value will not change) and check whether the tuple satisfies the constraint. The tuples are generated until all the tuples are generated or a tuple which satisfies the constraint is generated. If there is no tuple which satisfies the constraint for the selected variable value, then that variable value is inconsistent and removed from the variable domain. The process is repeated for all the domain values of the selected variable, then for all the variables in the constraint and for all the constraints in the constraint network.

To illustrate the idea discussed above, let us consider the following CSP: set of variables $X = \{m, n, o, p, q\}$, domain of the variables D(m) = $\{1, 2\}$, D(n) = $\{2, 3\}$, D(o) = $\{1, 2\}$, D(p) = $\{1, 3\}$, D(q) = $\{2, 3\}$ and the constraints $C1 : m + n + o + p = 7$ and $C2 : m + o + q = 9$. The consistency search (for conjunction of constraints) for $m = 1$ has to go through 16 tuples (because each of the remaining variables (n, o, p, q) has two variables in the domain) to find out that value is not consistent.

3.3   Intuitive Idea of GACCC-op

In consistency check, if any one constraint is not satisfied, then the tuple generated is inconsistent with the conjunction set. We can reduce the number of tuples generated during consistency search by using this property. Initially for a given variable, we consider the constraint with lowest number of variables and contain the specified variable. We generate tuples for the above constraint and search for consistency. If the tuple generated for the smallest constraint is not consistent then all the tuples generated for the conjunction of constraints are also not consistent. If the tuple generated for the smallest constraint is consistent, then only we need to generate the tuples for the conjunction of constraints (tuple generated for conjunction of constraints should contain the tuple which is consistent with the smallest constraint). Since the number of variables in the smallest constraint is less when compared to tuple for conjunction of constraints, consistency can be checked in less number of iterations.

In the above CSP, $C2$ is the smallest constraint in the set, which has 3 variables and the variable m. Consistency check is first performed on this constraint. In 4 (because each of the remaining variables (o, q) has two variables in the domain) iterations we can find that $m = 1$ is inconsistent with the constraint $C2$. Hence m = 1 is inconsistent for the conjunction of constraints. The tuples for a variable in conjunction of constraints is generated only if the smallest constraint containing the variable is satisfied by the tuple. Consider another set of constraints $C3 : m + n + o + p = 8$ and $C4 : m + o + q = 6$. By GACCC we have to generate 8 tuples to find a consistent tuple. By using the new algorithm we need only 5 (4 iterations for $C3$ and 1 for conjunction of $C3$ and $C4$) iterations to find the tuple which satisfies the constraints. So by using the proposed algorithm consistency check can be completed in less number of iteration when compared to GACCC.

So the difference between GACCC and GACCC-op are as follows:

1. In GACCC the support list is made by using some existing variable order scheme. In GACCC-op we propose a new variable ordering scheme in which the consistency search starts with the variable, which is present in the constraint with the lowest arity and has the largest number of domain values.
2. In GACCC during consistency search of a domain value of a variable, the tuples generated will contain all the variable in the conjunction set. In GACCC-op the consistency search for a variable x will begin with tuples which contain only variables from the smallest constraint (C_s) $(C_s$ should contain the variable x). If there is a tuple which satisfies the constraint C_s, only then GACCC-op generates tuples with the entire variable in the conjunction set.

3.4 GACCC-op

Let us start the discussion of the proposed GACCC-op algorithm with the main program (Algorithm 1). First the data structures (lastSc, supportlist, deletionlist and Sclast) must be created and initialized. Sclast, supportlist, deletionlist and lastSc are initialized in such a way that:

1. Sclast contains the last tuple returned by the function **SeekValidSupport-Set** as a support for variable value.
2. supportlist contains all tuples that are support for variable value.
3. deletionlist contains all variable values that are inconsistent.
4. lastSc is the last tuple returned by the function **SeekValidSupport** as a support for variable value.

Then for each set of constraints, for each variable present in the constraints, all the domain values of the variable are put in supportlist. The domain values of the variables in a conjunction set are added to supportlist using the following heuristics:

1. Find the lowest arity constraint (C_l) in the conjunction set.
2. Find a variable (x_l) where the variable and C_l is not added to the list, the variable is in C_l and has the highest number of domain values.
3. Add all the domain values of the selected variable (x_l), variable and the constraint to the list.
4. Repeat step 2 until all the variables in the constraint C_l are considered.
5. If there is any variable or constraint set to be added to the list from the conjunction set, then find the next highest arity constraint and repeat steps 2–4.

Algorithm 1. GACCC-op Algorithm

```
 1: for each constraint set (S) do
 2:    for each variable in set (y) do
 3:        for each domain value of variable (b) do
 4:            Add to support stream(S,y,b)
 5:        end for
 6:    end for
 7: end for
 8: while support stream ≠ nil do
 9:    σ = SeekInferableSupport(S,y,b)
10:    if σ = nil then
11:        c = smallest constraint containing variable y
12:        while found soln ‖ checked all tuples do
13:            σ∗ = lastSc(C,y,b)
14:            if σ∗ = nil then
15:                LOOP2: σ∗ = SeekValidSupport (C,y,b,σ∗)
16:                if σ∗ = nil then
17:                    DeletionStream (y,b)
18:                else
19:                    if variables in all the constraints are same then
20:                        Add to Sclast(S,y,b)
21:                    else
22:                        Add to lastSc(C,y,b)
23:                        go to LOOP1
24:                    end if
25:                end if
26:            else
27:                if Sclast(S,y,b)≠ nil then
28:                    σ∗∗ = Sclast(S,y,b)
29:                    go to LOOP1
30:                else
31:                    σ∗∗ = nil
32:                end if
33:            end if
34:            LOOP1: λ∗ = SeekValidSupportSet(S,y,b,σ∗∗)
35:            if λ∗≠ nil then
36:                Add to Sclast(S,y,b)
37:            else
38:                go to LOOP2
39:            end if
40:        end while
41:    end if
42: end while
```

This supportlist is used to find the support (support is a tuple which satisfies the constraint) for each variable value in the constraint set. For each value in supportlist the algorithm will try to find a valid support by using the function **SeekInferableSupport**. Function **SeekInferableSupport** checks whether an already checked tuple is a support for (y,b). If there is no valid support to be inferred then we will search for a valid support.

For every value 'b', for a variable 'y' in X(C), lastSc(C,y,b) is the last tuple returned by **SeekValidSupport** as a support for (y,b) if **SeekValid-Support**(C,y,b) has already been called or empty otherwise. The above two functions help to avoid checking several times whether the same tuple is a support for the constraint or not. If the search is new we look for support from the first valid tuple.

If no valid tuple is found then the variable value is not consistent with the constraint. Hence it is not consistent with constraint set. This variable value will be deleted from the domain of the variable by the function **DeletionStream**(y,b).

Algorithm 2. SeekInferableSupport

1: SeekInferableSupport (**in** S:constraint; **in** y:variable; **in** b:value):tuple
2: **while** support stream ≠ nil **do**
3:     **if** Sclast(var(S,y),τ[y]) = b **then**
4:         zigma = Sclast(S,y,b)
5:     **else**
6:         zigma = **nil**
7:     **end if**
8:     **return** zigma
9: **end while**

If a tuple is returned by lastSc(C,y,b), we will check for Sclast(S,y,b). Sclast (S,y,b) is the last tuple returned by **SeekValidSupportSet** as a support for (S,y,b) if **SeekValidSupportSet** has already been called or empty otherwise. If a tuple is returned we start the search for support for conjunction constraint set from that tuple, else we will start search from the first valid tuple for the conjunction set, with variables in constraint C has the values of the tuple from lastSc(C,y,b). If the **SeekValidSupportSet** returns empty then we will call function **SeekValidSupport** and repeat the process until a valid tuple for the for conjunction constraint set is found or the lastSc(C,y,b) returns empty. If the lastSc(C,y,b) returns empty then the variable value is deleted the function **DeletionStream**(y,b). The above processes will be repeated until both the dele-tionlist and supportlist are empty.

The function **SeekInferableSupport** (Algorithm 2) ensures that the algorithm will never look for a support for a value when a tuple supporting this value has already been checked. The idea is to exploit the property: "If (y,b) belongs to a tuple supporting another value, then this tuple also supports (y,b)".

After the function **SeekInferableSupport** fails to find any previously checked tuple as a support for (y,b) on the constraint C, the function **SeekValidSupport** (Algorithm 3) is called to find a new support for (y,b). But the function has to avoid checking tuples which are already checked. This is taken care by using the function **SeekCandidateTuple**. The function **NextTuple** will generate new tuples

---

Algorithm 3. SeekValidSupport

1: SeekValidSupport (**in** C:constraint; **in** y:variable; **in** b:value; **in** τ:tuple):tuple
2: **if** $\tau \neq$ nil **then**
3:     zigma = **NextTuple**(C,y,b,τ)
4: **else**
5:     zigma = **FirstTuple**(C,y,b)
6: **end if**
7: zigma1 = **SeekCandidateTuple**(C,y,b,τ)
8: solution found = **false**
9: **while** (zigma1 $\neq$ nil) **and** (not solution found) **do**
10:     **if** zigma1 satisfies constraint C **then**
11:         solution found = **true**
12:     **else**
13:         zigma1= **NextTuple**(C,y,b,zigma1)
14:         zigma1 = **SeekCandidateTuple**(C,y,b,zigma1)
15:     **end if**
16:     **return** zigma1
17: **end while**

---

in a lexicographical order which can be a valid support for the constraint variable value.

Function **SeekCandidateTuple**(C,y,b,τ) (Algorithm 4) returns the smallest candidate greater than or equal to τ. For each index from 1 to $|X(C)|$ **SeekCandidateTuple** verifies whether τ is greater than lastSc (λ). If τ is smaller than λ, the search moves forward to the smallest valid tuple following τ, else to the valid tuple following λ. When the search moves to the next valid tuple greater than τ or λ, some values before the index may have changed. In those cases we again repeats the previous process to make sure that we are not repeating a previously checked tuple.

The function **SeekValidSupportSet** (Algorithm 5) is called to find a new support for (y,b) on the conjunction of constraints. But the function has to avoid checking tuples which are already checked. This is taken care by using the function **SeekCandidateSet**. This function is similar to the function **SeekCandidateTuple**. The function **SeekCandidateSet** returns the smallest tuple which is a support of the conjunction of constraints.

If there is no support for a variable value, then that variable value is deleted from the variable domain by the function **DeletionStream** (Algorithm 6). The function also checks whether any tuple in Sclast contains the variable value. If there is such a tuple, then all the variable values in the tuple are added to supportlist to find new support.

3.5 Heuristic for Generating Conjunction Set

The CSPs associated with the verification scenarios have large number of constraints, large domain for each input variables and many of the constraints have the same variables. The pruning capability by consistency search can be

Algorithm 4. SeekCandidateTuple

```
1: SeekCandidateTuple (in C:constraint; in y:variable; in b:value; in τ:tuple):tuple
2:   k = 1
3:   while (τ ≠ nil) and (k≤X(C)) do
4:     if lastc(var(C,k),τ[k])≠ nil then
5:       λ = lastSc(var(C,k),τ[k])
6:       split = 1
7:       while τ[split] = λ[split] do
8:         split = split+1
9:       end while
10:      if τ[split] < λ[split] then
11:        if split < k then
12:          (τ,k')= NextTuple(C,y,b,λ)
13:          k = k'+1
14:        else
15:          (τ,k')= NextTuple(C,y,b,λ)
16:          k = min(k'-1, k)
17:        end if
18:      end if
19:    end if
20:    k = k+1
21:  end while
22:  return τ
```

Algorithm 5. SeekValidSupportSet

```
1: SeekCandidateTuple (in S:constraint set; in y:variable; in b:value; in τ:tuple):tuple
2:   if τ≠ nil then
3:     zigma = NextTuple(S,y,b,τ,θ)
4:   else
5:     zigma = FirstTuple(S,y,b)
6:   end if
7:   zigma1 = SeekCandidateSet(S,y,b,τ,θ)
8:   solution found = false
9:   while (zigma1 ≠ nil) and (not solution found) do
10:    if zigma1 satisfies constraint set S then
11:      solution found = true
12:    else
13:      zigma1= NextTuple(S,y,b,zigma1,θ)
14:      zigma1 = SeekCandidateSet(D,y,b,zigma1,θ)
15:    end if
16:    return zigma1
17:  end while
```

Algorithm 6. DeletionStream

```
1: SeekCandidateTuple (in y:variable; in b:value)
2:   if Sclast(var(C,y),τ[y])= b then
3:     Add to supportlist (S,(var(C,x)),a) where x≠ y and τ[x]=a
4:     delete λ from Sclast
5:   end if
```

increased, by combining/conjuncting a large number of constraints together. If a large number of constraints are conjuncted, the variables in the tuple increases and the number of tuples that has to be generated also increases. So there should be a limit to the number of constraints conjuncted together. Similarly the number of variables in the tuple has to be regulated to prevent the tuple from becoming very large. For conjunction of constraints to be effective in reducing the domain values, the constraints in the conjunction set should have a certain number of variables in common. The number of constraints (k), number of variable in the conjunction set (j) and the number of variable common to all the constraints in the conjunction set (i) depends on the CSP and the machine capacity. So there should be a heuristic based on the parameters i, j and k to determine which constraints can be combined together to make the conjunction set.

The heuristic for grouping constraints into conjunctive sets is as follows:

1. Initially there will be 'n' conjunctive sets (S), each containing a single constraint (where n is the total number of constraints in the CSP).
2. If there exists two conjunctive sets S1, S2 such that variables in S1 is equal to variables in S2, then remove S1 and S2 and add a new set which is conjunction of all the constraints in S1 and S2.
3. If there exist two conjunctive sets S1, S2 such that (a) S1, S2 share at least i variables (b) the number of variables in S1 ∪ S2 is less than j (c) the total number of constraints in S1 and S2 is less than k then remove S1 and S2 and add a new set which is conjunction of all the constraints in S1 and S2.
4. Repeat 2 and 3 until no more such pairs exist.

The Table 1 shows how constraints can be conjuncted using the above heuristic. During step 3 the constraints $S5$ and $S3$ are conjuncted to form constraint $S6$. The constraint $S4$ cannot be conjuncted with $S6$ because the total number of constraints in the conjunction set should be less than 4 (since k = 4).

3.6   Correctness of the Algorithm

To show the correctness of the algorithm it is necessary to prove that every inconsistent value is removed (completeness) and that no consistent value is removed by the algorithm (soundness) when the algorithm terminates. Moreover, we need to prove that the algorithm terminates.

Lemma 1. *Algorithm will terminate.*

Table 1. Conjunction of constraints

Constraints in CSP	After step1	After step2	After step3 (i = 1, j = 5, k = 4)
$C1 : a * b > 20$	$S1 : a * b > 20$	$S5 : S1 \wedge S2 :$	$S6 : S5 \wedge S3 :$
$C2 : a > b$	$S2 : a > b$	$a * b > 20 \wedge a > b$	$a * b > 20 \wedge a > b \wedge a + c = 25$
$C3 : a + c = 25$	$S3 : a + c = 25$	$S3 : a + c = 25$	
$C4 : c + d = 19$	$S4 : c + d = 19$	$S4 : c + d = 19$	$S4 : c + d = 19$

Proof. The algorithm consists of a for loop and two while loops. The generation of elements for the list called *support stream(S,y,b)* uses a for loop. The number of domain values, variable and constraints are finite. Hence the elements generated for the list is finite and the for loop will terminate. The pruning process for the domain values uses a while loop. During each cycle, one element is removed from the list. The elements are added to this list only when a value is removed from some domain. Thus, it is possible to add only a finite number of elements to the list (some elements can be added repeatedly). Hence the while loop will terminate. The algorithm uses a while loop to find support for a variable value in a constraint. The algorithm generates tuples in lexicographic order starting for the smallest one. Since the number of possible tuples for a constraint is finite, the while loop will terminate when it finds a valid support tuple or when all the tuples are generated.

Lemma 2. *SeekCandidateTuple will not miss any valid tuple during the generation of next tuple.*

Proof. Consider that there is a candidate tuple σ' between σ and the tuple returned by the function NextTuple. This implies that $\sigma'[1...k] = \sigma[1...k]$ else σ' will the tuple returned by NextTuple. Hence σ' should be smaller than λ (lines 10–11). If σ' is smaller than λ then that tuple is already generated and checked for consistency. So σ' cannot be a tuple between σ and the tuple returned by the function NextTuple.

Another possibility is that there can be a candidate tuple σ' between σ and λ. Then $\sigma'[1...k]$ should be equal to $\lambda[1...k]$ (lines 7–11). This is not possible candidate since λ is not a valid support tuple.

Lemma 3. *The algorithm does not remove any consistent value from the domain of variables.*

Proof. A value is removed from the domain of a variable only if the value is not arc consistent i.e. there is no valid support tuple for the variable value. Thus, the algorithm does not remove any consistent value from the variables' domains so the algorithm is sound.

Lemma 4. *When the algorithm terminates, then the domain of variables contain only arc consistent values (or some domain is empty).*

Proof. Every value in the domain has to pass the consistency test and inconsistent values will be deleted. When an inconsistent value is deleted and if the deleted value is part of a valid support tuple, then all variable values in that tuple are checked for consistency again. Hence when the algorithm terminates only consistent values remain in the domain.

3.7   Complexity of the Algorithm

Lemma 5. *The worst case time complexity of the algorithm is $O(en^2d^n)$.*

Proof. The worst-case time complexity of GACCC-op depends on the arity of the constraints involved in the constraint network. The greater the number of variables involved in a constraint, the higher the cost to propagate it. Let us first limit our analysis to the cost of enforcing GAC on a single conjunction constraint, S_i of arity n (n = $|X(S_i)|$) and d = size of the domain of the variable. For each variable $x_i \in X(S_i)$, for each value a$\in D(x_i)$, we look for supports in the search space where $x_i = a$, which can contain up to d^{n-1} tuples. If the cost to check whether a tuple satisfies the constraint is in $\mathbf{O}(n)$, then the cost for checking consistency of a value is in $\mathbf{O}(nd^{n-1})$. Since we have to find support for nd values, the cost of enforcing GAC on S_i is in $O(n^2d^n)$. If we enforce GAC on the whole constraint network, values can be pruned by other constraints, and each time a value is pruned from the domain of a variable involved in S_i, we have to call **SeekValidSupportSet** on S_i. So, S_i can be revised up to nd times. Fortunately, additional calls to **SeekValidSupportSet** do not increase its complexity since, $last(S_i, y, b)$ ensures that the search for support for (x_i, a) on S_i will never check twice the same tuple. Therefore, in a network involving e number of constraints with arity bounded by n, the total time complexity of GACCC-op is in $\mathbf{O}(en^2d^n)$.

Lemma 6. *The worst case space complexity of the algorithm is* $\boldsymbol{O}(en^2d)$.

Proof. Consistency search generates at most one valid support tuple for each variable value. Then there are at most nd tuples in memory for a constraint. One tuple will contain n elements. Then the set of all tuples which are a valid support for a constraint can be represented in $\mathbf{O}(n^2d)$. Therefore, in a network involving e constraints with arity bounded by n, the total space complexity of GACCC-op is in $\mathbf{O}(en^2d)$.

4   Experimental Results

We implemented the proposed algorithm in C++. The tool will take SystemVerilog constraints and the domain of the input variables as input and generates the reduced domain as output. For our purpose we considered a subset of SystemVerilog constraints which can be given as input to the tool. Our tool can handle unary constraint, binary constraints and some high order constraints. The high order constraints considered includes arithmetic, logical, mutex and implication constraints. The proposed consistency search algorithm is used along with existing CDTG as shown in Fig. 2. As shown earlier in Fig. 1, the different verification scenarios are converted to constraints. We used SystemVerilog to model the scenarios as constraints. The domain of the input variables are also specified as constraints. These constraints and the domain of the variables are given to the consistency check tool (based on GACCC-op). The reduced domain obtained from the tool and the SystemVerilog constraints are then given to the constraint solver of CDTG tool. The output of the solver is the input stimulus required for verification of the DUV.

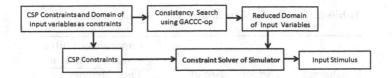

Fig. 2. CDTG with consistency

We report on experiments we performed with different CSP models. The first is a model for the 3-SAT problems [1] with different number of variables. The SAT problems with a set of clauses are converted into CSPs containing the same set of variables. In our case, we set i = 2, k = 2 and j = 5 (i, j and k are the values from the heuristic for generating conjunction set) and generated the conjunction set. Hence the model contained some conjunction set which has 2 variables shared between member constraints. The results are shown in Table 2. For each problem the experiment is repeated for 20 instances. We implemented the GAC-scheme on conjunction of constraints and the proposed algorithm using the C++ language. The result shows that the proposed algorithm attains consistency faster than the existing algorithm. In order to show the effect of consistency check on constraint solvers associated with CDTG, we took three different CSP benchmark problems, Langford Series, Magic Sequence and Golomb Ruler. The three CSPs are modeled using SystemVerilog (modeling language used by CDTGs). The SystemVerilog constraints are then used for consistency search. The reduced input variable domain are generated by the consistency search. This reduced domain is then used by the CDTG tool VCS to generate the CSP solutions. From Table 3, we can see that the time to solve the three CSPs are reduced after giving the reduced domain. In the cases of Magic Sequence the time is significantly reduced, because, after the domain reduction the number of domain values in most of the variables is reduced to one. Since the domain of input variables are reduced, the search space which has to be covered by the solver is reduced. This helps the solver to generate the solutions for CSP in less time and with reduced memory consumption.

Table 2. Time for consistency search for 3-SAT problem instances

No: of variables	No: of constraints	No: of tuples with GACCC	No: of tuples with GACCC-op	%Improvement in time
10	14	98	76	12.34
12	14	96	70	10.66
14	14	103	82	11.46
18	30	168	120	19.86
20	30	170	131	17.96
20	40	256	216	17.43

Table 3. Results for benchmark CSP problems using VCS

Benchmark problem	No: of variables	No: of domain values	Improvement after domain reduction	
			Time (%)	Memory (%)
Langford Series	6	3	10.0	23.5
	8	4	21.4	27.7
	14	7	25.0	40.8
Golomb Ruler	3	4	8.3	23.2
	4	7	7.1	28.2
	5	12	9.5	39.1
	6	18	13.8	73.1
Magic Sequence	4	4	30.0	50.0
	5	5	40.0	71.6
	7	7	55.0	73.3
	8	8	62.5	81.5

5   Conclusions

Existing CDTG tools take large amount of time and memory to generate the required test cases. In this paper, we presented a consistency check algorithm which helps CDTG tools to reduce the memory consumption and time required to generate the test cases. The results showed that the proposed algorithm helps in getting solution faster and with reduced memory consumption. For illustration purposes, we provided the analysis of the Magic Sequence, Langford Series, Golomb Ruler and 3-SAT problem. The requirements of constraint solvers associated with CDTG open the doors to many interesting and novel directions of research. Some worth mentioning are, generating all possible solutions and uniformity in randomization. In future we would like to propose a methodology based on consistency search which will be able to attain 100 % coverage at a faster rate with fewer iterations.

References

1. http://www.satlib.org. Accessed 16 Sept 2013
2. Arangú, M., Salido, M.: A fine-grained arc-consistency algorithm for non-normalized constraint satisfaction problems. Int. J. Appl. Math. Comput. Sci. **21**, 733–744 (2011)
3. Arangu, M., Salido, M.A., Barber, F.: Ac3-op: an arc-consistency algorithm for arithmetic constraints. In: Proceedings of the 2009 Conference on Artificial Intelligence Research and Development, pp. 293–300 (2009)
4. Arangú, M., Salido, M.A., Barber, F.: Ac2001-op: an arc-consistency algorithm for constraint satisfaction problems. In: García-Pedrajas, N., Herrera, F., Fyfe, C., Benítez, J.M., Ali, M. (eds.) IEA/AIE 2010, Part III. LNCS, vol. 6098, pp. 219–228. Springer, Heidelberg (2010)

5. Bessière, C.: Arc-consistency and arc-consistency again. In: Artifical Intelligence, pp. 179–190, January 1994
6. Bessière, C., Freuder, E.C., Regin, J.-C.: Using inference to reduce arc consistency computation. In: Proceedings of the 14th International Joint Conference on Artificial Intelligence, vol. 1, pp. 592–598 (1995)
7. Bessière, C., Régin, J.-C.: Local consistency on conjunctions of constraints. In: Proceedings of the ECAI'98 Workshop on Non-binary Constraints, pp. 53–59 (1998)
8. Bessire, C.: Refining the basic constraint propagation algorithm. In: Proceedings IJCAI01, pp. 309–315 (2001)
9. Bin, E., Emek, R., Shurek, G., Ziv, A.: Using a constraint satisfaction formulation and solution techniques for random test program generation. IBM Syst. J. **41**(3), 386–402 (2002)
10. van Dongen, M.R.C.: AC − 3_d an efficient arc-consistency algorithm with a low space-complexity. In: Van Hentenryck, P. (ed.) CP 2002. LNCS, vol. 2470, pp. 755–760. Springer, Heidelberg (2002)
11. Fine, S., Ziv, A.: Coverage directed test generation for functional verification using bayesian networks. In: Proceedings of the Design Automation Conference, pp. 286–291, June 2003
12. Fournier, L., Arbetman, Y., Levinger, M.: Functional verification methodology for microprocessors using the genesys test-program generator. application to the x86 microprocessors family. In: Design, Automation and Test in Europe Conference and Exhibition, pp. 434–441 (1999)
13. Iyer, M.: Race a word-level atpg-based constraints solver system for smart random simulation. In: International Test Conference, pp. 299–308 (2003)
14. Mackworth, A.: Consistency in networks of relations. In: Artificial Intelligence, pp. 99–118 (1977)
15. Moss, A.: Constraint patterns and search procedures for CP-based random test generation. In: Yorav, K. (ed.) HVC 2007. LNCS, vol. 4899, pp. 86–103. Springer, Heidelberg (2008)
16. Roger, M., Thomas, H.: Arc and path consistency revisited. In: Artificial Intelligence, pp. 225–233 (1986)
17. Van Hentenryck, P., Deville, Y., Teng, C.: A generic arc consistency algorithm and its specializations. Technical report, Providence, RI, USA (1991)

A Datalog Engine for GPUs

Carlos Alberto Martínez-Angeles[1], Inês Dutra[2], Vítor Santos Costa[2],
and Jorge Buenabad-Chávez[1](✉)

[1] Departamento de Computación, CINVESTAV-IPN,
Av. Instituto Politécnico Nacional 2508, 07360 Mexico, D.F., Mexico
camartinez@cinvestav.mx, jbuenabad@cs.cinvestav.mx
[2] Departmento de Ciência de Computadores, Universidade do Porto,
Rua do Campo Alegre, 1021, 4169-007 Porto, Portugal
{ines,vsc}@dcc.fc.up.pt

Abstract. We present the design and evaluation of a Datalog engine
for execution in Graphics Processing Units (GPUs). The engine evaluates recursive and non-recursive Datalog queries using a bottom-up
approach based on typical relational operators. It includes a memory
management scheme that automatically swaps data between memory in
the host platform (a multicore) and memory in the GPU in order to
reduce the number of memory transfers. To evaluate the performance of
the engine, four Datalog queries were run on the engine and on a single
CPU in the multicore host. One query runs up to 200 times faster on the
(GPU) engine than on the CPU.

Keywords: Logic programming · Datalog · Parallel computing · GPUs ·
Relational databases

1 Introduction

The traditional view of Datalog as a query language for deductive databases
is changing as a result of the new applications where Datalog has been in
use recently [18], including declarative networking [19], program analysis [9],
information extraction [23] and security [20] — datalog recursive queries are
at the core of these applications. This renewed interest in Datalog has in turn
prompted new designs of Datalog targeting computing architectures such as
GPUs, Field-programmable Gate Arrays (FPGAs) [18] and cloud computing
based on Google's Mapreduce programming model [7]. This paper presents a
Datalog engine for GPUs.

GPUs can substantially improve application performance and are thus now
being used for general purpose computing in addition to game applications.
GPUs are single-instruction-multiple-data (SIMD) [2] machines, particularly
suitable for compute-intensive, highly parallel applications. They fit scientific
applications that model physical phenomena over time and space, wherein the
"compute-intensive" aspect corresponds to the modelling over time, while the
"highly parallel" aspect corresponds to the modelling at different points in space.

M. Hanus and R. Rocha (Eds.): KDPD 2013, LNAI 8439, pp. 152–168, 2014.
DOI: 10.1007/978-3-319-08909-6_10, © Springer International Publishing Switzerland 2014

Data-intensive, highly parallel applications such as database relational operations can also benefit from the SIMD model, substantially in many cases [11, 16,17]. However, the communication-to-computation ratio must be relatively low for applications to show good performance, i.e.: the cost of moving data from host memory to GPU memory and vice versa must be low relative to the cost of the computation performed by the GPU on that data.

The Datalog engine presented here was designed considering various optimisations aimed to reduce the communication-to-computation ratio. Data is preprocessed in the host (a multicore) in order that: (i) data transfers between the host and the GPU take less time, and (ii) data can be processed more efficiently by the GPU. Also, a memory management scheme swaps data between host memory and GPU memory seeking to reduce the number of transfers.

Datalog queries, recursive and non-recursive, are evaluated using typical relational operators, *select, join* and *project*, which are also optimised in various ways in order to capitalise better on the GPU architecture.

Sections 2 and 3 present background material to the GPU architecture and the Datalog language. Section 4 presents the design and implementation of our Datalog Engine as a whole, and Sect. 5 of its relational operators. Section 6 presents an experimental evaluation of our Datalog engine. Section 7 presents related work. We conclude in Sect. 8.

2 GPU Architecture and Programming

GPUs are SIMD machines: they consist of many processing elements (PEs) that run the *same program* but on distinct data items. This same program, referred to as the *kernel*, can be quite complex including control statements such as *if* and *for* statements. However, a kernel is run in *bulk-synchronous parallelism* [28] by the GPU hardware, i.e.: each instruction within a kernel is executed across all PEs running the kernel. Thus, if a kernel compares strings, PEs that compare longer strings will take longer and the other PEs will wait for them.

Scheduling GPU work is usually as follows. A thread in the host platform (e.g., a multicore) first copies the data to be processed from host memory to GPU memory, and then invokes GPU threads to run the *kernel* to process the data. Each GPU thread has a unique id which is used by each thread to identify what part of the data set it will process. When all GPU threads finish their work, the GPU signals the host thread which will copy the results back from GPU memory to host memory and schedule new work.

GPU memory is organised hierarchically as shown in Fig. 1. Each (GPU) thread has its own *per-thread local* memory. Threads are grouped into *blocks*, each block having a memory *shared* by all threads in the block. Finally, thread blocks are grouped into a single *grid* to execute a kernel — different grids can be used to run different kernels. All grids share the *global memory*.

The global memory is the GPU "main memory". All data transfers between the host (CPU) and the GPU are made through reading and writing global memory. It is the slowest memory. A common technique to reducing the number

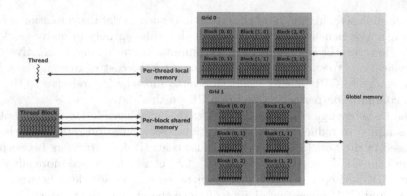

Fig. 1. GPU memory organization.

of global memory reads is *coalesced memory access*, which takes place when consecutive threads read consecutive memory locations allowing the hardware to coalesce the reads into a single one.

Nvidia GPUs are mostly programmed using the CUDA toolkit, a set of developing tools and a compiler that allow programmers to develop GPU applications using a version of the C language extended with keywords to specify GPU code. CUDA also includes various libraries with algorithms for GPUs such as the Thrust library [5] which resembles the C++ Standard Template Library (STL) [21]. We use the functions in this library to perform sorting, prefix sums [15] and duplicate elimination as their implementation is very efficient.

CUDA provides the following reserved words, each with three components x,y and z, to identify each thread and each block running a kernel: **threadIdx** is the index of a thread in its block; **blockIdx** is the index of a block in its grid; **blockDim** is the size of a block in number of threads; and **gridDim** is the size of a grid in number of blocks. With these identifiers, new identifiers can be derived with simple arithmetic operations. For example, the global identifier of a thread in a three-dimensional block would be:

```
unsigned int GID = threadIdx.x + threadIdx.y * blockDim.x +
                   threadIdx.z * blockDim.x * blockDim.z;
```

x,y and z are initialised by CUDA according to the *shape* with which a kernel is invoked, either as a 1D *Vector* (y=z=0), a 2D *Matrix* (z=0), or a 3D *Volume*.

3   Datalog

As is well known, Datalog is a language based on first order logic that has been used as a data model for relational databases [26,27]. A Datalog program consists of *facts* about a subject of interest and *rules* to deduce new facts. Facts can be seen as rows in a relational database table, while rules can be used to specify complex queries. Datalog recursive rules facilitate specifying (querying for) the transitive closure of relations, which is a key concept to many applications [18].

3.1   Datalog Programs

A Datalog program consists of a finite number of facts and rules. Facts and rules are specified using atomic formulas, which consist of predicate symbols with arguments [26], e.g.:

```
FACTS                             father relational table
                                  -----------------------
father(harry, john).              harry    john
father(john, david).              john     david
...                               ...
```

```
RULE
grandfather(Z, X) :- father(Y, X), father(Z, Y).
```

Traditionally, names beginning with lower case letters are used for predicate names and constants, while names beginning with upper case letters are used for variables; numbers are considered constants. Facts consist of a single atomic formula, and their arguments are constants; facts that have the same name must also have the same arity. Rules consist of two or more atomic formulas with the first one from left to right, the rule *head*, separated from the other atomic formulas by the implication symbol ':-'; the other atomic formulas are *subgoals* separated by ',', which means a logical AND. We will refer to all the subgoals of a rule as the *body* of the rule. Rules, in order to be general, are specified with variables as arguments, but can also have constants.

3.2   Evaluation of Datalog Programs

Datalog programs can be evaluated through a top-down approach or a bottom-up approach. The top-down approach (used by Prolog) starts with the goal which is reduced to subgoals, or simpler problems, until a trivial problem is reached. It is tuple-oriented: each tuple is processed through the goal and subgoals using all relevant facts. It is not suitable for GPU bulk-synchronous parallelism (BSP) because the processing time of distinct tuples may vary significantly.

The bottom-up approach first applies the rules to the given facts, thereby deriving new facts, and repeating this process with the new facts until no more facts are derived. The query is considered only at the end, to select the facts matching the query. Based on relational operations (as described shortly), this approach is suitable for GPU BSP because such operations are set-oriented and relatively simple overall; hence show similar processing time for distinct tuples. Also, rules can be evaluated in any order. This approach can be improved using the magic sets transformation [8] or the subsumptive tabling transformation [25]. Basically, with these transformations the set of facts that can be inferred contains only facts that would be inferred during a top-down evaluation.

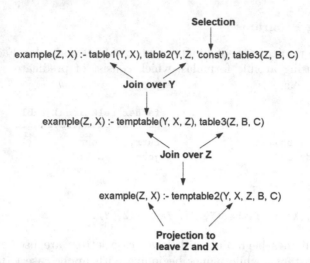

Fig. 2. Evaluation of a Datalog rule based on relational algebra operations.

3.3   Evaluation Based on Relational Algebra Operators

Evaluation of Datalog rules can be implemented using the typical relational algebra operators *select*, *join* and *projection*, as outlined in Fig. 2. *Selections* are made when constants appear in a rule body. Then a *join* is made between two or more subgoals in the rule body using the variables as reference. The result of a join becomes a temporary subgoal that must be joined to the other subgoals in the body. Finally, a *projection* is made over the variables in the rule head.

For recursive rules, fixed-point evaluation is used. The basic idea is to iterate through the rules in order to derive new facts, and using these new facts to derive even more new facts until no new facts are derived.

4   Our Datalog Engine for GPUs

This section presents the design of our Datalog engine for GPUs.

4.1   Architecture

Figure 3 shows the main components of our Datalog engine. There is a single *host* thread that runs in the host platform (a multi-core in our evaluation). The host thread schedules GPU work as outlined in Sect. 2, and also preprocesses the data to send to the GPU for efficiency, as described in Sect. 4.2.

The data sent to the GPU is organized into arrays that are stored in global memory. The results of rule evaluations are also stored in global memory.

Our Datalog (GPU) engine is organised into various GPU kernels. When evaluating rules, for each pair of subgoals in a rule, selection and selfjoin kernels are applied first in order to eliminate irrelevant tuples as soon as possible, followed

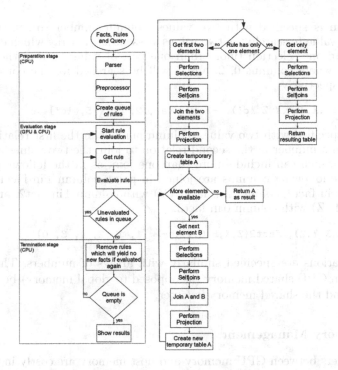

Fig. 3. GPU Datalog engine organisation.

by join and projection kernels. At the end of each rule evaluation, duplicate elimination kernels are applied. Figure 3 (on the right) shows these steps.

A memory management module helps identifying most recently used data by the GPU in order to keep it in global memory and to discard other data instead.

4.2 Host Thread Tasks

Parsing. To capitalise on the GPU capacity to process numbers and to have short and constant processing time for each tuple (the variable size of strings entails varying processing time), we identify and use facts and rules with/as numbers, keeping their corresponding strings in a hashed dictionary. Each unique string is assigned a unique id, equal strings are assigned the same id. The GPU thus works with numbers only; the dictionary is used at the very end when the final results are to be displayed.

Preprocessing. A key factor for good performance is preprocessing data before sending it to the GPU. As mentioned before, Datalog rules are evaluated through a series of relational algebra operations: selections, joins and projections. For the evaluation of each rule, the specification of what operations to perform, including constants, variables, facts and other rules involved, is carried out in the host (as opposed to be carried out in the GPU by each kernel thread), and sent to the GPU for all GPU threads to use. Examples:

- **Selection** is specified with two values, column number to search and the constant value to search; the two values are sent as an array which can include more than one selection (more than one pair of values), as in the following example, where columns 0, 2, and 5 will be searched for the constants a, b and c, respectively:

    ```
    fact1('a',X,'b',Y,Z,'c'). -> [0, 'a', 2, 'b', 5, 'c']
    ```

- **Join** is specified with two values, column number in the first relation to join and column number in the second relation to join; the two values are sent as an array which can include more than one join, as in the following example where the following columns are joined in pairs: column 1 in fact1 (X) with column 1 in fact2, column 2 in fact1 (Y) with column 4 in fact2, and column 3 in fact1 (Z) with column 0 in fact2.

    ```
    fact1(A,X,Y,Z), fact2(Z,X,B,C,Y). -> [1, 1, 2, 4, 3, 0]
    ```

Other operations are specified similarly with arrays of numbers. These arrays are stored in GPU shared memory (as opposed to global memory) because they are small and the shared memory is faster.

4.3   Memory Management

Data transfers between GPU memory and host memory are costly in all CUDA applications [1]. We designed a memory management scheme that tries to minimize the number of such transfers. Its purpose is to maintain facts and rule results in GPU memory for as long as possible so that, if they are used more than once, they may often be reused from GPU memory. To do so, we keep track of GPU memory available and GPU memory used, and maintain a list with information about each fact and rule result that is resident in GPU memory. When data (facts or rule results) is requested to be loaded into GPU memory, it is first looked up in that list. If found, its entry in the list is moved to the beginning of the list; otherwise, memory is allocated for the data and a list entry is created at the beginning of the list for it. In either case, its address in memory is returned. If allocating memory for the data requires deallocating other facts and rule results, those at the end of the list are deallocated first until enough memory is obtained — rule results are written to CPU memory before deallocating them. By so doing, most recently used fact and rule results are kept in GPU memory.

5   GPU Relational Algebra Operators

This section presents the design decisions we made for the relational algebra operations we use in our Datalog engine: select, join and project operations for GPUs. The GPU kernels that implement these operations access (read/write) tables from GPU global memory.

5.1   Selection

Selection has two main issues when designed for running in GPUs. The first issue is that the size of the result is not known beforehand, and increasing the size of the results buffer is not convenient performance-wise because it may involve reallocating its contents. The other issue is that, for efficiency, each GPU thread must know onto which global memory location it will write its result without communicating with other GPU threads.

To avoid those issues, our selection uses three different kernel executions. The first kernel marks all the rows that satisfy the selection predicate with a value one. The second kernel performs a prefix sum on the marks to determine the size of the results buffer and the location where each GPU thread must write the results. The last kernel writes the results.

5.2   Projection

Projection requires little computation, as it simply involves taking all the elements of each required column and storing them in a new memory location. While it may seem pointless to use the GPU to move memory, the higher memory bandwidth of the GPU, compared to that of the host CPUs, and the fact that the results remain in GPU memory for further processing, make projection a suitable operation for GPU processing.

5.3   Join

Our Datalog engine uses these types of join: Single join, Multijoin and Selfjoin. A single join is used when only two columns are to be joined, e.g.: $table_1(X, Y) \bowtie table_2(Y, Z)$. A multijoin is used when more than two columns are to be joined: $table_1(X, Y) \bowtie table_2(X, Y)$. A selfjoin is used when two columns have the same variable in the same predicate: $table_1(X, X)$.

Single join. We use a modified version of the Indexed Nested Loop Join described in [16], which is as follows:

```
Make an array for each of the two columns to be joined
Sort one of them
Create a CSS-Tree for the sorted column
Search the tree to determine the join positions
Do a first join~to determine the size of the result
Do a second join~to write the result
```

The CSS-Tree [22] (Cache Sensitive Search Tree) is very adequate for GPUs because it can be quickly constructed in parallel and because tree traversal is performed via address arithmetic instead of the traditional memory pointers.

While the tree allows us to know the location of an element, it does not tell us how many times each element is going to be joined with other elements nor in which memory location must each thread write the result, so we must perform a "preliminary" join. This join counts the number of times each element has to be joined and returns an array that, as in the select operation, allows us to determine the size of the result and write locations when a prefix sum is applied to it. With the size and write locations known, a second join writes the results.

Multijoin. To perform a join over more than two columns, e.g., $table_1(X, Y) \bowtie table_2(X, Y)$, first we take a pair of columns, say (X, X), to create and search on the CSS-Tree as described in the single join algorithm. Then, as we are doing the first join, we also check if the values of the remaining columns are equal (in our example we check if $Y = Y$) and discard the rows that do not comply.

Selfjoin. The selfjoin operation is similar to the selection operation. The main difference is that, instead of checking for a constant value on the corresponding row, it checks if the values of the columns involved by the self join match.

5.4   Optimisations

Our relational algebra operations make use of the following optimisations in order to improve performance. The purpose of these optimisations is to reduce memory use and in principle processing time — the cost of the optimisations themselves is not yet evaluated.

Duplicate Elimination. Duplicate elimination uses the *unique* function of the Thrust library. It takes an array and a function to compare two elements in the array, and returns the same array with the unique elements at the beginning. We apply duplicate elimination to the result of each rule: when a rule is finished, its result is sorted and the *unique* function is applied.

Optimising Projections. Running a *projection* at the end of each join, as described below, allows us to discard unnecessary columns earlier in the computation of a rule. For example, consider the following rule:

```
rule1(Y, W) :- fact1(X, Y), fact2(Y, Z), fact3(Z,W).
```

The evaluation of the first join, $fact_1 \bowtie_Y fact_2$, generates a temporary table with columns (X, Y, Y, Z), not all of which are necessary. One of the two Y columns can be discarded; and column X can also be discarded because it is not used again in the body nor in the head of the rule.

Fusing Operations. Fusing operations consists of applying two or more operations to a data set in a single read of the data set, as opposed to applying only one operation, which involves as many reads of the data set as the number of operations to be applied. We fuse the following operations.

- All selections required by constant arguments in a subgoal of a rule are performed at the same time.
- All selfjoins are also performed at the same time.

– Join and projection are always performed together at the same time.

To illustrate these fusings consider the following rule:

`rule1(X,Z):- fact1(X,'const1',Y,'const2'),fact2(Y,'const3',Y,Z,Z).`

This rule will be evaluated as follows. $fact_1$ is processed first: the selections required by $const_1$ and $const_2$ are performed at the same time — $fact_1$ does not require selfjoins. $fact_2$ is processed second: (a) the selection required by $const_3$ is performed, and then (b) the selfjoins between Ys and Zs are performed at the same time. Finally, a join is performed between the third column of $fact_1$ and the first column of $fact_2$ and, at the same time, a projection is made (as required by the arguments in the rule head) to leave only the first column of $fact_1$ and the fourth column of $fact_2$.

6   Experimental Evaluation

This section describes our platform, applications and experiments to evaluate the performance of our Datalog engine. We are at this stage interested in the performance benefit of using GPUs for evaluating Datalog queries, compared to using a CPU only. Hence we present results that show the performance of 4 Datalog queries running on our engine compared to the performance of the same queries running on a single CPU in the host platform. (We plan to compare our Datalog engine to similar GPU work discussed in Sect. 7, Related Work, in another paper).

On a single CPU in the host platform, the 4 queries were run with the Prolog systems YAP [10] and XSB [24], and the Datalog system from the MITRE Corporation [3]. As the 4 queries showed the best performance with YAP, our results show the performance of the queries with YAP and with our Datalog engine only. YAP is a high-performance Prolog compiler developed at LIACC/Universidade do Porto and at COPPE Sistemas/UFRJ. Its Prolog engine is based on the WAM (Warren Abstract Machine) [10], extended with some optimizations to improve performance. The queries were run on this platform:

Hardware. *Host platform*: Intel Core 2 Quad CPU Q9400 2.66GHz (4 cores in total), Kingston RAM DDR2 6GB 800 MHz. *GPU platform*: Fermi GeForce GTX 580 - 512 cores - 1536 MB GDDR5 memory.

Software. Ubuntu 12.04.1 LTS 64bits. CUDA 5.0 Production Release, gcc 4.5, g++ 4.5. YAP 6.3.3 Development Version, Datalog 2.4, XSB 3.4.0.

For each query, in each subsection below, we describe first the query, and then discuss the results. Our results show the evaluation of each query once all data has been preprocessed and in CPU memory, i.e.: I/O, parsing and preprocessing costs are not included in the evaluation.

6.1   Join over Four Big Tables

Four tables, all with the same number of rows filled with random numbers, are
joined together to test all the different operations of our Datalog engine. The
rule and query used are:

```
join(X,Z) :- table1(X), table2(X,4,Y), table3(Y,Z,Z), table4(Y,Z).
join(X,Z)?
```

Figure 4 shows the performance of the join with YAP and our engine, in
both normal and logarithmic scales to better appreciate details. Our engine is
clearly faster, roughly 200 times. Both YAP and our engine take proportionally
more time as the size of tables grows. Our engine took just above two seconds to
process tables with five million rows each, while YAP took about two minutes
to process tables with one million rows each. Joins were the most costly operations with multijoin alone taking more than 70 % of the total time; duplicate
elimination and sorting were also time consuming but within acceptable values;
prefix sums and selections were the fastest operations.

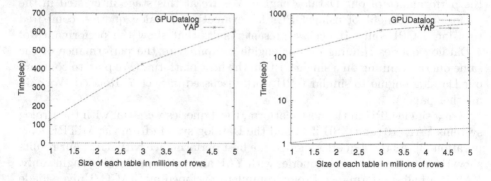

Fig. 4. Performance of join over four big tables (NB: log. scale on the right).

6.2   Transitive Closure of a Graph

The transitive closure of a graph (TCG) is a recursive query. We use a table with
two columns filled with random numbers that represent the edges of a graph [13].
The idea is to find all the nodes that can be reached if we start from a particular
node. This query is very demanding because recursive queries involve various
iterations over the relational operations that solve the query. The rules and the
query are:

```
path(X,Y) :- edge(X,Y).
path(X,Z) :- edge(X,Y), path(Y,Z).
path(X,Y)?
```

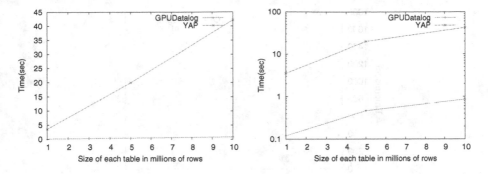

Fig. 5. Performance of transitive closure of a graph (NB: log. scale on the right).

Figure 5 shows the performance of TCG with YAP and our engine. Similar observations can be made as for the previous experiment. Our engine is 40x times faster than YAP for TCG. Our engine took less than a second to process a table of 10 million rows while YAP took 3.5 s to process 1 million rows.

For the first few iterations, duplicate elimination was the most costly operation of each iteration, and the join second but closely. As the number of rows to process in each iteration decreased, the join became by far the most costly operation.

6.3 Same-Generation Program

This is a well-known program in the Datalog literature, and there are various versions. We use the version described in [6]. Because of the initial tables and the way the rules are written, it generates lots of new tuples in each iteration. The three required tables are created with the following equations:

$$up = \{(a, b_i)|i\epsilon[1, n]\} \cup \{(b_i, c_j)|i, j\epsilon[1, n]\}. \tag{1}$$
$$flat = \{(c_i, d_j)|i, j\epsilon[1, n]\}. \tag{2}$$
$$down = \{(d_i, e_j)|i, j\epsilon[1, n]\} \cup \{(e_i, f)|i\epsilon[1, n]\}. \tag{3}$$

Where a and f are two known numbers and b, c, d and e are series of n random numbers. The rules and query are as follows:

```
sg(X,Y) :- flat(X,Y).
sg(X,Y) :- up(X,X1), sg(X1,Y1), down(Y1,Y).
sg(a,Y)?
```

The results show (Fig. 6) very little gain in performance, with our engine taking an average of 827 ms and YAP 1600 ms for $n = 75$. Furthermore, our engine cannot process this application for n > 90 due to lack of memory.

The analysis of each operation revealed that duplicate elimination takes more than 80 % of the total time and is also the cause of the memory problem. The

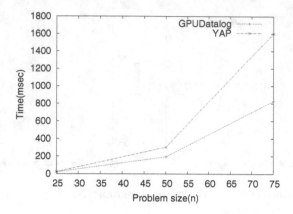

Fig. 6. Same-generation program.

reason of this behaviour is that the join creates far too many new tuples, but most of these tuples are duplicates (as an example, for $n = 75$ the first join creates some 30 million rows and, after duplicate elimination, less than 10 thousand rows remain).

6.4 Tumour Detection

Correctly determining whether or not a tumour is malignant requires analysing and comparing a great amount of information from medical studies. Considering each characteristic of a tumour as a fact, the rules and query below can be used to determine, for each patient, if his/her tumour is malignant or not:

```
is_malignant(A):-
    same_study(A,B),                   'HO_BreastCA'(B,hxDCorLC),
    'MassPAO'(B,present),              'ArchDistortion'(A,notPresent),
    'Calc_Round'(A,notPresent),        'Sp_AsymmetricDensity'(A,notPresent),
    'SkinRetraction'(B,notPresent),    'Calc_Popcorn'(A,notPresent),
    'FH_DCNOS'(B,none).
same_study(Id,OldId) :-
    'IDnum'(Id,X),                     'MammoStudyDate'(Id,DO),
    'IDnum'(OldId,X),                  'MammoStudyDate'(OldId,DO),
    OldId \= Id.
is_malignant(A)?
```

The query asks for those studies which detect a malignant tumour. Some tumour characteristics are taken from the most recent study, while others must be taken from past studies. This restriction requires defining an additional rule (`same_study`) to determine if two studies belong to the same person and if they have different dates. The last subgoal of `same_study` (`OldId \= Id`) prevents a study from referencing to itself, thus avoiding incorrect results. We evaluated this program with 65800 studies, i.e., each table is composed of 65800 rows.

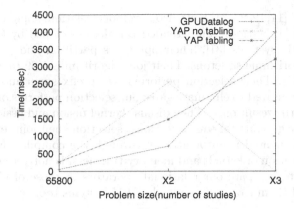

Fig. 7. Performance of tumour detection.

Figure 7 shows the performance of tumour detection with YAP and our engine. We used different sizes of input data through duplicating and triplicating each table, i.e.: each table had 65800 rows for the first test, and 131600 and 197400 rows for the second and third tests. We could thus increase processing time while obtaining the same results thanks to duplicate elimination.

Our engine performs best for the first and second tests, but is surpassed in the third test by YAP with tabling [25]. A detailed analysis showed that the multijoin required by `same_study` consumed almost 90 % of the total execution time, and mostly in duplicate elimination, as follows. In the third test, `same_study` generated 4095036 rows, which were reduced to 50556 after duplicate elimination. For `is_malignant` the results were similar: 4881384 rows were generated but only 550 remained after duplicate elimination.

YAP with no tabling in the third test is so affected by duplicates that it simply terminates after throwing an error — shown in the figure as zero execution time. In contrast, YAP with tabling avoids performing duplicate work and thus performs rather well.

7   Related Work

He *et al.* [17] have designed, implemented and evaluated GDB, an in-memory relational query coprocessing system for execution on both CPUs and GPUs. GDB consists of various primitive operations (scan, sort, prefix sum, etc.) and relational algebra operators built upon those primitives.

We modified the Indexed Nested Loop Join (INLJ) of GDB for our single join and multijoin, so that more than two columns can be joined, and a projection performed, at the same time. Their selection operation and ours are similar too; ours takes advantage of GPU shared memory and uses the Prefix Sum of the Thrust Library. Our projection is fused into the join and does not perform duplicate elimination, while they do not use fusion at all.

Diamos *et al.* [11, 12, 29–31] have also developed relational operators for GPUs for the Red Fox [4] platform, an extended Datalog developed by LogicBlox [14] for multiple-GPU systems [31]. Their operators partition and process data in blocks using algorithmic skeletons. Their join algorithm is 1.69 times faster than that of GDB [11]. Their selection performs two prefix sums and the result is written and then moved to eliminate gaps; our selection performs only one prefix sum and writes the result once. They discuss kernel fusion and fission in [30]. We applied fusion (e.g., simultaneous selections, selection then join, etc.) at source code; they implemented it automatically through the compiler. Kernel fission, the parallel execution of kernels and memory transfers, is not yet adopted in our work. We plan to compare our relational operators to those of GDB and Red Fox, and extend them to run on multiple-GPU systems too.

8   Conclusions

Our Datalog engine for GPUs evaluates queries based on the relational operators select, join and projection. Our evaluation using 4 queries shows a dramatic performance improvement for two queries, up to 200 times for one of them. The other two queries did not perform that well, but we are working on the following extensions to our engine in order to improve its performance further.

- Evaluation based on tabling [25] or magic sets [8] methods.
- Managing tables larger than the total amount of GPU memory.
- Mixed processing of rules both on the GPU and on the host multicore.
- Improved join operations to eliminate duplicates before writing final results.
- Extended syntax to accept built-in predicates and negation [6].

Acknowledgments. CMA thanks the support during his MSc studies from: the University of Porto, the Centre for Research and Postgraduate Studies of the National Polytechnic Institute (CINVESTAV-IPN) of Mexico, and the Mexican Council of Science and Technology (CONACyT). ICD and VSC were partially supported by: the European Regional Development Fund (ERDF), COMPETE Programme; the Portuguese Foundation for Science and Technology (FCT), project LEAP FCOMP-01-0124-FEDER-015008; and project ABLe PTDC/EEI-SII/2094/2012.

References

1. CUDA, C Best Practices Guide. http://docs.nvidia.com/cuda/cuda-c-best-practices-guide/index.html
2. CUDA, C Programming Guide. http://docs.nvidia.com/cuda/cuda-c-programming-guide/index.html
3. Datalog by the MITRE Corporation. http://datalog.sourceforge.net/
4. Red Fox: A Compilation Environment for Data Warehousing. http://gpuocelot.gatech.edu/projects/red-fox-a-compilation-environment-for-data-warehousing/

5. Thrust: A Parallel Template Library. http://thrust.github.io/
6. Abiteboul, S., et al.: Foundations of Databases. Addison-Wesley, Boston (1995)
7. Afrati, F.N., Borkar, V., Carey, M., Polyzotis, N., Ullman, J.D.: Cluster Computing, Recursion and Datalog. In: de Moor, O., Gottlob, G., Furche, T., Sellers, A. (eds.) Datalog 2010. LNCS, vol. 6702, pp. 120–144. Springer, Heidelberg (2011)
8. Beeri, C., Ramakrishnan, R.: On the power of magic. J. Log. Program. **10**(3–4), 255–299 (1991)
9. Bravenboer, M., Smaragdakis, Y.: Strictly declarative specification of sophisticated points-to analyses. In: OOPSLA, pp. 243–262 (2009)
10. Costa, V.S., et al.: The YAP prolog system. TPLP **12**(1–2), 5–34 (2012)
11. Diamos, G., et al.: Efficient relational algebra algorithms and data structures for GPU. Technical report, Georgia Institute of Technology (2012)
12. Diamos G. et al.: Relational algorithms for multi-bulk-synchronous processors. In: 18th Symposium on Principles and Practice of Parallel Programming (2013)
13. Dong, G., Jianwen, S., Topor, R.W.: Nonrecursive incremental evaluation of datalog queries. Ann. Math. Artif. Intell. **14**(2–4), 187–223 (1995)
14. Green, T.J., Aref, M., Karvounarakis, G.: LogicBlox, Platform and Language: A Tutorial. In: Barceló, P., Pichler, R. (eds.) Datalog 2.0 2012. LNCS, vol. 7494, pp. 1–8. Springer, Heidelberg (2012)
15. Harris, M., et al.: Parallel prefix sum (scan) with CUDA. In: Nguyen, H. (ed.) GPU Gems 3, pp. 851–876. Addison Wesley, Boston (2007)
16. He, B., et al.: Relational joins on graphics processors. In: SIGMOD Conference, pp. 511–524 (2008)
17. He, B., et al.: Relational query coprocessing on graphics processors. ACM Trans. Database Syst. (TODS) **34**(4), 21:1–21:39 (2009)
18. Huang, S.S., et al.: Datalog and emerging applications: an interactive tutorial. In: SIGMOD Conference. pp. 1213–1216 (2011)
19. Loo, B.T., et al.: Declarative networking: language, execution and optimization. In: SIGMOD Conference, pp. 97–108 (2006)
20. Marczak W.R., et al.: Secureblox: customizable secure distributed data processing. In: SIGMOD Conference, pp. 723–734 (2010)
21. Musser, D.R., Derge, G.J., Saini, A.: STL Tutorial and Reference Guide: C++ Programming With The Standard Template Library, 2nd edn. Addison-Wesley Longman Publishing Co. Inc., Boston (2001)
22. Rao, J., Ross, K.A.: Cache conscious indexing for decision-support in main memory. In: 25th VLDB Conference, San Francisco., CA, USA, pp. 78–89 (1999)
23. Shen, W., et al.: Declarative information extraction using datalog with embedded extraction predicates. In: VLDB, pp. 1033–1044 (2007)
24. Swift, T., Warren, D.S.: Xsb: Extending prolog with tabled logic programming. TPLP **12**(1–2), 157–187 (2012)
25. Tekle, K.T., Liu, Y.A.: More efficient datalog queries: subsumptive tabling beats magic sets. In: SIGMOD Conference, pp. 661–672 (2011)
26. Ullman, J.D.: Principles of Database and Knowledge-Base Systems, vol. 1. Computer Science Press, Beijing (1988)
27. Ullman, J.D.: Principles of Database and Knowledge-Base Systems, vol. 2. Computer Science Press, Beijing (1989)
28. Valiant, L.G.: A bridging model for parallel computation. Commun. ACM **33**(8), 103–111 (1990)

29. Wu, H., et al.: Kernel weaver: Automatically fusing database primitives for efficient GPU computation. In: 45th International Symposium on Microarchitecture (2012)
30. Wu, H., et al.: Optimizing data warehousing applications for GPUs using kernel fusion/fission. In: IEEE 26th International Parallel and Distributed Processing Symposium Workshops and PhD Forum (2012)
31. Young, J., et al.: Satisfying data-intensive queries using GPU clusters. In: 2nd Annual Workshop on High-Performance Computing meets Databases (2012)

Towards Parallel Constraint-Based Local Search
with the X10 Language

Danny Munera[1]([✉]), Daniel Diaz[1], and Salvador Abreu[2]

[1] University of Paris 1-Sorbonne, Paris, France
Danny.Munera@malix.univ-paris1.fr, Daniel.Diaz@univ-paris1.fr
[2] Universidade de Évora and CENTRIA, Évora, Portugal
spa@di.uevora.pt

Abstract. In this study, we started to investigate how the Partitioned Global Address Space (PGAS) programming language X10 would suit the implementation of a Constraint-Based Local Search solver. We wanted to code in this language because we expect to gain from its ease of use and independence from specific parallel architectures. We present our implementation strategy, and quest for different sources of parallelism. We discuss the algorithms, their implementations and present a performance evaluation on a representative set of benchmarks.

1   Introduction

Constraint Programming has been successfully used to model and solve many real-life problems in diverse areas such as planning, resource allocation, scheduling and product line modeling [18,19]. Classically, constraint satisfaction problems (CSPs) may be solved exhaustively by complete methods which are capable of finding all solutions, and therefore to determine whether any solutions exist. However efficient these solvers may be, a significant class of problems remains out of reach because of exponential growth of the search space, which must be exhaustively explored. Another approach to solving CSPs entails giving up completeness and resorting to (meta-) heuristics which will guide the process of searching for solutions to the problem. Solvers in this class make choices which limit the search space which actually gets visited, enough so to make problems tractable. For instance a complete solver for the *magic squares* benchmark will fail for problems larger than 15×15 whereas a local search method will easily solve a 100×100 problem instance, within similar memory and CPU time bounds. On the other hand, a local search procedure may not be able to find a solution, even when one exists.

However, it is unquestionable that the more computational resources are available, the more complex the problems that may be solved. We would therefore like to be able to tap into the forms of augmented computational power which are actually available, as conveniently as feasible. This requires taming various forms of explicitly parallel architectures.

M. Hanus and R. Rocha (Eds.): KDPD 2013, LNAI 8439, pp. 169–184, 2014.
DOI: 10.1007/978-3-319-08909-6_11, © Springer International Publishing Switzerland 2014

Present-day parallel computational resources include increasingly multi-core processors, General Purpose Graphic Processing Units (GPGPUs), computer clusters and grid computing platforms. Each of these forms requires a different programming model and the use of specific software tools, the combination of which makes software development even more difficult.

The foremost software platforms used for parallel programming include POSIX Threads [2] and OpenMP [17] for shared-memory multiprocessors and multicore CPUs, MPI [23] for distributed-memory clusters or CUDA [16] and OpenCL [12] for massively parallel architectures such as GPGPUs. This diversity is a challenge from the programming language design standpoint, and a few proposals have emerged that try to simultaneously address the multiplicity of parallel computational architectures.

Several modern language designs are built around the Partitioned Global Address Space (PGAS) memory model, as is the case with X10 [21], Unified Parallel C [7] or Chapel [10]. Many of these languages propose abstractions which capture the several forms in which multiprocessors can be organized. Other, less radical, approaches consist in supplying a library of inter-process communication which relies on and uses a PGAS model [14].

In our quest to find a scalable and architecture-independent implementation platform for our exploration of high-performance parallel constraint-based local search methods, we decided to experiment with one of the most promising new-generation languages, X10 [21].

The remainder of this article is organized as follows: Section 2 discusses the PGAS Model and briefly introduces the X10 programming language. Section 3 introduces native X10 implementations exploiting different sources of parallelism of the Adaptive Search algorithm. Section 4 presents an evaluation of these implementations. A short conclusion ends the paper.

2   X10 and the Partitioned Global Address Space (PGAS) Model

The current arrangement of tools to exploit parallelism in machines are strongly linked to the platform used. Two broad programming models stand out in this matter: *distributed* and *shared memory* models. For large distributed memory systems, like clusters and grid computing, the Message Passing Interface (MPI) [23] is a de-facto programming standard. The key idea in MPI is to decompose the computation over a collection of processes, each with its private memory space. These processes can communicate with each other through message passing, generally over a communication network.

With the recent growth of many-core architectures, the shared memory approach has grown in popularity. This model decomposes the computation in multiple threads of execution which share a common address space, communicating with each other by reading and writing shared variables. Actually, this is the model used by traditional programming tools like Fortran or C through libraries like *pthreads* [2] or OpenMP [17].

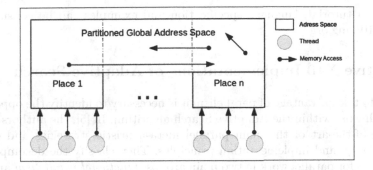

Fig. 1. PGAS model

The PGAS model tries to combine the advantages of these two approaches: it extends shared memory to a distributed memory setting. The execution model allows having multiple processes (like MPI), multiple threads in a process (like OpenMP), or a combination thereof (see Fig. 1). Ideally, the user would be allowed to decide how tasks get mapped to physical resources. X10 [21], Unified Parallel C [25] and Chapel [10] are examples of PGAS-enabled languages, but there are also PGAS-based IPC libraries such as GPI [14], for use in traditional programming languages. For the experiments described herein, we used the X10 language.

X10 [21] is a general-purpose language developed by IBM, which provides a PGAS variation: Asynchronous PGAS (APGAS). APGAS extends the PGAS model making it flexible, even in non-HPC platforms [20]. With this model X10 supports different levels of concurrency with simple language constructs.

There are two main abstractions in X10: *places* and *activities*. A *place* is the abstraction of a virtual shared-memory process, it has a coherent portion of the address space together with threads (activities) that operate on that memory. The X10 construct for creating a place in X10 is *at*, and is commonly used to create a place for each processing unit in the platform. An *activity* is the mechanism to abstract the single threads that perform computation within a place. Multiple activities may be simultaneously active in one place.

X10 implements the major components of the PGAS model, by the use of places and activities. However, the language includes other interesting tools with the goal of improving the abstraction level of the language. Synchronization is supported thanks to various operations such as *finish*, *atomic* and *clock*. The *finish* operation is used to wait for the termination of a set of activities, it behaves like a traditional barrier. The constructs *atomic* ensures exclusive access to a critical portion of code. Finally, the construct *clock* is the standard way to ensure the synchronization between activities or places. X10 supports the distributed array construct, which makes it possible to divide an array into sub-arrays which are mapped to available places. Doing this ensures a local access from each place to the related assigned sub-array. A detailed discussion of X10, including a tutorial, language specification and examples can be consulted at http://x10-lang.org/.

3   Native X10 Implementations of Adaptive Search

In order to take advantage of parallelism it is necessary to identify the opportunities which exist within the Adaptive Search algorithm. In [5], the authors survey the state-of-the-art of the main parallel meta-heuristic strategies and discuss general design and implementation principles. They classify the decomposition of activities for parallel work in two main groups: *functional parallelism* and *data parallelism*.[1]

On the one hand, in *functional parallelism* different tasks run on multiple compute instances across the same or different data-sets. On the other hand, *data parallelism* refers to the methods in which the problem domain or the associated search space is decomposed. A particular solution methodology is used to address the problem on each of the resulting components of the search space. This article reports on our experiments concerning both kinds of parallelism applied to the Adaptive Search method.

3.1   Sequential Implementation

Our first experiment with AS in X10 was to develop a sequential implementation corresponding to a specialized version of the Adaptive Search for permutation problems [15].[2]

Figure 2 shows the class diagram of the basic X10 project. The class *ASPermutSolver* contains the Adaptive Search permutation specialized method implementation. This class inherits the basic functionality from a general implementation of the Adaptive Search solver (in class *AdaptiveSearchSolver*), which in turn inherits a very simple Local Search method implementation from the class *LocalSearchSolver*. This class is then specialized for different parallel approaches, which we experimented with. As we will see below, we experimented with two versions of Functional Parallelism (FP1 and FP2) and a Data Parallelism version (called Independent Multi-Walk, i.e. IMW).

Moreover, a simple CSP model is described in the class *CSPModel*, and specialized implementations of each CSP benchmark problem are contained in the classes *PartitModel, MagicSquareModel, AllIntervallModel* and *CostasModel*, which have all data structures and methods to implement the error function of each problem.

Listing 1.1 shows a simplified skeleton code of our X10 sequential implementation, based on Algorithm 1. The core of the Adaptive Search algorithm

---

[1] Their relation is similar to that of AND- and OR-parallelism in the Logic Programming community.

[2] In a permutation problem, all N variables have the same initial domain of size N and are subject to an implicit *all-different* constraint. The associated algorithm is reported in the appendix.

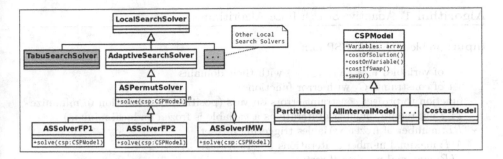

Fig. 2. X10 Class diagram basic project

is implemented in the method *solve*. The *solve* method receives a *CSPModel* instance as parameter. On line 8, the CSP variables of the model are initialized with a random permutation. On the next line the total cost of the current configuration is computed. The *while* instruction on line 10 corresponds to the main loop of the algorithm. The *selectVarHighCost* function (Line 12) selects the variable with the maximal error and saves the result in the *maxI* variable. The *selectVarMinConflict* function (Line 13) selects the best neighbor move from the highest cost variable *maxI*, and saves the result in the *minJ* variable. Finally, if no local minimum is detected, the algorithm swaps the variables *maxI* and *minJ* (permutation problem) and computes the total cost of the resulting new configuration (Line 16). The solver function ends if the *totalCost* variable equals 0 or when the maximum number of iterations is reached.

Listing 1.1. Simplified AS X10 Sequential Implementation

```
1   class ASPermutSolver {
2     var totalCost: Int;
3     var maxI: Int;
4     var minJ: Int;
5
6     public def solve (csp: CSPModel): Int {
7       <local variables>
8       csp.initialize();
9       totalCost = csp.costOfSolution();
10      while (totalCost != 0) {
11        <restart code>
12        maxI = selectVarHighCost (csp);
13        minJ = selectVarMinConflict (csp);
14        <local min tabu list, reset code>
15        csp.swapVariables (maxI, minJ);
16        totalCost = csp.costOfSolution ();
17      }
18      return totalCost;
19    }
20  }
```

Algorithm 1. Adaptive Search Base Algorithm

Input: problem given in CSP format:

- set of variables $V = \{X_1, X_2 \cdots\}$ with their domains
- set of constraints C_j with error functions
- function to project constraint errors on vars (positive) cost function to minimize
- T: Tabu tenure (number of iterations a variable is frozen on local minima)
- RL: number of frozen variables triggering a reset
- MI: maximal number of iterations before restart
- MR: maximal number of restarts

Output: a solution if the CSP is satisfied or a quasi-solution of minimal cost otherwise.

1: $Restart \leftarrow 0$
2: **repeat**
3:     $Restart \leftarrow Restart + 1$
4:     $Iteration \leftarrow 0$
5:     Compute a random assignment A of variables in V
6:     $Opt\_Sol \leftarrow A$
7:     $Opt\_Cost \leftarrow cost(A)$
8:     **repeat**
9:         $Iteration \leftarrow Iteration + 1$
10:        Compute errors constraints in C and project on relevant variables
11:        Select variable X with highest error: $MaxV$
12:   ▷ not marked Tabu
13:        Select the move with best cost from X: $MinConflictV$
14:        **if** no improvement move exists **then**
15:            mark X as Tabu for T iterations
16:            **if** number of variables marked Tabu $\geq RL$ **then**
17:                randomly reset some variables in V
18:   ▷ and unmark those Tabu
19:            **end if**
20:        **else**
21:            swap($MaxV$,$MinConflictV$),
22:   ▷ modifying the configuration A
23:            **if** $cost(A) < Opt\_Cost$ **then**
24:                $Opt\_Sol \leftarrow A$
25:                $Opt\_Cost \leftarrow costs(A)$
26:            **end if**
27:        **end if**
28:     **until** $Opt\_Cost = 0$ (solution found) or $Iteration \geq MI$
29: **until** $Opt\_Cost = 0$ (solution found) or $Restart \geq MR$
30: $output(Opt\_Sol, Opt\_Cost)$

3.2   Functional Parallel Implementation

Functional parallelism is our first attempt to parallelize the Adaptive Search algorithm. The key aim for this implementation is to decompose the problem into different tasks, each task working in parallel on the same data. To achieve this objective it is necessary to change the inner loop of the sequential Adaptive Search algorithm.

In this experiment, we decided to change the structure of the *selectVarHigh-Cost* function, because therein lies the most costly activities performed in the inner loop. The most important task performed by this function is to go through the variable array of the CSP model to compute the cost of each variable (in order to select the variable with the highest cost). A X10 skeleton implementation of *selectVarHighCost* function is presented in Listing 1.2.

Listing 1.2. Function selVarHighCost in X10

```
1   public def selectVarHighCost( csp : CSPModel ) : Int {
2     <local variables>
3     // main loop: go through each variable in the CSP
4     for (i = 0; i < size; i++) {
5       <count marked variables>
6       cost = csp.costOnVariable (i);
7       <select the highest cost>
8     }
9     return maxI; // (index of the highest cost)
10  }
```

Since this function must process the entire variable vector at each iteration, it is natural to try to parallelize the task. For problems with many variables (e.g. the magic square problem involves N^2 variables) the gain could be very interesting. We developed a *first approach* (called FP1), in which **n** single activities are created at each iteration. Each activity processes a portion of the variables array and performs the required computations. The X10 construct *async* was chosen to create individual *activities* sharing the global array. Listing 1.3 shows the X10 skeleton code for the *first approach* of the *functional parallelism* in the function *selectVarHighCost*.

Listing 1.3. First approach to *functional parallelism*

```
1   public def selectVarHighCost (csp : CSPModel) : Int {
2     // Initialization of Global variables
3     var partition : Int = csp.size/THNUM;
4     finish for(th in 1..THNUM){
5       async{
6         for (i = ((th−1)*partition); i < th*partition; i++){
7           <calculate individual cost of each variable>
8           <save variable with higher cost>
9         }
10      }
11    }
12    <terminate function: merge solutions>
13    return maxI; //(Index of the higher cost)
14  }
```

In this implementation the constant THNUM on line 4 represents the number of concurrent activities that are deployed by the program. On the same line, the keyword *finish* ensures the termination of all spawned activities. Finally, the construct *async* on line 5 spawns independent individual tasks to cross over a portion of the variable array (sentence *for* on line 6). With this strategy we face up with a well known problem of functional parallelism: the overhead due to the management of fine-grained activities. As expected results are not good enough (see Sect. 4 for detailed results).

In order to limit the overhead due to activity creation, we implemented a *second approach* (called FP2). Here the n working activities are created at the very beginning of the solving process, just before the main loop of the algorithm. These activities are thus available for all subsequent iterations. However, it is necessary to develop a synchronization mechanism to assign tasks to the working

Listing 1.4. Second approach to *functional parallelism*

```
1   public class ASSolverFP1 extends ASPermutSolver{
2     val computeInst : Array[ComputePlace];
3     var startBarrier : ActivityBarrier;
4     var doneBarrier : ActivityBarrier;
5
6     public def solve(csp : CSPModel):Int{
7       for(var th : Int = 1; th <= THNUM ; th++)
8         computeInst(th) = new ComputePlace(th , csp);
9
10      for(id in computeInst)
11        async computeInst(id).run();
12
13      while(total_cost!=0){
14        <restart code>
15        for(id in computeInst)
16          computeInst(id).activityToDo = SELECVARHIGHCOST;
17
18        startBarrier.wait(); // send start signal
19        // activities working...
20        doneBarrier.wait(); // work ready
21        maxI=terminateSelVarHighCost();
22        <local min tabu list, reset code>
23      }
24      // Finish activities
25      for(id in computeInst)
26        computeInst(id).activityToDo = FINISH;
27
28      startBarrier.wait();
29      doneBarrier.wait();
30      return totalCost;
31    }
32  }
```

activities and to wait for their termination. For this purpose we created two new classes: *ComputePlace* and *ActivityBarrier*. *ComputePlace* is a compute instance, which contains the functionality of the working activities. *ActivityBarrier* is a very simple barrier developed with X10 monitors (X10 concurrent package).

Listing 1.4 shows the X10 implementation[3] of the *second approach*.

This code begins with the definition of three global variables on lines 2–4: *computeInst*, *startBarrier* and *doneBarrier*; *computeInst* is an array of *ComputePlace* objects, one for each working activity desired. *startBarrier* and *doneBarrier* are *ActivityBarrier* instances created to signalize the starting and ending of the task in the compute place. On lines 7–11, before the main loop THNUM working activities are created and started over an independent X10 activity. When the algorithm needs to execute the *selectVarHighCost* functionality, the main activity assigns this task putting a specific value into the variable *activityToDo* in the corresponding instance of the *ComputePlace* class (lines 15 and 16), then the function *wait()* is executed over the barrier *startBarrier* to notify all working activities to start (line 18). Finally, the function *wait()* is executed over the barrier *doneBarrier* to wait the termination of the working activities (line 20). Then on line 21 the main activity can process the data with the function *terminateSelVarHighCost*. When the main loop ends, all the working activities are notified to end and the *solve* function returns (lines 25–30). Unfortunately, as we will see below, the improvement of this *second approach* is not important enough (and, in addition, it has its own overhead due to synchronization mechanisms).

3.3   Data Parallel Implementation

A straightforward implementation of data parallelism in the Adaptive Search algorithm is the multiple independent Independent Multi-Walks (IMW) approach. The idea is to use isolated sequential Adaptive Search solver instances dividing the search space of the problem through different random starting points. This strategy is also known as Multi Search (MPSS, Multiple initial Points, Same search Strategies) [5] and has proven to be very efficient [6,13].

The key of this implementation is to have several independent and isolated instances of the Adaptive Search Solver applied to the same problem model. The problem is distributed to the available processing resources in the computer platform. Each solver runs independently (starting with a random assignment of values). When one instance finds a solution it is necessary to stop all other running instances. This is achieved using a termination detection communication strategy. This simple parallel version has no inter-process communication, making it *Embarrassingly* or *Pleasantly Parallel*. The skeleton code of the algorithm is shown in the Listing 1.5.

---

[3] Remark: in X10 the array notation is *table(index)* instead of *table[index]* as in C.

Listing 1.5. Adaptive Search *data parallel* X10 implementation

```
1   public class ASSolverIMW{
2      val solDist : DistArray[ASPermutSolver];
3      val cspDist : DistArray[CSPModel];
4      def this( ){
5        solDist=DistArray.make[ASPermutSolver](Dist.makeUnique());
6        cspDist=DistArray.make[CSPModel](Dist.makeUnique());
7      }
8      public def solve(){
9        val random = new Random();
10       finish for(p in Place.places()){
11         val seed = random.nextLong();
12         at(p) async {
13           cspDist(here.id) = new CSPModel(seed);
14           solDist(here.id) = new ASPermutSolver(seed);
15           cost = solDist(here.id).solve(cspDist(here.id));
16           if (cost==0){
17             for (k in Place.places())
18               if (here.id != k.id)
19                 at(k) async{
20                   solDist(here.id).kill = true;
21                 }
22           }
23         }
24       }
25       return cost;
26     }
27   }
```

For this implementation the *ASSolverIMW* class was created. The algorithm has two global distributed arrays: *solDist* and *cspDist* (lines 2 and 3). As
explained in Sect. 2, the *DistArray* class creates an array which is spread across
multiple X10 places. In this case, an instance of *ASPermutSolver* and *CSPModel*
are stored at each available place in the program. On lines 5 and 6 function *make*
creates and initializes the distributed vector in the region created by the function *Dist.makeUnique()*. The *makeUnique* function creates a distribution over a
region that maps every point in the region to a distinct place, and also maps some
point in the region to every place. On line 10 a *finish* operation is executed over a
for loop that goes through all the places in the program (*Place.places()*). Then,
an activity is created in each place with the sentence *at(p) async* on line 12. Into
the *async* block, a new instance of the solver (*new ASPermutSolver(seed)*) and
the problem (*new CSPModel(seed)*) are created (lines 13 and 14) and a random
seed is passed. On line 15, the solving process is executed and the returned cost
is assigned to the *cost* variable. If this cost is equal to 0, the solver in a place has
reached a valid solution, it is then necessary to send a termination signal to the
remaining places (lines 16–22). For this, every place (i.e. every solver), checks
the value of a *kill* variable at each iteration. When it becomes equal to *true*
the main loop of the solver is broken and the activity is finished. To set a *kill*

remote variable from any X10 place it was necessary to create a new activity into each remaining place (sentence *at(k) async* on line 19) and into the *async* block to change the value of the *kill* variable. On line 18, the sentence *if (here.id !=k.id)* filters all places which are not the winning one (*here*). Finally, the function returns the solution of the fastest place on line 25.

4   Performance Analysis

We now present and discuss the experimental results of our X10 implementations of the Adaptive Search algorithm. The testing environment used was a non-uniform memory access (NUMA) computer, with 2 Intel Xeon W5580 CPUs each one with 4 hyper-threaded cores running at 3.2 GHz as well as a system based on 4 16-core AMD Opteron 6272 CPUs running at 2.1 GHz.

We used a set of benchmarks composed of four classical problems in constraint programming: the magic square problem (MSP), the number partitioning problem (NPP) and the all-interval problem (AIP), all three taken from the CSPLib [8]; also we include the Costas Arrays Problem (CAP) introduced in [11], which is a very challenging real problem. The problems were all tested on significantly large instances which are generally out of reach of the traditional complete solvers like Gecode [22]. The interested reader may find more information on these benchmarks in [15].

It is worth noting, at the software level, that the X10 runtime system can be deployed in two different backends: a Java or a C++ backend; they differ in the native language used to implement the X10 program (Java or C++), also they present different trade-offs on different machines. Currently, the C++ backend seems relatively more mature and faster for scientific computation, and therefore became our choice for this experimentation.

Regarding the stochastic nature of the Adaptive Search behavior, several executions of the same problem were done and the times averaged. We ran 100 samples for each experimental case in the benchmark.

In this presentation, all tables report raw times in seconds (average of 100 runs) and relative speed-ups. These tables respect the same format: the first column identifies the problem instance, the second column is the execution time of the problem in the sequential implementation, the next group of columns contains the corresponding speed-up obtained with a varying number of cores (places), and the last column presents the execution time of the problem with the highest number of places.

4.1   Sequential Performance

Even if our first goal in using X10 is parallelism, it is interesting to compare the sequential X10 implementation with a reference implementation: our low-level and highly optimized C version initially used in [3,4] and continuously improved since then. With the latest versions, the X10 implementation appears to be 2 to 3 times slower than the C version, a consequence of the more complex runtime and the lack of optimization in the compiler. We feel that this is not a prohibitive price to pay, if one takes into account the possibilities promised by X10 for future experimentation.

A possible explanation of the difference between the performances of both implementations is probably the richness of the X10 language (OOP, architecture abstractions, communication abstractions, etc.). Also, maybe it is necessary to improve our X10 language skills good enough to get the best performance of this tool.

4.2   Functional Parallel Performance

Table 1 shows the results of the *first version* of the *functional parallelism* X10 implementation. Only two benchmarks (2 instances of MSP and CAP) are presented. Indeed, we did not investigate this approach any further since the results are clearly not good. Each problem instance was executed with a variable number of activities (THNUM = 2, 4 and 8). It is worth noting, that the environmental X10 variable *X10_NTHREADS* was passed to the program with an appropriate value to each execution. This variable controls the number of initial working threads per place in the X10 runtime system.

As seen in Table 1, for all the treated cases the obtained speed-up is less than 1 (i.e. a slowdown factor), showing a deterioration of the execution time due to this parallel implementation. So, it is possible to conclude that no gain time is obtainable in this approach. To analyze this behavior it is important to return to the description of the Listing 1.3. As already noted, the parallel function *selVarHighCost* in this implementation are located into the main loop of the algorithm, so THNUM activities are created, scheduled and synchronized at each iteration in the program execution, being a very important source of overhead. The results we obtained suggest that this overhead is larger than the improvement obtained by the implementation of this parallel strategy.

Turning to the *second approach*, Table 2 shows the results obtained with this strategy. Equally, the number of activities spawn, in this case at the beginning, was varied from 2 to 8.

Even if the results are slightly better, there is no noticeable speed-up. This is due to a new form of overhead tied to the synchronization mechanism which

Table 1. Functional parallelism – first approach (timings and speed-ups)

Problem instance	Time (s) seq.	Speed-up with k places			Time (s) 8 places
		2	4	8	
MSP-100	11.98	0.86	0.95	0.77	15.49
MSP-120	24.17	1.04	0.97	0.98	24.65
CAP-17	1.56	0.43	0.28	0.24	6.53
CAP-18	12.84	0.51	0.45	0.22	57.16

Table 2. Functional parallelism – second approach (timings and speed-ups)

Problem instance	Time (s) seq.	Speed-up with k places			Time (s) 8 places
		2	4	8	
MSP-100	11.98	1.15	0.80	0.86	13.87
MSP-120	24.17	1.23	0.94	0.63	38.34
CAP-17	1.56	0.56	0.30	0.25	6.35
CAP-18	12.84	0.74	0.39	0.27	46.84

is used in the inner loop of the algorithm, to assign tasks and wait for their termination (see Listing 1.4).

The performance model in X10 [9] specifies that the current implementation of async tasks in X10 has a considerable amount of overhead. Large number of fine grained async tasks are likely to decrease the performance of the application. Actually an implementation of Adaptive Search on GPU using CUDA language, reported by [1] shows that some performance improvement can be achieved, but activities have to be fine-tuned at a low level.

4.3   Data Parallel Performance

Table 3 and Fig. 3 document the speedups we obtained when resorting to data parallelism. Observe that, for this particular set of runs, we used a different hardware platform, with more cores than for the other runs.

The performance of data parallel version is clearly above the performance of the functional parallel version. The resulting average runtime and the speed-ups obtained in the entire experimental test performed seems to lie within the predictable bounds proposed by [24]. The Costas Arrays Problem displays remarkable performance with this strategy, e.g. the CAP reaches a speed-up of 20.5 with 32 places. It can be seen that the speed-up increases almost linearly with the number of used places. However, for other problems (e.g. MSP), the curve clearly tends to flatten out when the number of places increases.

Table 3. Data parallelism (timings and speed-ups)

Problem instance	Time (s) seq.	Speed-up with k places				Time (s) 32 places
		8	16	24	32	
AIP-300	56.7	4.7	7.1	9.9	10.0	5.6
NPP-2300	6.6	6.1	9.8	10.5	12.0	0.5
M3P-200	365	8.3	12.2	13.6	14.6	24.9
CAP-20	731	5.6	12.0	16.1	20.5	35.7

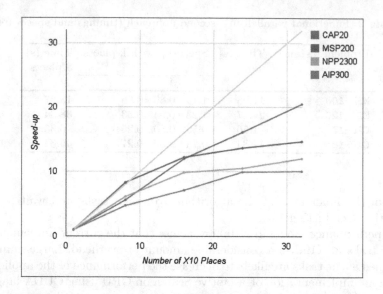

Fig. 3. Speed-ups for the most difficult instance of each problem

5   Conclusion and Future Work

We presented different parallel X10 implementations of an effective Local Search algorithm, Adaptive Search in order to exploit various sources of parallelism. We first experimented two *functional parallelism* versions, i.e. trying to divide the inner loop of the algorithm into various concurrent tasks. This turned out not to yield any speed-up at all, most likely because of the bookkeeping overhead (creation, scheduling and synchronization) that is incompatible with such a fine-grained level of parallelism.

We then proceeded with a data parallel implementation, in which the search space is decomposed into possible different random initial configurations of the problem and getting isolated solver instances to work on each point concurrently. We got a good level of performance for the X10 data-parallel implementation with monotonously increasing speed-ups in all the problems we studied, although they taper off after some point.

The main result we draw from this experiment, is that X10 has proved a suitable platform to exploit parallelism in different ways for constraint-based local search solvers. These entail experimenting with different forms of parallelism, ranging from single shared memory inter-process communication to a distributed memory programming model. Moreover, the use of the X10 implicit communication mechanisms allowed us to abstract away from the complexity of the parallel architecture with a very simple and consistent device: the *distributed arrays* and the *termination detection system* in our data parallel implementation.

Considering that straightforward forms of parallelism seem to get lower gains as we increase the number of cores, we want to look for ways of improving on this situation. Future work will focus on the implementation of a cooperative Local

Search parallel solver based on data parallelism. The key idea being to take advantage of the many communications tools available in the APGAS model, to exchange information between different solver instances in order to obtain a more efficient and, most importantly, scalable solver implementation. We also plan to test the behavior of a cooperative implementation under different HPC architectures, such as the many-core Xeon Phi, GPGPU accelerators and grid computing platforms.

References

1. Arbelaez, A., Codognet, P.: A GPU implementation of parallel constraint-based local search. In: 22nd Euromicro International Conference on Parallel Distributed and Network-Based Processing, Turin, Italy (2014)
2. Butenhof, D.: Programming with Posix Threads. Addison-Wesley Professional, Reading (1997)
3. Codognet, P., Díaz, D.: Yet another local search method for constraint solving. In: Steinhöfel, K. (ed.) SAGA 2001. LNCS, vol. 2264, pp. 73–90. Springer, Heidelberg (2001)
4. Codognet, P., Diaz, D.: An efficient library for solving CSP with local search. In: 5th International Conference on Metaheuristics, Kyoto, Japan, pp. 1–6 (2003)
5. Crainic, T.G., Toulouse, M.: Parallel Meta-Heuristics. In: Gendreau, M., Potvin, J.-Y. (eds.) Handbook of Metaheuristics, vol. 146, pp. 497–541. Springer, New York (2010)
6. Diaz, D., Abreu, S., Codognet, P.: Targeting the cell broadband engine for constraint-based local search. Concurrency Comput.: Pract. Exper. (CCP&E) **24**(6), 647–660 (2011)
7. El-Ghazawi, T., Carlson, W., Sterling, T., Yelick, K.: UPC: Distributed Shared Memory Programming. Wiley, New York (2005)
8. Gent, I.P., Walsh, T.: CSPLib: a benchmark library for constraints. Technical report (1999)
9. Grove, D., Tardieu, O., Cunningham, D., Herta, B., Peshansky, I., Saraswat, V.: A performance model for X10 applications: what's going on under the hood? In: 2011 ACM SIGPLAN X10 Workshop, San Jose, California, pp. 1–8. ACM (2011)
10. Cray Inc. Chapel Language Specification (2012). http://chapel.cray.com
11. Kadioglu, S., Sellmann, M.: Dialectic search. In: Gent, I.P. (ed.) CP 2009. LNCS, vol. 5732, pp. 486–500. Springer, Heidelberg (2009)
12. Khronos OpenCL Working Group. OpenCL Specification (2008). https://www.khronos.org/opencl
13. Machado, R., Abreu, S., Diaz, D.: Parallel local search: experiments with a PGAS-based programming model. In: 12th International Colloquium on Implementation of Constraint and Logic Programming Systems, Budapest, Hungary, pp. 1–17 (2012)
14. Machado, R., Lojewski, C.: The Fraunhofer virtual machine: a communication library and runtime system based on the RDMA model. Comput. Sci. Res. Dev (CSRD) **23**(3–4), 125–132 (2009)
15. Múnera, D., Diaz, D., Abreu, S.: Experimenting with X10 for Parallel constraint-based local search. In: Rocha, R., Theil Have, C. (eds.) Proceedings of the 13th International Colloquium on Implementation of Constraint and LOgic Programming Systems (CICLOPS 2013), August 2013 (2013)

16. NVIDIA. CUDA C Programming Guide (2013). http://docs.nvidia.com/cuda/cuda-c-programming-guide
17. OpenMP. The OpenMP API specification for parallel programming. http://openmp.org
18. Rossi, F., Van Beek, P., Walsh, T. (eds.): Handbook of Constraint Programming. Foundations of Artificial Intelligence, vol. 2. Elsevier Science, New York (2006)
19. Salinesi, C., Mazo, R., Djebbi, O., Diaz, D., Lora-michiels, A.: Constraints: the core of product line engineering. In: Conference on Research Challenges in Information Science (RCIS), Guadeloupe, French West Indies, France, number ii, pp. 1–10 (2011)
20. Saraswat, V., Almasi, G., Bikshandi, G., Cascaval, C., Cunningham, D., Grove, D., Kodali, S., Peshansky, I., Tardieu, O.: The asynchronous partitioned global address space model. In: The First Workshop on Advances in Message Passing, Toronto, Canada, pp. 1–8 (2010)
21. Saraswat, V., Bloom, B., Peshansky, I., Tardieu, O., Grove. D.: X10 language specification - Version 2.3. Technical report (2012). http://x10.sourceforge.net/documentation/languagespec/x10-latest.pdf
22. Schulte, C., Tack, G., Lagerkvist, M.: Modeling and programming with gecode (2013). http://www.gecode.org/
23. Snir, M., Otto, S., Huss-Lederman, S., Walker, D., Dongarra, J.: MPI: The Complete Reference. The MIT Press, Cambridge (1996)
24. Truchet, C., Richoux, F., Codognet, P.: Prediction of parallel speed-ups for las vegas algorithms. In: ICPP'2013, 42nd International Conference on Parallel Processing, Lyon, France, 1–4 October 2013. IEEE Computer Society (2013)
25. UPC Consortium, editor. UPC Language Specifications (2005). http://upc.gwu.edu/docs/upc_specs_1.2.pdf

Debate Games in Logic Programming

Chiaki Sakama(✉)

Department of Computer and Communication Sciences,
Wakayama University, Sakaedani, Wakayama 640-8510, Japan
sakama@sys.wakayama-u.ac.jp

Abstract. A *debate game* provides an abstract model of debates between two players based on the formal argumentation framework. This paper presents a method of realizing debate games in logic programming. Two players have their knowledge bases represented by extended logic programs, and build claims using arguments associated with those programs. A player updates its knowledge base with arguments posed by the opponent player, and tries to refute claims by the opponent. During a debate game, a player may claim false or incorrect arguments as a tactic to win the game. The result of this paper provides a new formulation of debate games in a non-abstract argumentation framework associated with logic programming. Moreover, it provides a novel application of logic programming to modelling social debates which involve argumentative reasoning, belief update and dishonest reasoning.

1 Introduction

Logic programming and argumentation are two different frameworks for knowledge representation and reasoning in artificial intelligence (AI). In his seminal paper, Dung [4] points out a close connection between the two frameworks and shows that a logic program can be considered as a schema for generating arguments. Since then, several attempts have been made for integrating the two frameworks ([1,8,13,21]; see [9] for an overview). A line of research of formal argumentation is concerned with the dialectical process of two or more players who are involved in a discussion [3,11,12,14]. Along this line, Sakama [19] introduces a *debate game* between two players based on the formal argumentation framework. In a debate game, a player makes the initial claim, then the opponent player tries to refute it by building a counter-claim. A debate continues until one cannot refute the other, and the player who makes the last claim wins the game. A debate game has unique features such that (i) each player has its own argumentation framework as its background knowledge, (ii) during a debate each player updates its argumentation framework by new arguments provided by the opponent player, and (iii) a player may claim inaccurate or even false arguments as a tactic to win a debate. The study [19] formulates debate games using the abstract argumentation theory of [4].

The abstract argumentation theory has an advantage that it is not bound to any particular representation for arguments on the one hand, but on the

M. Hanus and R. Rocha (Eds.): KDPD 2013, LNAI 8439, pp. 185–201, 2014.
DOI: 10.1007/978-3-319-08909-6_12, © Springer International Publishing Switzerland 2014

other hand it does not specify how arguments are generated from the underlying knowledge base and what conclusions are yielded by those arguments. In [2] the authors argue that "Argumentation, as it happens in the world around us, is almost never completely abstract.... Instead, the arguments one encounters in daily life consist of *reasons* that support particular *claims*. These reasons can formally be modelled in the form of *rules*, that are instances of underlying *argumentation schemes* [15]." In this respect, debate games based on the abstract argumentation theory need yet another formulation based on non-abstract argumentation frameworks.

With this motivation, this paper uses logic programming as an underlying representation language and formulates debate games in a non-abstract argumentation framework. In this framework, each player has a knowledge base represented by an extended logic program, and builds claims using arguments which can contain information brought by the opponent as well as information in the player's program. During a game, a player may use *dishonest* claims to refute the opponent, while a player must be self-consistent in its claims. The proposed framework provides an abstraction of real-life debates and realizes a formal dialogue system in logic programming. The rest of this paper is organized as follows. Section 2 reviews a framework of argument-based logic programming. Section 3 introduces debate games in logic programming and investigates formal properties. Section 4 discusses related issues and Sect. 5 concludes the paper.

2   Arguments in Logic Programming

In this paper we consider the class of extended logic programs [10]. An *objective literal* is a ground atom B or its explicit negation $\neg B$. We define $\neg\neg B = B$. A *default literal* is of the form $not\,L$ where L is an objective literal and not is *negation as failure* (NAF). An *extended logic program* (or simply a *program*) P is a finite set of *rules* of the form:

$$L_0 \leftarrow L_1, \ldots, L_m, not\,L_{m+1}, \ldots, not\,L_n$$

where each L_i $(0 \leq i \leq n)$ is an objective literal. The literal L_0 is the *head* of the rule and the conjunction $L_1, \ldots, L_m, not\,L_{m+1}, \ldots, not\,L_n$ is the *body* of the rule. A rule r is *believed-true* in P if $r \in P$. A rule containing default literals is called a *default rule*. A rule $L \leftarrow$ with the empty body is also called a *fact* and is identified with a literal L.

Let Lit be the set of all objective literals in the language of a program. A set $S\,(\subset Lit)$ is *consistent* if $L \in S$ implies $\neg L \notin S$ for any $L \in Lit$. The semantics of a program is given by its *answer sets* [10]. First, let P be a program containing no default literal and $S \subset Lit$. Then, S is an *answer set* of P if S is a consistent minimal set satisfying the condition that for each rule of the form $L_0 \leftarrow L_1, \ldots, L_m$ in P, $\{L_1, \ldots, L_m\} \subseteq S$ implies $L_0 \in S$. Second, given any program P (possibly containing default literals) and $S \subset Lit$, a *reduct* of P with respect to S (written P^S) is defined as follows: a rule $L_0 \leftarrow L_1, \ldots, L_m$ is in P^S iff there is a rule of the form $L_0 \leftarrow L_1, \ldots, L_m, not\,L_{m+1}, \ldots, not\,L_n$ in P such that

$\{L_{m+1}, \ldots, L_n\} \cap S = \emptyset$. Then, S is an *answer set* of P if S is an answer set of P^S. A program may have none, one or multiple answer sets in general. A program is *consistent* if it has an answer set; otherwise, it is *inconsistent*.

Definition 2.1. ([13,21]) An *argument* associated with a program P is a finite sequence[1] $A = [r_1; \cdots; r_n]$ of rules $r_i \in P$ such that for every $1 \le i \le n$, for every objective literal L_j in the body of r_i there is a rule r_k $(k > i)$ such that the head of r_k is L_j.

The head of a rule in an argument A is called a *conclusion* of A, and a default literal *not* L in the body of a rule in A is called an *assumption* of A. We write $assum(A)$ for the set of assumptions and $concl(A)$ for the set of conclusions of an argument A. By Definition 2.1, every objective literal in the body of a rule r_i is justified by the consequence of a rule that appears later in the sequence. For instance, $[p \leftarrow q,\ not\ r;\ q \leftarrow]$ is an argument but $[p \leftarrow q;\ q \leftarrow p]$ is not. A *subargument* of A is a subsequence of A which is an argument. An argument A with a conclusion L is a *minimal argument for* L if there is no subargument of A with the conclusion L. An argument is *minimal* if it is minimal for some literal L. The minimality condition presents that an argument does not include rules which do not contribute to concluding some particular literal L.

Remark: In this paper, we slightly abuse the notation and use the same capital letter A to denote the *set* of rules included in an argument A. Thus, $P \cup A$ means the set of rules included either in a program P or in an argument A.

Example 2.1. Let P be the program:

$$p \leftarrow q,$$
$$\neg p \leftarrow not\ q,$$
$$q \leftarrow,$$
$$r \leftarrow s.$$

Then, the following facts hold.

- The minimal argument for p is $A_1 = [p \leftarrow q;\ q \leftarrow]$, $concl(A_1) = \{p, q\}$, and $assum(A_1) = \emptyset$.
- The minimal argument for $\neg p$ is $A_2 = [\neg p \leftarrow not\ q]$, $concl(A_2) = \{\neg p\}$ and $assum(A_2) = \{not\ q\}$.
- The minimal argument for q is $A_3 = [q \leftarrow]$, $concl(A_3) = \{q\}$ and $assum(A_3) = \emptyset$.
- r and s have no minimal arguments.

Proposition 2.1. *Let P be a consistent program containing no default literal. Then, for any argument A associated with Γ, $concl(A) \subseteq S$ holds for the answer set S of P.*

---

[1] Rules are separated by semicolons in a sequence A of rules, while they are separated by commas in a set P of rules.

Proof. Let P' be the program which is obtained by replacing every negative literal $\neg L$ in P with a new atom L' that is uniquely associated with $\neg L$. As P is consistent, P' has the least model S' iff P has the answer set S where $\neg L$ in S is replaced by the atom L' in S'. Let A' be an argument associated with P'. Viewing A' as the set of rules included in it, it holds that $A' \subseteq P'$ which implies $concl(A') \subseteq S'$ by the monotonicity of deduction. By replacing L' with $\neg L$, $A \subseteq P$ implies $concl(A) \subseteq S$. □

Definition 2.2. ([13, 21]) Let A_1 and A_2 be two arguments.

- A_1 *undercuts* A_2 if there is an objective literal L such that L is a conclusion of A_1 and *not L* is an assumption of A_2.
- A_1 *rebuts* A_2 if there is an objective literal L such that L is a conclusion of A_1 and $\neg L$ is a conclusion of A_2.
- A_1 *attacks* A_2 if A_1 undercuts or rebuts A_2.
- A_1 *defeats* A_2 if A_1 undercuts A_2, or A_1 rebuts A_2 and A_2 does not undercut A_1.

An argument A is *coherent* if it does not attack itself; otherwise, A is *incoherent*. A set S of arguments is *conflict-free* if no argument in S attacks an argument in S. Given a program P, we denote the set of minimal and coherent arguments associated with P by $Args(P)$.

If an argument A_1 undercuts another argument A_2, then A_1 denies an assumption of A_2. This means that the assumption conflicts with the evidence to the contrary, and A_1 defeats A_2 in this case. If A_1 rebuts A_2, on the other hand, two arguments support contradictory conclusions. In this case, the attack relation is symmetric and A_1 defeats A_2 under the condition that A_2 does not undercut A_1. The coherency condition presents self-consistency of an argument. By definition, if $A \in Args(P)$ then the set A of rules is consistent, that is, A has an answer set.

Example 2.2. In the program P of Example 2.1, the following facts hold.

- $Args(P) = \{A_1, A_2, A_3\}$.
- A_1 and A_3 undercut (and also defeat) A_2.
- A_1 rebuts A_2 and A_2 rebuts A_1.
- $\{A_1, A_3\}$ is conflict-free, but $\{A_1, A_2\}$ and $\{A_2, A_3\}$ are not.
- The argument $A_4 = [\neg p \leftarrow not\, q\,;\, q \leftarrow]$ is incoherent.

Proposition 2.2. *Let P be a consistent program. For any argument $A \in Args(P)$, let A^+ be the set of rules obtained from A by removing every default literal in A. Then, $concl(A) = concl(A^+)$.*

Proof. Since A is coherent, $not\, L \in assum(A)$ implies $L \notin concl(A)$. Then, $concl(A) = concl(A^+)$. □

Proposition 2.3. *Let P be a consistent program. For any argument $A \in Args(P)$, if A is not defeated by any argument associated with P, then $concl(A) \subseteq S$ for any answer set S of P.*

Proof. Let A^+ be the set of rules obtained from A by removing every default literal in A. When A is not defeated by any argument associated with P, $A^+ \subseteq A^S$ for any answer set S of P, where A^S is the reduct of A wrt S.[2] By Proposition 2.1, for any argument A^S associated with P^S, $concl(A^S) \subseteq S$ for any answer set S of P^S. Since $A^+ \subseteq A^S$ implies $concl(A^+) \subseteq concl(A^S)$, it holds that $concl(A^+) \subseteq S$. As $concl(A) = concl(A^+)$ by Proposition 2.2, the result holds. $\qquad\square$

In Example 2.1, A_1 and A_3 are defeated by no argument, then $concl(A_1)$ and $concl(A_3)$ are subsets of the answer set $\{p, q\}$ of P.

3   Debate Games in Logic Programming

3.1   Debate Games

A debate game involves two players. Each player has its knowledge base defined as follows.

Definition 3.1 (player). A *player* has a knowledge base $K = (P, O)$ where P is a consistent program representing the player's belief and O is a set of rules brought by another player. In particular, the initial knowledge base of a player is $K = (P, \emptyset)$.

In this paper, we identify a player with its knowledge base. We represent two players by K_1 and K_2. For a player K_1 (resp. K_2), the player K_2 (resp. K_1) is called the *opponent*.

Definition 3.2 (update). Let $K = (P, O)$ be a player and A an argument. Then, the *update* of K with A is defined as[3]

$$u(K, A) = (P, O \cup A).$$

The function u is iteratively applied to a player. We represent the result of the i-th update of K by $K^i = (P, O^i)$ $(i \geq 0)$, that is, $K^i = (P, O^i) = u(K^{i-1}, A_i)$ $(i \geq 1)$ for arguments $A_1, \ldots, A_i$ and $K^0 = (P, O^0) = (P, \emptyset)$.

During a debate, a player incrementally obtains new information from the opponent player. The situation is realized by a series of updates. By definition, update adds rules A to O while it does not change P. The reason of separating P and O is to distinguish belief originated in a player's program from information brought by the opponent player. A player having a knowledge base after the i-th update is represented by K^i, but we often omit the superscript i when it is unimportant in the context.

---

[2] Here, A is viewed as a set of rules.
[3] Note that the sequence A is treated as a set here.

Definition 3.3 (claim). Let $K_1 = (P_1, O_1)$ and $K_2 = (P_2, O_2)$ be two players.

1. The *initial claim* is a pair of the form: $(\text{in}(A), \_)$ where $A \in Args(P_1)$. It is read that "the player K_1 claims the argument A."
2. A *counter-claim* is a pair of the form: $(\text{out}(B), \text{in}(A))$ where $A \in Args(P_k \cup O_k)$ and $B \in Args(P_l \cup O_l)$ $(k, l = 1, 2; k \neq l)$. It is read that "the argument B by the player K_l does not hold because the player K_k claims the argument A".

The initial claim or counter-claims are simply called *claims*. A claim $(\text{in}(A), \_)$ or $(\text{out}(B), \text{in}(A))$ by a player is *refuted* by the claim $(\text{out}(A), \text{in}(C))$ with some argument C by the opponent player.

Definition 3.4 (debate game). Let $K_1^0 = (P_1, O_1^0)$ and $K_2^0 = (P_2, O_2^0)$ be two players. Then, an *admissible debate* Δ is a sequence of claims: $[(\text{in}(X_0), \_), (\text{out}(X_0), \text{in}(Y_1)), (\text{out}(Y_1), \text{in}(X_1)), \ldots, (\text{out}(X_i), \text{in}(Y_{i+1})), (\text{out}(Y_{i+1}), \text{in}(X_{i+1})), \ldots]$ such that

(a) $(\text{in}(X_0), \_)$ is the initial claim by K_1^0 where $X_0 \in Args(P_1)$.
(b) $(\text{out}(X_0), \text{in}(Y_1))$ is a claim by K_2^1 where $K_2^1 = u(K_2^0, X_0) = (P_2, O_2^1)$ and $Y_1 \in Args(P_2 \cup O_2^1)$.
(c) $(\text{out}(Y_{i+1}), \text{in}(X_{i+1}))$ is a claim by K_1^{i+1} where $K_1^{i+1} = u(K_1^i, Y_{i+1}) = (P_1, O_1^{i+1})$ and $X_{i+1} \in Args(P_1 \cup O_1^{i+1})$ $(i \geq 0)$.
(d) $(\text{out}(X_i), \text{in}(Y_{i+1}))$ is a claim by K_2^{i+1} where $K_2^{i+1} = u(K_2^i, X_i) = (P_2, O_2^{i+1})$ and $Y_{i+1} \in Args(P_2 \cup O_2^{i+1})$ $(i \geq 0)$.
(e) for each $(\text{out}(U), \text{in}(V))$, V defeats U.
(f) for each $\text{out}(Z)$ in a claim by K_1^i (resp. K_2^i), there is $\text{in}(Z)$ in a claim by K_2^j such that $j \leq i$ (resp. K_1^j such that $j < i$).
(g) both $\bigcup_{i \geq 0} \{ X_i \mid X_i \subseteq P_1 \}$ and $\bigcup_{j \geq 1} \{ Y_j \mid Y_j \subseteq P_2 \}$ are conflict-free.

Let Γ_n $(n \geq 0)$ be any claim. A *debate game* Δ *(for an argument X_0)* is an admissible debate between two players $[\Gamma_0, \Gamma_1, \ldots]$ where the initial claim is $\Gamma_0 = (\text{in}(X_0), \_)$. A debate game Δ for an argument X_0 *terminates* with Γ_n if $\Delta = [\Gamma_0, \Gamma_1, \ldots, \Gamma_n]$ is an admissible debate and there is no claim Γ_{n+1} such that $[\Gamma_0, \Gamma_1, \ldots, \Gamma_n, \Gamma_{n+1}]$ is an admissible debate. In this case, the player who makes the last claim Γ_n *wins* the game.

By definition, (a) the player K_1^0 starts a debate with the claim $\Gamma_0 = (\text{in}(X_0), \_)$. (b) The player K_2^0 then updates its knowledge base with X_0, and responds to the player K_1^0 with a counter-claim $\Gamma_1 = (\text{out}(X_0), \text{in}(Y_1))$ based on the updated knowledge base K_2^1. In response to Γ_1, the player K_1^1 updates its knowledge base and builds a counter-claim $\Gamma_2 = (\text{out}(Y_1), \text{in}(X_1))$. A debate continues by iterating updates and claims $((c),(d))$. (e) In each claim an argument V of $\text{in}(V)$ defeats an argument U of $\text{out}(U)$. (f) A player can refute not only the last claim of the opponent player, but any previous claim of the opponent. (g) During a debate game, arguments which come from a player's own program must be conflict-free, that is, each player must be self-consistent in its claims. Note that a player

K_l^i ($l = 1, 2$; $i \geq 1$) can construct arguments using rules included in arguments O_l^i posed by the opponent player as well as rules in its own program P_l. This means that conclusions of arguments claimed by a player may change nonmonotonically during a game. If a player K_l^i claims $(\text{out}(A), \text{in}(B))$ which is refuted by a counter-claim $(\text{out}(B), \text{in}(C))$ by the opponent, then the player K_l^j ($i < j$) can use rules in the argument C for building a claim. Once the player K_l^j uses rules in C, it may be the case that K_l^j withdraws some conclusions of the argument B previously made by K_l^i. Thus, two different claims by the same player may conflict during a game. The condition (g) states that such a conflict is not allowed among arguments which consist of rules from a player's original program P.

Example 3.1. Suppose a dispute between a prosecutor and a defense. First, the prosecutor and the defense have knowledge bases $K_1^0 = (P_1, \emptyset)$ and $K_2^0 = (P_2, \emptyset)$, respectively, where

$$P_1 : \; guilty \leftarrow suspect, motive,$$
$$evidence \leftarrow witness, not \neg credible,$$
$$suspect \leftarrow, \quad motive \leftarrow, \quad witness \leftarrow .$$
$$P_2 : \; \neg guilty \leftarrow suspect, not \; evidence,$$
$$\neg credible \leftarrow witness, dark,$$
$$suspect \leftarrow, \quad dark \leftarrow .$$

A debate game proceeds as follows.

- First, the prosecutor K_1^0 makes the initial claim $\Gamma_0 = (\text{in}(X_0), \_)$ with

  $$X_0 = [\, guilty \leftarrow suspect, motive; \;\; suspect \leftarrow; \;\; motive \leftarrow \,]$$
  ("The suspect is guilty because he has a motive for the crime.")

  where $X_0 \subseteq Args(P_1)$.

- The defense updates K_2^0 to $K_2^1 = u(K_2^0, X_0) = (P_2, O_2^1)$ where $O_2^1 = X_0$, and makes the counter-claim $\Gamma_1 = (\text{out}(X_0), \text{in}(Y_1))$ with

  $$Y_1 = [\, \neg guilty \leftarrow suspect, not \; evidence; \;\; suspect \leftarrow \,]$$
  ("The suspect is not guilty as there is no evidence.")

  where $Y_1 \in Args(P_2 \cup O_2^1)$ and Y_1 rebuts X_0.

- The prosecutor updates K_1^0 to $K_1^1 = u(K_1^0, Y_1) = (P_1, O_1^1)$ where $O_1^1 = Y_1$, and makes the counter-claim $\Gamma_2 = (\text{out}(Y_1), \text{in}(X_1))$ with

  $$X_1 = [\, evidence \leftarrow witness, not \neg credible; \;\; witness \leftarrow \,]$$
  ("There is an eyewitness who saw the suspect on the night of the crime.")

  where $X_1 \in Args(P_1 \cup O_1^1)$ and X_1 undercuts Y_1.

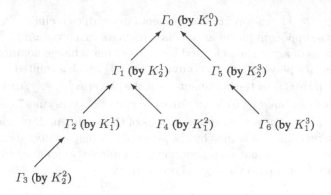

Fig. 1. Debate game

- The defense updates K_2^1 to $K_2^2 = u(K_2^1, X_1) = (P_2, O_2^2)$ where $O_2^2 = X_0 \cup X_1$, and makes the counter-claim $\Gamma_3 = (\text{out}(X_1), \text{in}(Y_2))$ with

$$Y_2 = [\neg credible \leftarrow witness, dark; \; witness \leftarrow; \; dark \leftarrow]$$

("The testimony is incredible because it was dark at night.")

where $Y_2 \in Args(P_2 \cup O_2^2)$ and Y_2 undercuts X_1.

- The prosecutor updates K_1^1 to $K_1^2 = u(K_1^1, Y_2) = (P_1, O_1^2)$ where $O_1^2 = Y_1 \cup Y_2$, but cannot refute the claim by K_2^2. As a result, the defense is a winner of the game $\Delta = [\Gamma_0, \Gamma_1, \Gamma_2, \Gamma_3]$.

Suppose another game in which the defense makes the initial claim $\Gamma_1' = (\text{in}(Y_1), \_)$. In this case, a debate proceeds as $\Delta' = [\Gamma_1', \Gamma_2, \Gamma_3]$, and the defense also wins the game Δ'. On the other hand, a sequence $[(\text{in}(X_1), \_), (\text{out}(X_1), \text{in}(Y_2)), (\text{out}(Y_2), \text{in}(X_0)), (\text{out}(X_0), \text{in}(Y_1))]$ is not a debate game because the argument X_0 does not defeat the argument Y_2 (i.e., $(\text{out}(Y_2), \text{in}(X_0))$ does not satisfy the condition of Definition 3.4(e)). Thus, a player may have options for submitting arguments, but the player needs to select the most effective ones to win a debate.

A debate game is represented as a directed tree in which the root node represents the initial claim, each node represents a claim, and there is a directed edge between two nodes Γ_i and Γ_j if the former refutes the latter. Figure 1 represents a debate game $\Delta = [\Gamma_0, \Gamma_1, \dots, \Gamma_6]$ in which the player K_1^0 makes the initial claim Γ_0, the player K_2^1 makes a counter-claim Γ_1, the player K_1^1 refutes Γ_1 by Γ_2, and the player K_1^2 refutes Γ_2 by Γ_3. At this stage, K_1^2 cannot refute Γ_3 but refutes Γ_1 by Γ_4. The player K_2^3 cannot refute Γ_4 but refutes Γ_0 by Γ_5. Then, K_1^3 refutes Γ_5 by Γ_6. The player K_2^4 cannot refute Γ_6 and other claims by the opponent. As a result, the player K_1^3 wins the game. In what follows, we simply say "a debate game" instead of "a debate game for an argument X_0" when the argument X_0 in the initial claim is clear or unimportant in the context.

Proposition 3.1. *Let Γ be a claim of either $(\text{in}(U), \_)$ or $(\text{out}(V), \text{in}(U))$ in a debate game. Then, U has a single answer set S such that $concl(U) = S$ and $concl(V) \not\subseteq S$.*

Proof. Since U is minimal and coherent, U has a single answer set S such that $concl(U) = S$. As U defeats V, there is a rule $r \in V$ such that the head of r is not included in S. □

Proposition 3.2. *A debate game Δ terminates if $\Gamma_i \neq \Gamma_{i+2k}$ $(k = 1, 2, \ldots)$ for any Γ_i $(i \geq 1)$ in Δ.*

Proof. In case of $\Gamma_i \neq \Gamma_{i+2k}$, each player cannot repeat the same claim in a debate game. Since the number of minimal and coherent arguments associated with a propositional program is finite, the result holds. □

A debate may not terminate when arguments go round in circles.

Example 3.2. Suppose a debate game between two players such that

Γ_0: $(\text{in}([p \leftarrow not\, q]), \_)$
Γ_1: $(\text{out}([p \leftarrow not\, q]), \text{in}([q \leftarrow not\, p]))$
Γ_2: $(\text{out}([q \leftarrow not\, p]), \text{in}([p \leftarrow not\, q]))$
$\Gamma_{k+1} = \Gamma_{k-1}$ $(k \geq 2)$.

Such a debate does not terminate and there is no winner of the game.[4]

3.2 Dishonest Player

In debate games, each player constructs claims using rules included in its program or rules brought by the opponent. To defeat a claim by the opponent, a player may claim an argument which the player does not believe its conclusion.

Example 3.3. Suppose that the prosecutor in Example 3.1 has the program

$$P_1' = P_1 \cup \{\neg dark \leftarrow light, not\, broken, \quad light \leftarrow, \quad broken \leftarrow \}.$$

In response to the last claim $\Gamma_3 = (\text{out}(X_1), \text{in}(Y_2))$ by the defense K_2^2, suppose that the prosecutor $K_1^2 = (P_1', O_1^2)$ makes the counter-claim $(\text{out}(Y_2), \text{in}(X_2))$ with

$$X_2 = [\neg dark \leftarrow light, not\, broken; \quad light \leftarrow].$$

("It was not dark because the witness saw the suspect under the light of the victim's apartment"). Then, X_2 defeats Y_2.

In Example 3.3, the prosecutor K_1^2 claims the argument X_2 but he/she does not believe its conclusion $concl(X_2)$. In fact, $\neg dark$ is included in no answer set of the program $P_1' \cup Q$ for any $Q \subseteq O_1^2$. Generally, a player may behave dishonestly by concealing believed facts to justify another fact which the player wants to conclude. We classify different types of claims which may appear in a debate game.

---

[4] A well-known example of this type is: "which came first, the chicken or the egg?"

Definition 3.5 (credible, misleading, incredible, incorrect, false claims).
Let Γ be a claim of either $(\text{in}(U), \_)$ or $(\text{out}(V), \text{in}(U))$ by a player $K_l^i = (P_l, O_l^i)$ $(l = 1, 2; i \geq 0)$. Also, let U^S be an argument which consists of rules in the reduct of U with respect to a set S.

- Γ is *credible* if $concl(U) \subseteq S$ for every answer set S of $P_l \cup Q$ for some $Q \subseteq O_l^i$ such that $P_l \cup Q$ is consistent and $concl(U) = concl(U^S)$.
- Γ is *misleading* if $concl(U) \subseteq S$ for every answer set S of $P_l \cup Q$ for some $Q \subseteq O_l^i$ such that $P_l \cup Q$ is consistent but $concl(U) \neq concl(U^S)$.
- Γ is *incredible* if $concl(U) \subseteq S$ for some (but not every) answer set S of $P_l \cup Q$ for any $Q \subseteq O_l^i$ such that $P_l \cup Q$ is consistent.
- Γ is *incorrect* if $concl(U) \not\subseteq S$ for any answer set S of $P_l \cup Q$ for any $Q \subseteq O_l^i$ such that $P_l \cup Q$ is consistent, and $concl(U) \cup S$ is consistent for some answer set S of $P_l \cup Q$ for some $Q \subseteq O_l^i$ such that $P_l \cup Q$ is consistent.
- Γ is *false* if $concl(U) \cup S$ is inconsistent for any answer set S of $P_l \cup Q$ for any $Q \subseteq O_l^i$ such that $P_l \cup Q$ is consistent.

A claim is called *dishonest* if it is not credible. A player K_l is *honest* in a debate game Δ if every claim made by K_l^i $(i \geq 0)$ in Δ is credible. Otherwise, K_l is *dishonest*.

During a game, a player K_l^i constructs an argument U using some rules $Q \subseteq O_l^i$. Then, U has the answer set which coincides with $concl(U)$ (Proposition 3.1), but this does not always imply that $concl(U)$ is a subset of an answer set of P_l^i. A honest player makes a claim in which rules are properly used in its argument and the player believes the conclusions of his/her own claim. By contrast, a dishonest player may make a claim in which rules are misused in its argument or the player does not believe conclusions of his/her own claim.

Proposition 3.3. *Every claim in a debate game is classified as one of the five types of claims of Definition 3.5.*

Example 3.4.

- Given $K_1 = (\{p \leftarrow not\, q\}, \emptyset)$, the claim $\Gamma_1 = (\text{in}([p \leftarrow not\, q]), \_)$ is credible.
- Given $K_2 = (\{p \leftarrow not\, q, \quad p \leftarrow q, \quad q \leftarrow\}, \emptyset)$, the claim $\Gamma_2 = (\text{in}([p \leftarrow not\, q]), \_)$ is misleading.
- Given $K_3 = (\{p \leftarrow not\, q, \quad q \leftarrow not\, p\}, \emptyset)$, the claim $\Gamma_3 = (\text{in}([p \leftarrow not\, q]), \_)$ is incredible.
- Given $K_4 = (\{p \leftarrow not\, q, \quad q \leftarrow\}, \emptyset)$, the claim $\Gamma_4 = (\text{in}([p \leftarrow not\, q]), \_)$ is incorrect.
- Given $K_5 = (\{p \leftarrow not\, \neg p, \quad \neg p \leftarrow\}, \emptyset)$, the claim $\Gamma_5 = (\text{in}([p \leftarrow not\, \neg p]), \_)$ is false.

In Example 3.4, Γ_1 is credible because $concl([p \leftarrow not\, q]) = \{p\}$ coincides with the answer set of the program $\{p \leftarrow not\, q\}$ in K_1. By contrast, Γ_2 is misleading because for $U = [p \leftarrow not\, q]$ it becomes $U^S = \emptyset$ by the answer set $S = \{p, q\}$ of the program in K_2, so that $concl(U) \neq concl(U^S)$. That is, a misleading claim does not use rules in a proper manner to reach conclusions.[5] Γ_3 is incredible because p is included in some but not in every answer set of the program in K_3. Γ_4 is incorrect because p is included in no answer set of the program in K_4. Γ_5 is false because $\neg p$ is included in every answer set of the program in K_5, which is inconsistent with $concl([p \leftarrow not\ \neg p]) = \{p\}$.

The existence of dishonest claims is due to the nonmonotonic nature of a program. A player $K = (P, O)$ is *monotonic* if P contains no default literal. In this case, the following result holds.

Proposition 3.4. *Let Δ be a debate game between two monotonic players. Then, every claim in Δ is credible.*

Proof. Let Γ be a claim of either $(\text{in}(U), \_)$ or $(\text{out}(V), \text{in}(U))$ by a player $K = (P, O)$. By $U \in Args(P \cup O)$, $U \subseteq P \cup Q$ for some $Q \subseteq O$ such that $P \cup Q$ is consistent. Since U is an argument associated with $P \cup Q$, $concl(U) \subseteq S$ holds for the answer set S of $P \cup Q$ by Proposition 2.1. By $U^S = U$, $concl(U) = concl(U^S)$. Hence, Γ is credible.   □

In other words, dishonest behavior requires intelligence of performing non-monotonic (commonsense) reasoning.

Generally, it is unknown which player wins a debate game. In real life, a player who is more knowledgeable than another player is likely to win a debate. The situation is formulated as follows.

Proposition 3.5. *Let Δ be a debate game between two players $K_1^0 = (P_1, \emptyset)$ and $K_2^0 = (P_2, \emptyset)$.*

1. *Suppose that K_1 is honest and $P_2 \subset P_1$. If Δ terminates, K_1^i ($i \geq 1$) wins the game.*
2. *Suppose that K_2 is honest, $P_1 \subset P_2$ and the claim $(\text{out}(X_0), \text{in}(Y_1))$ by K_2^1 refutes the initial claim $(\text{in}(X_0), \_)$ by K_1^0. If Δ terminates, K_2^i ($i \geq 1$) wins the game.*

Proof. (1) Suppose that K_1 is honest and $P_2 \subset P_1$. Let Γ_m be a honest claim of either $(\text{in}(X_0), \_)$ or $(\text{out}(Y_i), \text{in}(X_i))$ by $K_1^i = (P_1, O_1^i)$ ($i \geq 0$) in Δ. Then, $concl(X_i) \subseteq S$ for every answer set S of P_1. Suppose that K_2^{i+1} makes a counter-claim $\Gamma_{m+1} = (\text{out}(X_i), \text{in}(Y_{i+1}))$, and K_1^{i+1} cannot refute Γ_{m+1} by any honest claim. In this case, P_1 has no rule to defeat Y_{i+1}. By $P_2 \subset P_1$, it holds that $Y_{i+1} \subset P_1$. Since Y_{i+1} defeats X_i, either (i) Y_{i+1} undercuts X_i or (ii) Y_{i+1} rebuts X_i but X_i does not undercut Y_{i+1}. In either case, $concl(X_i) \nsubseteq S$ for any answer set S of P_1. This contradicts the fact that $concl(X_i) \subseteq S$. Hence, K_1^{i+1} can refute Γ_{m+1} by a honest claim in Δ. As such, every claim by K_2^i is honestly refuted by K_1^i. Then, if Δ terminates, K_1^i wins the game.

(2) Suppose that K_2 is honest and $P_1 \subset P_2$. By the assumption, the honest claim $(\text{out}(X_0), \text{in}(Y_1))$ by K_2^1 refutes the initial claim $(\text{in}(X_0), \_)$ by K_1^0. In

---

[5] An example of K_2 is found in the famous speech by John F. Kennedy in September 12, 1962. "We choose to go to the moon in this decade and do the other things, not because they are easy, but because they are hard." Put $p = gotoMoon$ and $q = hard$.

this case, $concl(Y_1) \subseteq S$ for every answer set S of P_2. Then, we can show that K_2^i ($i \geq 1$) wins the game whenever Δ terminates in a way similar to (1).   □

Proposition 3.5 presents that if a player K_1 has information more than K_2, then K_1 has no reason to play dishonestly to win a game. By contrast, if K_2 has information more than K_1, then K_2 may have reason to play dishonestly to win a game.

Example 3.5. Consider two players $K_1^0 = (P_1, \emptyset)$ and $K_2^0 = (P_2, \emptyset)$ where $P_1 = \{p \leftarrow not\, q\}$ and $P_2 = \{p \leftarrow not\, q,\ q \leftarrow not\, r,\ r \leftarrow\}$. Then, $P_1 \subset P_2$. If K_1^0 claims $\Gamma_0 = (\text{in}([p \leftarrow not\, q]), \_)$, then K_2^1 has no honest counter-claim against Γ_0. The only possibility to win the game for the player K_2^1 is making an incorrect claim $\Gamma_1 = (\text{out}([p \leftarrow not\, q]), \text{in}([q \leftarrow not\, r]))$ against Γ_0.

In Example 3.5, the player K_2 agrees with the initial claim Γ_0 made by the opponent K_1. Given Γ_0, the player K_2 has two options: (i) accept the claim Γ_0 and stop the debate, or (ii) make an incorrect counter-claim Γ_1 and continue the debate. In case of (i), K_2 just loses the game. In case of (ii), K_2 can test how knowledgeable the opponent K_1 is. If K_1^1 makes a counter-claim $\Gamma_2 = (\text{out}([q \leftarrow not\, r]), \text{in}([r \leftarrow]))$ against Γ_1, then K_2 realizes that K_1 is at least as knowledgeable as K_2. Else if there is no counter-claim by K_1^1, then K_2 realizes that K_1 is less knowledgeable than K_2 and K_2 wins the game. As such, a dishonest claim is used for testing whether the opponent player is knowledgeable enough to refute the claim successfully. A similar situation happens in real life when one agrees with the other's argument while just challenges the argument to see justification.

A player has an incentive to build a dishonest claim if the player cannot build a honest counter-claim in response to the claim by the opponent. Then, our next question is how a player effectively uses dishonest claims as a tactic to win a debate. We first show that among different types of dishonest claims, misleading claims are useless for the purpose of winning a debate.

Proposition 3.6. *Let Δ be a debate game between two players $K_1^0 = (P_1, \emptyset)$ and $K_2^0 = (P_2, \emptyset)$.*

1. *If the initial claim $\Gamma_0 = (\text{in}(X_0), \_)$ by K_1^0 is misleading, there is a credible claim $\Gamma_0' = (\text{in}(X), \_)$ by K_1^0 such that $concl(X) = concl(X_0)$.*
2. *If a claim $\Gamma_k = (\text{out}(V), \text{in}(U))$ by a player K_l^i ($l = 1, 2$; $i \geq 1$) is misleading, there is a credible claim $\Gamma_k' = (\text{out}(V), \text{in}(W))$ by K_l^i such that $concl(W) = concl(U)$.*

Proof. (1) Since $concl(X_0) \subseteq S$ for every answer set S of P_1, there is a set $X \subseteq P_1$ of rules such that $concl(X_0) = concl(X) = concl(X^S)$. By selecting a minimal set X of rules satisfying the conditions of Definition 2.1, the result holds. (2) Since $concl(U) \subseteq S$ for every answer set S of $P_l \cup Q$ for some $Q \subseteq O_l^i$ such that $P_l \cup Q$ is consistent, there is a set $W \subseteq P_l \cup Q$ of rules such that $concl(U) = concl(W) = concl(W^S)$. By selecting a minimal set W of rules satisfying the conditions of Definition 2.1, the result holds.   □

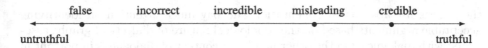

Fig. 2. Degree of truthfulness

Thus, dishonest claims which are effectively used for the purpose of winning a debate game are either incredible, incorrect or false claims. Once a player makes a dishonest claim in a game, however, it will restrict what the player can claim later in the game.

Example 3.6. Consider two players $K_1^0 = (P_1, \emptyset)$ and $K_2^0 = (P_2, \emptyset)$ where $P_1 = \{p \leftarrow not\, q, \quad r \leftarrow s, \quad s \leftarrow not\, t\}$ and $P_2 = \{q \leftarrow not\, r, not\, t, \quad t \leftarrow\}$. Suppose the following debate between K_1 and K_2:

Γ_0: $(\text{in}([p \leftarrow not\, q]), \_)$
Γ_1: $(\text{out}([p \leftarrow not\, q]), \text{in}([q \leftarrow not\, r, not\, t]))$
Γ_2: $(\text{out}([q \leftarrow not\, r, not\, t]), \text{in}([r \leftarrow s; s \leftarrow not\, t]))$.

Note that the claim Γ_1 by K_2^1 is incorrect. The player K_2^2 cannot refute Γ_2 because the counter-claim $\Gamma_3 = (\text{out}([r \leftarrow s; s \leftarrow not\, t]), \text{in}([t \leftarrow]))$ against Γ_2 conflicts with the claim Γ_1 by K_2^1 (i.e., "$t \leftarrow$" attacks "$q \leftarrow not\, r, not\, t$"). Thus, Γ_3 violates the condition of Definition 3.4(g).

To keep conflict-freeness of a player's claims in a game, dishonest claims would restrict the use of believed-true rules in later claims and may result in a net loss of freedom in playing the game. With this reason, it seems reasonable to select a dishonest claim only if there is no choice among honest claims. Comparing different types of dishonest claims, it is considered that incredible claims are preferred to incorrect claims, and incorrect claims are preferred to false claims. If a claim $\Gamma = (\text{out}(V), \text{in}(U))$ is incredible, the player does not *skeptically* believe the conclusion of U but *credulously* believes the conclusion of U. If Γ is incorrect, the player does not credulously believe the conclusion of U but the conclusion is consistent with the player's belief. If Γ is false, on the other hand, the conclusion of U is inconsistent with the player's belief. Thus, the degree of truthfulness (against the belief state of a player) decreases from incredible claims to incorrect claims, and from incorrect claims to false claims (Fig. 2). Generally, a dishonest claim deviates from the reality as believed by a player, and a claim which increases such deviation is undesirable for a player because it increases a chance of making the player's claims conflict. A player wants to keep claims close to its own belief as much as possible, so the best-practice strategy for a debate game is to firstly use credible claims, secondly use incredible ones, thirdly use incorrect ones, and finally use false ones to refute the opponent.

4   Discussion

A formal argumentation framework has been used for modelling dialogue games or discussion games ([3,11,12,14]; and references therein). However, most of the studies use abstract argumentation and pay much attention on identifying acceptable arguments based on the topological nature of dialectical graphs associated with dialogues. On the other hand, the content of dialogue is important in human communication. Participants in debates are interested in why one's argument is defeated by the opponent, whether arguments made by the opponent are logically consistent, which arguments made by the opponent are unacceptable, and so on. In debate games proposed in this paper, each player can see the *inside* of the arguments in claims made by the opponent. As a result, a player can judge whether a counter-claim made by the opponent is grounded on evidences, and whether claims made by the opponent are consistent throughout a debate. Moreover, a player can obtain new information from arguments posed by the opponent.

In AI agents are usually assumed to be honest and little attention has been paid for representing and reasoning with dishonesty. In real-life debates, however, it is a common practice for one to misstate their beliefs or opinions [20]. In formal argumentation, [16] characterizes dishonest agents in a game-theoretic argumentation mechanism design and [19] introduces dishonest arguments in a debate game. These studies use the abstract argumentation framework and do not show how to construct dishonest arguments from the underlying knowledge base. In this paper, we show how to build dishonest arguments from a knowledge base represented by a logic program. Using arguments associated with logic programs, we argue that at least four different types of dishonest claims exist. In building dishonest claims, default literals play an important role—concealing known rules or facts could produce conclusions which are not believed by a player. Proposition 3.4 shows an interesting observation that players cannot behave dishonestly without default assumption. Dishonest reasoning in logic programs is studied by [17] in which the notion of *logic programs with disinformation* is introduced and its computation by abductive logic programming is provided. An application of dishonest reasoning to multiagent negotiation is provided by [18] in which agents represented by abductive logic programs misstate their bargaining positions to gain one's advantage over the other. The current study shows yet another application of dishonest reasoning in argumentation-based logic programming.

Prakken and Sartor [13] introduce *dialogue trees* in order to provide a proof theory of argumentation-based extended logic programs. A dialogue tree consists of nodes representing arguments by the proponent and the opponent, and edges representing attack relations between arguments. Given the initial argument of the proponent at the root node of a dialogue tree, the opponent attacks the argument by a counterargument if any (called a *move*). Two players move in turn and one player wins a dialogue if the other player run out of moves in a tree. Comparing dialogue trees with debate games, a dialogue tree is constructed by arguments associated with a *single* extended logic program. In debate games, on the other hand, two players have different knowledge bases and build arguments associated with them. Dialogue trees are introduced to provide a proof theory of argumentation-based logic programs, and they do not intend to provide a formal theory of dialogues between two players. As a result, dialogue trees do not have mechanisms of update and dishonest reasoning. Fan and Toni [6] propose a formal model for argumentation-based dialogues between agents. They use *assumption-based argumentation* (ABA) [5] in which arguments are built from rules and supported by assumptions. In ABA attacks against arguments are directed at the assumptions supporting the arguments and are provided by arguments for the contrary of assumptions. In their dialogue model, agents can utter claims to be debated, rules, assumptions, and contraries. A dialogue between the proponent and the opponent constructs a dialectical tree which represents moves by agents during a dialogue and outcomes. In their framework, two agents share a common ABA framework and have common background knowledge. With this setting, an agent cannot behave dishonestly as one cannot keep some information from the other.

5   Conclusion

The contributions of this paper are mainly twofold. First, we developed debate games using a non-abstract argumentation framework associated with logic programming. We applied argumentation-based extended logic programs to formal modelling of dialogue games. Second, we showed an application of dishonest reasoning in argumentation-based logic programming. Debate games introduced in this paper realize dishonest reasoning by players using nonmonotonic nature of logic programs. To the best of our knowledge, there is no formal dialogical system which can deal with argumentative reasoning, belief update and dishonest reasoning in a uniform and concrete manner. The current study contributes to a step toward integrating logic programming and formal argumentation.

The proposed framework will be extended in several ways. In real-life debates, players may use *assumptions* in their arguments. Assumptions are also used for constructing arguments in an assumption-based argumentation framework [5]. Arguments considered in this paper use assumptions in the form of default literals. To realize debate games in which players can also use objective literals as assumptions, we can consider a non-abstract assumption-based argumentation framework associated with *abductive logic programs*. In this framework, an argument associated with an abductive logic program can contain *abducibles* as well as rules in a program. A player can claim an argument containing abducibles whose truthfulness are unknown. This is another type of dishonest claims called *bullshit* [7]. To realize debate games, we are now implementing a prototype system of debate games based on the abstract argumentation framework [17]. We plan to extend the system to handle non-abstract arguments associated with extended logic programs.

Acknowledgement. We thank Martin Caminada for useful discussion on the subject of this paper.

References

1. Bondarenko, A., Dung, P.M., Kowalski, R., Toni, F.: An abstract, argumentation-theoretic approach to default reasoning. Artif. Intell. **93**(1–2), 63–101 (1997)
2. Caminada, M., Wu, Y.: On the limitation of abstract argumentation. In: Proceedings of the 23rd Benelux Conference on Artificial Intelligence (BNAIC), Gent, Belgium (2011)
3. Caminada, M.: Grounded semantics as persuasion dialogue. In: Proceedings of the 4th International Conference on Computational Models of Argument (COMMA), Frontiers in Artificial Intelligence and Applications, vol. 245, pp. 478–485. IOS Press (2012)
4. Dung, P.M.: On the acceptability of arguments and its fundamental role in non-monotonic reasoning, logic programming and n-person games. Artif. Intell. **77**(2), 321–357 (1995)
5. Dung, P.M., Kowalski, R.A., Toni, F.: Assumption-based argumentation. In: Rahwan, I., Simari, G.R. (eds.) Argumentation in Artificial Intelligence, pp. 199–218. Springer, New York (2009)
6. Fan, X., Toni, F.: Assumption-based argumentation dialogues. In: Proceedings of the 22nd International Joint Conference on Artificial Intelligence (IJCAI), pp. 198–203 (2011)
7. Frankfurt, H.G.: On Bullshit. Princeton University Press, Princeton (2005)
8. García, A.J., Simari, G.R.: Defeasible logic programming: an argumentative approach. Theor. Pract. Logic Program. **4**(1–2), 95–138 (2004)
9. García, A.J., Dix, J., Simari, G.R.: Argumentation-based logic programming. In: Rahwan, I., Simari, G.R. (eds.) Argumentation in Artificial Intelligence, pp. 153–171. Springer, New York (2009)
10. Gelfond, M., Lifschitz, V.: Logic programs with classical negation. In: Proceedings of the 7th International Conference on Logic Programming (ICLP), pp. 579–597. MIT Press (1990)
11. Jakobovits, H., Vermeir, D.: Dialectic semantics for argumentation frameworks. In: Proceedings of the 7th International Conference on Artificial Intelligence and Law (ICAIL), pp. 53–62. ACM Press (1999)
12. Parsons, S., Wooldridge, M., Amgoud, L.: Properties and complexity of some formal inter-agent dialogues. J. Log. Comput. **13**(3), 347–376 (2003)
13. Prakken, H., Sartor, G.: Argument-based extended logic programming with defeasible priorities. J. Appl. Non-Class. Log. **7**(1), 25–75 (1997)
14. Prakken, H.: Coherence and flexibility in dialogue games for argumentation. J. Log. Comput. **15**(6), 1009–1040 (2005)
15. Prakken, H.: On the nature of argument schemes. In: Reed, C.A., Tindale, C. (eds.) Dialectics, Dialogue and Argumentation, An Examination of Douglas Walton's Theories of Reasoning and Argument, pp. 167–185. College Publications, London (2010)
16. Rahwan, I., Larson, K., Tohmé, F.: A characterisation of strategy-proofness for grounded argumentation semantics. In: Proceedings of the 21st International Joint Conference on Artificial Intelligence (IJCAI), pp. 251–256 (2009)
17. Sakama, C.: Dishonest reasoning by abduction. In: Proceedings of the 22nd International Joint Conference on Artificial Intelligence (IJCAI), pp. 1063–1068 (2011)
18. Sakama, C., Son, T.C., Pontelli, E.: A logical formulation for negotiation among dishonest agents. In: Proceedings of the 22nd International Joint Conference on Artificial Intelligence (IJCAI), pp. 1069–1074 (2011)

19. Sakama, C.: Dishonest arguments in debate games. In: Proceedings of the 4th International Conference on Computational Models of Argument (COMMA), Frontiers in Artificial Intelligence and Applications, vol. 245, pp. 177–184. IOS Press (2012)
20. Schopenhauer, A.: The Art of Controversy. Originally published in 1896 and is translated by T. Bailey Saunders. Cosimo Classics, New York (2007)
21. Schweimeier, R., Schroeder, M.: A parameterized hierarchy of argumentation semantics for extended logic programming and its application to the well-founded semantics. Theor. Pract. Log. Program. 5(1–2), 207–242 (2005)

A Descriptive Handling of Directly Conflicting Rules in Preferred Answer Sets

Alexander Šimko[✉]

Comenius University in Bratislava, 842 48 Bratislava, Slovakia
simko@fmph.uniba.sk
http://dai.fmph.uniba.sk/~simko

Abstract. We consider the problem of extending the answer set semantics for logic programs with preferences on rules. Many interesting semantics have been proposed. In this paper we develop a descriptive semantics that ignores preferences between non-conflicting rules. It is based on the Gelfond-Lifschitz reduction extended by the condition: a rule cannot be removed because of a less preferred conflicting rule. It turns out that the semantics continues to be in the hierarchy of the approaches by Delgrande et al., Wang et al., and Brewka and Eiter, and guarantees existence of a preferred answer set for the class of call-consistent head-consistent extended logic programs. The semantics can be also characterized by a transformation from logic programs with preferences to logic programs without preferences such that the preferred answer sets of an original program correspond to the answer sets of the transformed program. We have also developed a prototypical solver for preferred answer sets using the meta-interpretation technique.

Keywords: Knowledge representation · Logic programming · Preferred answer sets

1 Introduction

A knowledge base of a logic program possibly contains conflicting rules – rules that state mutually exclusive things. Having such rules, we often want to specify which of the rules to apply if both the rules can be applied.

Many interesting extensions of the answer set semantics for logic programs with preferences on rules have been proposed, e.g., [2,4,10,13,17,18]. Among the ones that stay in the NP complexity class ([2,4,17]), there is none that guarantees existence of a preferred answer set for the subclass of stratified [1,3] normal logic programs. This is the result of the fact, that the semantics do not ignore preferences between non-conflicting rules. Therefore the semantics are not usable in situations where we need to automatically induce preferences between rules. Such situations can be found, e.g., in [8,19]. An example is a system that allows a user to write his/her own rules, in order to override the rules defined by developers. Since the user's rules are know at run time, all the user's rules must be

M. Hanus and R. Rocha (Eds.): KDPD 2013, LNAI 8439, pp. 202–217, 2014.
DOI: 10.1007/978-3-319-08909-6_13, © Springer International Publishing Switzerland 2014

preferred. It is then important that the preferences between the non-conflicting rules do not cause side effects, e.g., non existence of a preferred answer set.

In this paper we propose a semantics, staying in the NP complexity class, with a very simple and elegant definition that ignores preferences between non-conflicting rules. The semantics guarantees existence of a preferred answer set for the class of call-consistent [9,11] head-consistent [14] extended logic programs, which is a superclass of stratified normal logic programs. The semantics is defined using a modified version of the Gelfond-Lifschitz reduction. An additional principle is incorporated: a rule cannot be removed because of a less preferred conflicting rule. It turns out that the semantics continues to be in the hierarchy of the semantics [2,4,17] discovered in [12]. It preserves the preferred answer sets of these semantics and admits additional ones, which were rejected because of preferences between non-conflicting rules. We also present a simple and natural transformation from logic programs with preferences to logic programs without preferences such that the answer sets of the transformed program (modulo new special-purpose literals) are exactly the preferred answer sets of an original one.

The rest of the paper is organized as follows. Section 2 recapitulates preliminaries from logic programming and answer set semantics. Section 3 informally describes the approach. Section 4 develops an alternative definition of the answer set semantics that is extended to a preferred answer set semantics in Sect. 5. Section 6 presents a transformation from logic programs with preferences to logic programs without preferences. Section 7 analyses the properties of the semantics. In Sect. 8 we show the connection between the semantics and existing approaches. Section 9 summarizes the paper.

All the proofs not presented here can be found in a technical report [15].

2   Preliminaries

Let At be a set of all atoms. A *literal* is an atom or an expression $\neg a$, where a is an atom. A *rule* is an expression of the form $l_0 \leftarrow l_1, \ldots, l_m, not\ l_{m+1}, \ldots, not\ l_n$, where $0 \leq m \leq n$, and each l_i ($0 \leq i \leq n$) is a literal. Given a rule r of the above form we use $head(r) = l_0$ to denote the *head* of r, $body(r) = \{l_1 \ldots, not\ l_n\}$ the *body* of r. Moreover, $body^+(r) = \{l_1, \ldots, l_m\}$ denotes the *positive* body of r, and $body^-(r) = \{l_{m+1}, \ldots, l_n\}$ the *negative* body of r. For a set of rules R, $head(R) = \{head(r) : r \in R\}$. A *logic program* is a finite set of rules.

A set of literals S is *consistent* iff $a \in S$ and $\neg a \in S$ holds for no atom a. A set of literals S *satisfies*: (i) the body of a rule r iff $body^+(r) \subseteq S$, and $body^-(r) \cap S = \emptyset$, (ii) a rule r iff $head(r) \in S$ whenever S satisfies $body(r)$, (iii) a logic program P iff S satisfies each rule of P.

A logic program P is *head-consistent* [14] iff $head(P)$ is consistent. A logic program without *not* is called *positive*.

For a positive logic program P, an *answer set* is defined as the least consistent set of literals satisfying P, and we denote it by $\mathcal{M}(P)$.

The Gelfond-Lifschitz *reduct* of a program P w.r.t. a set of literals S, denoted P^S, is the set $\{head(r) \leftarrow body^+(r) : r \in P$ and $body^-(r) \cap S = \emptyset\}$.

Definition 1 (Answer set [7]). *A consistent set of literals S is an* answer set *of a logic program P iff S is an answer set of P^S.*

We will use $\mathcal{AS}(P)$ to denote the set of all the answer sets of a logic program P.

For a set of literals S, we also denote $\Gamma_P(S) = \{r \in P : body^+(r) \subseteq S \text{ and } body^-(r) \cap S \neq \emptyset\}$.

We will say that a set of literals S *defeats* a rule r iff $body^-(r) \cap S \neq \emptyset$. A set of rules R *defeats* a rule r iff $head(R)$ defeats r.

A *dependency graph* of a program P is an directed labelled graph where (i) the literals are the vertices, (ii) there is an edge labelled $+$, called *positive*, from a vertex a to a vertex b iff there is a rule $r \in P$ with $head(r) = a$ and $b \in body^+(r)$, (iii) there is an edge labelled $-$, called *negative*, from a vertex a to a vertex b iff there is a rule $r \in P$ with $head(r) = a$ and $b \in body^-(r)$.

A program P is called *call-consistent* [9,11] iff its dependency graph contains no cycle with an odd number of negative edges.

Definition 2. *A logic program with preferences is a pair $(P, <)$ where: (i) P is a logic program, and (ii) $<$ is a transitive and asymmetric relation on P. If $p < r$ for $p, r \in P$ we say that r is* preferred *over p.*

3   Informal Presentation

The logic programming way of handling conflicting rules is adding guards to rules in the form of default negated literals. They allow us to make a rule inapplicable when a conflicting rule is applied.

Consider the well known penguin-fly program P:

$r_1:$ $flies$ $\leftarrow$ $bird, not \neg flies$
$r_2:$ $\neg flies$ $\leftarrow$ $penguin, not flies$
$r_3:$ $bird$ $\leftarrow$
$r_4:$ $penguin$ $\leftarrow$

The rules r_1 and r_2 have contrary heads, and contain guards $not \neg flies$ and $not flies$. The rule r_1 is made inapplicable whenever r_2 is applied, and vice versa. We call this mutual exclusivity a conflict. Due to this conflict, the program has the two answer sets $A_1 = \{bird, penguin, flies\}$ and $A_2 = \{bird, penguin, \neg flies\}$.

Each answer set S can be associated with a unique set of rules that generate it: $\Gamma_P(S) = \{r \in P : body^+(r) \subseteq S \text{ and } body^-(r) \cap S = \emptyset\}$. Here we call it a generating set. The generating set of A_1 is $R_1 = \Gamma_P(A_1) = \{r_1, r_3, r_4\}$ and the generating set of A_2 is $R_2 = \Gamma_P(A_2) = \{r_2, r_3, r_4\}$.

Our intuition behind preference handling is that *a rule cannot be made inapplicable due to a less preferred conflicting rule*. If we say that the rule r_1 is preferred over r_2, we expect that r_2 cannot defeat r_1, and consequently A_1 is the only preferred answer set.

We can review generating set R_2 and check whether it is in line with the mentioned intuitive principle. We have that r_2 is the only rule from R_2 that defeats r_1. However r_2 is less preferred and conflicting. We also have that the positive body of r_1 can be supported by a fact r_3. This means that r_1's body is satisfied w.r.t. R_2 and r_1 should also be a generating rule, i.e., $r_1 \in R_2$. Since this is a contradiction, R_2 is not a preferred generating set and $A_2 = head(R_2)$ is not a preferred answer set. A_1 is the unique preferred answer set.

The following sections formalize this kind of reasoning. In the next section we make precise the alternative definitions of generating sets and answer sets, upon which the definitions of preferred answer sets and preferred generating sets are built.

4   Alternative Definition of Answer Sets

When working with preferences on rules, we need to work on the level of rules rather than on the level of literals. We need to check which rules are used to generate an answer set, compare the rules w.r.t. preference relation, make a rule inapplicable, etc. Therefore in order to keep the definition of preferred answer sets as simple as possible, we reformulate answer set semantics in the terms of sets of rules rather than sets of literals. Instead of guessing a set of literals and testing whether it is an answer set, we guess a set of rules, and test whether it is a generating set (of an answer set).

The first step in the standard definition of answer sets, is to define an answer set of a positive program as a minimal set of literals that satisfies the program. We alternatively define when a set of rules positively satisfies the program.

Definition 3 (Positive satisfaction). *Let P be a set of rules. A set of rules $R \subseteq P$ positively satisfies P iff for each rule $r \in P$ we have that: If $body^+(r) \subseteq head(R)$, then $r \in R$. We will use $\mathcal{Q}(P)$ to denote the least (w.r.t. $\subseteq$) set of rules that positively satisfies P.*

Informally, $\mathcal{Q}(P)$ is the set of the rules that are applied during the bottom-up evaluation of a program P.

Example 1. Consider the following program P

r_1:	a	$\leftarrow$	b	r_3:	a	$\leftarrow$
r_2:	b	$\leftarrow$		r_4:	d	$\leftarrow$ c

$\{a, b\}$ and $\{a, b, d\}$ satisfy the program. Alternatively $\{r_1, r_2, r_3\}$ and $\{r_1, r_2, r_3, r_4\}$ positively satisfy the program. We also have that $\mathcal{M}(P) = \{a, b\}$. Alternatively $\mathcal{Q}(P) = \{r_1, r_2, r_3\}$. Note that $head(\mathcal{Q}(P)) = \mathcal{M}(P)$.

The second step of the standard definition of answer sets is the Gelfond-Lifschitz reduction. We define an alternative version taking a guess of rules on its input.

Definition 4 (Reduct). *Let P be a logic program, and $R \subseteq P$ be a set of rules. The* reduct P^R *is obtained from P by removing each rule r with $head(R) \cap body^-(r) \neq \emptyset$.*

An answer set is defined as a stable set w.r.t. the operator $\mathcal{M}(P^S)$. A generating set is alternatively a stable set w.r.t. the operator $\mathcal{Q}(P^R)$.

Definition 5 (Generating set). *Let P be a logic program. A set of rules $R \subseteq P$ is a* generating set *of P iff $R = Q(P^R)$.*

The next propositions justify the name "generating set". It turns out that an answer set S of a program P is represented by a unique generating set, namely the set $\Gamma_P(S)$.

Proposition 1. *Let P be a logic program. Let R_1 and R_2 be generating sets such that $head(R_1) = head(R_2)$. Then $R_1 = R_2$.*

Proposition 2. *Let P be a logic program. Let R be a generating set, and S be a consistent set of literals such that $S = head(R)$.*
 Then $\Gamma_P(S) = R$.

Theorem 1. *Let P be a logic program. A consistent set of literals S is an answer set of P iff $\Gamma_P(S)$ is a generating set and $head(\Gamma_P(S)) = S$.*

From that we have an alternative definition of answer sets.

Theorem 2 (An alternative characterization of answer sets). *Let P be a logic program, and S be a consistent set of literals.*
 S is an answer set *of P iff there is a generating set R such that $head(R) = S$.*

The following example illustrates the alternative definition of answer sets alongside the original one.

Example 2. Consider the following program P:

r_1: a $\leftarrow$ $not\ b$
r_2: c $\leftarrow$ $d, not\ b$
r_3: b $\leftarrow$ $not\ a$

We will show that $S = \{a\}$ is an answer set.

	Gelfond-Lifschitz definition	Alternative definition
guess	$S = \{a\}$	$R = \{r_1\}$
reduct	$P^S = \{a \leftarrow, c \leftarrow d\}$	$P^R = \{r_1, r_2\}$
		r_3 is removed as $body^-(r_3) \cap head(R) \neq \emptyset$
"min model"	$\mathcal{M}(P^S) = \{a\}$	$Q(P^R) = \{r_1\}$
		r_2 is not included as $d \in body^+(r_2)$ cannot be derived
test	$\mathcal{M}(P^S) = S$	$Q(P^R) = R$
conclusion	S is an answer set	R is a generating set and $head(R) = S$ is an answer set

5   Preferred Answer Sets

In this section, we define preferred answer sets by formalizing the intuitions from Sect. 3.

In order to develop a definition of preferred answer sets we need to make clear what exactly a conflict is.

Definition 6 (Conflict). *Let r_1, r_2 be rules. We say that r_1 and r_2 are conflicting iff: (i) $head(r_1) \in body^-(r_2)$, and (ii) $head(r_2) \in body^-(r_1)$.*

In this paper we only consider direct conflicts. This in no way means we consider indirect conflicts unimportant. However there are good reasons why we consider only direct conflicts in this paper:

– We believe it is always a good practice to proceed from simple cases to complex ones.
– There are different ways in which a semantics can handle direct conflicts. Indirect conflicts only add additional variability to this. Restriction to direct conflicts allows us to better communicate the core idea behind the approach to other researchers while avoiding the different subjective views on indirect preference handling.
– Restriction to direct conflicts is essential in order to obtain the result presented by Theorem 13.

We obtain the definition of preferred answer sets by requiring in Definition 4 that less preferred conflicting rules cannot cause a rule to be removed.

Definition 7 (Override). *Let $\mathcal{P} = (P, <)$ be a logic program with preferences. Let r_1 and r_2 be rules. We say that r_1 overrides r_2 iff (i) r_1 and r_2 are conflicting, and (ii) $r_2 < r_1$.*

Definition 8 (Reduct). *Let $\mathcal{P} = (P, <)$ be a logic program with preferences, and $R \subseteq P$ be a set of rules. The reduct $\mathcal{P}^R$ is obtained from P by removing each rule $r \in P$ such that there is a rule $q \in R$ such that:*

– $head(q) \in body^-(r)$, *and*
– *r does not override q.*

Note that Definitions 4 and 8 differ only in the second condition. After removing it, the obtained definition is equivalent with Definition 4. Moreover, when a preference relation is empty, the reduct coincides with the reduct for logic programs without preferences as defined in Definition 4, which in turn corresponds to Gelfond-Lifschitz reduct.

Proposition 3. *Let $\mathcal{P} = (P, <)$ be a logic program with preferences. Let $R \subseteq P$ be a set of rules. If $<= \emptyset$, then $\mathcal{P}^R = P^R$.*

Definition 9 (Preferred generating set). *Let $\mathcal{P} = (P, <)$ be a logic program with preferences. A set of rules $R \subseteq P$ is a preferred generating set iff $R = Q(\mathcal{P}^R)$.*

Proposition 4. *Let* $\mathcal{P} = (P, <)$ *be a logic program with preferences, and* $R \subseteq P$ *be a set of rules. If* R *is a preferred generating set, then* R *is a generating set.*

Definition 10 (Preferred answer set). *Let* $\mathcal{P} = (P, <)$ *be a logic program with preferences. A consistent set of literals* S *is a* preferred answer set *of* $\mathcal{P}$ *iff there is a preferred generating set* R *such that* $head(R) = S$.

We will use $\mathcal{PAS}(\mathcal{P})$ to denote the set of all the preferred answer sets of a logic program with preferences $\mathcal{P}$.

From Proposition 4 we have that analogous versions of Proposition 1 and Theorem 1 hold also for preferred answer sets.

Notice that the definition of preferred answer sets is almost identical to the definition of answer sets. The only difference is the additional simple condition in the definition of reduct. Consider we would adapt the original definition of reduct rather than the alternative one. It would be as follows: *Given a set of literals* S *we remove from a program* P *each rule* r *such that there exists a rule* q *where* ... Since the guess S is a set of literals, it is not straightforward where to get q from, i.e., we would have to introduce additional conditions that q must meet. We would then have to justify those conditions. On the other hand, the use of the alternative definition of reduct has completely freed us from this job as no such conditions are needed. Exactly this elegance was the main motivation for the alternative definition of answer sets.

6   Transformation to Logic Programs Without Preferences

A logic program with preferences under the preferred answer set semantics as defined in Definition 10 can be transformed to a logic program without preferences such that the preferred answer sets of the original program are exactly the standard answer sets of the transformed program (modulo new special-purpose literals).

The basic idea of the transformation is to remove a default negated literal from the body of a rule if it is the head of a less preferred conflicting rule.

Example 3. Consider the program in the first column.

$$
\begin{array}{llll}
r_1: & a \leftarrow not\ b & & a \leftarrow \\
r_2: & b \leftarrow not\ a & \mapsto & b \leftarrow not\ a \\
& r_2 < r_1 & &
\end{array}
$$

We have that r_1 and r_2 are conflicting, and r_1 is preferred. Hence r_2 cannot defeat the rule r_1. Therefore we remove *not b* from r_1's body.

However, the situation is complicated if at least two rules have the same head. In general, we have to distinguish which rule can defeat a rule and which cannot. In order to do so, for each rule r we introduce a special-purpose literal n_r that denotes that r is applicable, and replace default negated literal *not x* by a collection of *not* n_r such that $head(r) = x$.

Example 4. Consider the program in the first column. First (the second column) we split a rule into two rules: (i) the first deriving n_r whenever r is applicable, and (ii) the second deriving $head(r)$ of the original rule. We also use n_t literals in the negative bodies. Next (the third column) we remove $not\ n_q$ from the negative body of r if q is a less preferred conflicting rule.

$$
\begin{array}{lll}
r_1: \quad a \;\leftarrow\; not\ b & \qquad n_{r_1} \;\leftarrow\; not\ n_{r_2},\ not\ n_{r_3} & \qquad n_{r_1} \;\leftarrow\; not\ n_{r_3} \\
r_2: \quad b \;\leftarrow\; not\ a & \qquad a \;\leftarrow\; n_{r_1} & \qquad a \;\leftarrow\; n_{r_1} \\
r_3: \quad b \;\leftarrow\; c \quad\longmapsto & \qquad n_{r_2} \;\leftarrow\; not\ n_{r_1} \quad\longmapsto & \qquad n_{r_2} \;\leftarrow\; not\ n_{r_1} \\
\quad\ r_2 < r_1 & \qquad b \;\leftarrow\; n_{r_2} & \qquad b \;\leftarrow\; n_{r_2} \\
& \qquad n_{r_3} \;\leftarrow\; c & \qquad n_{r_3} \;\leftarrow\; c \\
& \qquad b \;\leftarrow\; n_{r_3} & \qquad b \;\leftarrow\; n_{r_3}
\end{array}
$$

The next definition formalizes the transformation.

Definition 11 (Transformation). *Let $\mathcal{P} = (P, <)$ be a logic program with preferences, and $r \in P$ be a rule. The names of $r's$ potential blockers are*

$$B_{\mathcal{P}}(r) = \{n_q : q \in P, head(q) \in body^-(r),\ and\ r\ does\ not\ override\ q\}.$$

The transformation $t_{\mathcal{P}}(r)$ of a rule r is the set of the rules

$$head(r) \leftarrow n_r \tag{1}$$
$$n_r \leftarrow body^+(r),\ not\ B_{\mathcal{P}}(r). \tag{2}$$

The transformation $t(\mathcal{P})$ of a program $\mathcal{P}$ is given by $\bigcup_{r \in P} t_{\mathcal{P}}(r)$.

It can be easily seen from the definition that the transformation can be computed in polynomial time, and the size of the transformed program is polynomial in the size of an original one.

The transformation captures semantics of preferred answer sets.

Theorem 3. *Let $\mathcal{P} = (P, <)$ be a logic program with preferences. Let S be a consistent set of literals (of the program $\mathcal{P}$). Let T be a consistent set of literals (of the program $t(\mathcal{P})$). Let $N_{\mathcal{P}}(S) = \{n_r : r \in \Gamma_P(S)\}$ and $N(\mathcal{P}) = \{n_r : r \in P\}$.*

If S is a preferred answer set of $\mathcal{P}$, then $S \cup N_{\mathcal{P}}(S)$ is an answer set of $t(\mathcal{P})$. If T is an answer set of $t(\mathcal{P})$, then $T \setminus N(\mathcal{P})$ is a preferred answer set of $\mathcal{P}$.

Next we show that the transformation does not introduce cycles with odd number of negative edges.

Proposition 5. *Let $\mathcal{P} = (P, <)$ be a logic program with preferences. If P is call-consistent, then $t(\mathcal{P})$ is call-consistent.*

Proof. In the following we use the notation: $t(l) = l$ if l is a literal from P, and $t(n_r) = head(r)$ for $r \in P$.

Assume there is an odd cycle in the $t(\mathcal{P})$'s dependency graph: there is a sequence $l_1, s_1, \ldots, s_{n-1}, l_n = l_1$ such that (l_i, l_{i+1}, s_i) is a labelled edge for each $i < n$, and for some $i < n$ we have $s_i = -$. Assume a literal l_i and an edge (l_i, l_{i+1}, s).

If l_i is a literal from P, then the edge came from a rule of the form (1). Hence $l_{i+1} = n_r$ for some $r \in P$, $s = +$, and $head(r) = l_i$.

If $l_i = n_r$ for some $r \in P$, then: (i) If $s = +$ then the edge came from a rule of the form (2). Hence l_{i+1} is a literal from P, $l_{i+1} \in body^+(r)$, and $(t(l_i), t(l_{i+1}), +)$ is an edge in P's dependency graph. (ii) If $s = -$ then the edge came from a rule of the form (2). Hence $l_{i+1} = n_p$ for some $p \in P$, $head(p) \in body^-(r)$, and $(t(l_i), t(l_{i+1}), -)$ is an edge in P's dependency graph.

Now, we create a new sequence by iterating over the sequence $l_1, \ldots, l_n$: (i) If $i < n$ and l_i is a literal of the program P skip l_i and s_i. From the above analysis we have $t(l_{i+1}) = l_i$, and $s_i = +$, hence the literal will be added in the next step, and no negative edge is lost, (ii) If $i < n$ and $l_i = n_r$ for some $r \in P$, add $t(l_i), s_i$ to the end of the resulting sequence. (iii) If $i = n$ add $t(l_i)$ to the end of the resulting sequence. From the above analysis we have that the sequence forms a cycle in P's dependency graph, and the number of negative edges is preserved.□

If P is not call-consistent, then $t(\mathcal{P})$ can be call-consistent for some $<$ and not call-consistent for other $<$.

Example 5. Consider the program $\mathcal{P} = (P, <)$:

$$r_1: \quad a \quad \leftarrow \quad not\ b, not\ c$$
$$r_2: \quad b \quad \leftarrow \quad not\ a, not\ c$$
$$r_3: \quad c \quad \leftarrow \quad not\ b$$

If $r_3 < r_2 < r_1$, then $t(\mathcal{P})$:

$$n_{r_1} \leftarrow not\ n_{r_3} \qquad n_{r_2} \leftarrow not\ n_{r_1} \qquad n_{r_3} \leftarrow not\ n_{r_2}$$
$$a \leftarrow n_{r_1} \qquad\qquad b \leftarrow n_{r_2} \qquad\qquad c \leftarrow n_{r_3}$$

is not call-consistent.

If $r_1 < r_2 < r_3$, then $t(\mathcal{P})$:

$$n_{r_1} \leftarrow not\ n_{r_2}, not\ n_{r_3} \qquad n_{r_2} \leftarrow not\ n_{r_3} \qquad n_{r_3} \leftarrow$$
$$a \leftarrow n_{r_1} \qquad\qquad\qquad\qquad b \leftarrow n_{r_2} \qquad\qquad c \leftarrow n_{r_3}$$

is call-consistent.

7   Properties of Preferred Answer Sets

In this section we show that the semantics enjoys several nice properties. First of all, the semantics is selective, i.e., each preferred answer set is an answer set.

Theorem 4. $\mathcal{PAS}(\mathcal{P}) \subseteq \mathcal{AS}(P)$ *for each logic program with preferences* $\mathcal{P} = (P, <)$.

Proof. The theorem follows directly from Proposition 4.   □

Second, for the two simple classes of programs: (i) programs with an empty preference relation, and (ii) stratified programs, the semantics is equivalent to the answer set semantics.

Theorem 5. $\mathcal{PAS}((P, \emptyset)) = \mathcal{AS}(P)$ *for each logic program* P.

Theorem 6. $\mathcal{PAS}(\mathcal{P}) = \mathcal{AS}(P)$ *for each logic program with preferences* $\mathcal{P} = (P, <)$ *where* P *is stratified.*

Proof. If P is stratified, then there are no conflicting rules. Hence $P^R = \mathcal{P}^R$. $\square$

Next we show that our semantics satisfies both principles for preferential reasoning proposed in [2] by Brewka and Eiter.

Principle I tries to capture the meaning of preferences. If two answer sets are generated by the same rules except for two rules, the one generated by a less preferred rule is not preferred.

Principle I ([2]). *Let* $\mathcal{P} = (P, <)$ *be a logic program with preferences,* A_1, A_2 *be two answer sets of* P. *Let* $\Gamma_P(A_1) = R \cup \{d_1\}$ *and* $\Gamma_P(A_2) = R \cup \{d_2\}$ *for* $R \subset P$. *Let* $d_2 < d_1$. *Then* $A_2 \notin \mathcal{PAS}(\mathcal{P})$.

Theorem 7 (Principle I is satisfied). *Preferred answer sets as defined in Definition 10 satisfy Principle I.*

Proof. Assume that A_2 is a preferred answer set. Hence $\Gamma_P(A_2)$ is a preferred generating set, i.e., $\Gamma_P(A_2) = \mathcal{Q}(\mathcal{P}^{\Gamma_P(A_2)})$. We have that d_2 is the only rule $r \in \Gamma_P(A_2)$ with $head(r) \in body^-(d_1)$. We also have that d_1, and d_2 are conflicting, and $d_2 < d_1$. Hence $d_1 \in \mathcal{P}^{\Gamma_P(A_2)}$, and consequently $d_1 \in \Gamma_P(A_2)$. A contradiction. Therefore $\Gamma_P(A_2)$ is not a preferred generating set. $\square$

Principle II says that the preferences specified on a rule with the violated positive body are irrelevant.

Principle II ([2]). *Let* $\mathcal{P} = (P, <)$ *be a logic program with preferences,* $S \in \mathcal{PAS}(\mathcal{P})$ *and* r *be a rule such that* $body^+(r) \not\subseteq S$. *Let* $\mathcal{P}' = (P', <')$ *be a logic program with preferences, where* $P' = P \cup \{r\}$ *and* $<' \cap (P \times P) =<$. *Then* $S \in \mathcal{PAS}(\mathcal{P}')$.

Theorem 8 (Principle II is satisfied). *Preferred answer sets as defined in Definition 10 satisfy Principle II.*

Proof. Let S be a preferred answer set of $\mathcal{P}$, i.e., there is a set of rules $R \subseteq P$ such that $R = \mathcal{Q}(\mathcal{P}^R)$ and $head(R) = S$. We show that $R = \mathcal{Q}(\mathcal{P}'^R)$.

First we show that R positively satisfies $\mathcal{P}'^R$. Assume $p \in \mathcal{P}'^R$ such that $body^+(p) \subseteq head(R)$. Hence $p \in \mathcal{P}'$ and $p \neq r$. Hence $p \in P$ and $p \in \mathcal{P}^R$. As R positively satisfies $\mathcal{P}^R$ we have that $p \in R$.

Second, we show that no proper subset of R positively satisfies $\mathcal{P}'^R$. Assume that $V \subset R$ positively satisfies $\mathcal{P}'^R$. Since $\mathcal{P}^R \subseteq \mathcal{P}'^R$, we have that V positively satisfies $\mathcal{P}^R$. A contradiction with $R = \mathcal{Q}(\mathcal{P}^R)$. Hence such V does not exist.

Therefore $\mathcal{Q}(\mathcal{P}'^R) = R$. $\square$

On the other hand, the semantics violates the following Principle III[1]. It requires that a program has a preferred answer set whenever a standard answer set exists. It follows the view that the addition of preferences should not cause a consistent program to be inconsistent.

Principle III. *Let $\mathcal{P} = (P, <)$ be a logic program with preferences. If $\mathcal{AS}(P) \neq \emptyset$, then $\mathcal{PAS}(\mathcal{P}) \neq \emptyset$.*

Theorem 9. (Principle III is violated). *Preferred answer sets as defined in Definition 10 violate Principle III.*

Proof. Consider the following program $\mathcal{P} = (P, <)$:

$$r_1: \quad a \quad \leftarrow \quad not\ b \qquad\qquad r_3: \quad inc \quad \leftarrow \quad a, not\ inc$$
$$r_2: \quad b \quad \leftarrow \quad not\ a \qquad\qquad\qquad\qquad r_2 < r_1$$

P has the unique answer set $\{b\}$. However it is not a preferred one.   □

The semantics does not guarantee existence of a preferred answer set when a standard answer set exists for the class of all programs (which is believed to rise computational complexity). The reason is the strict character of preferences. In the view of the transformation, we can understand preferences as a handy way of encoding exceptions between rules – a rule does not define an exception to a preferred conflicting rule. An underlying logic program without preferences can be understood as a program schema that is transformed to the intended program using the preferences. Using this view, answer sets of the underlying program are of no interest.

On the other hand, the semantics guarantees existence of a preferred answer set for a subclass of programs.

Theorem 10. *Let $\mathcal{P} = (P, <)$ be a logic program with preferences such that P is call-consistent and head-consistent. Then $\mathcal{PAS}(\mathcal{P}) \neq \emptyset$.*

Proof. Since P is head-consistent, we can assume explicitly negated literals to be new literals and view $t(\mathcal{P})$ as a normal logic program. From Proposition 5 we have that $t(P)$ is call-consistent. Then from Theorem 5.8 from [5] we get that $\mathcal{AS}(t(\mathcal{P})) \neq \emptyset$. Finally we get $\mathcal{PAS}(\mathcal{P}) \neq \emptyset$ using Theorem 3.   □

Theorem 11. *Deciding whether $\mathcal{PAS}(\mathcal{P}) \neq \emptyset$ for a logic program with preferences $\mathcal{P}$ is NP-complete.*

Proof. Membership: $t(\mathcal{P})$ can be computed in polynomial time. Using Theorem 3 the decision problem whether $\mathcal{PAS}(\mathcal{P}) \neq \emptyset$ can be reduced to the decision whether $\mathcal{AS}(t(\mathcal{P})) \neq \emptyset$, which is in NP. *Hardness:* Deciding whether $\mathcal{AS}(P) \neq \emptyset$ for a logic program P is NP-complete. Using Theorem 5 we can reduce it to decision whether $\mathcal{PAS}((P, \emptyset)) \neq \emptyset$.   □

---

[1] It is an idea from Proposition 6.1 from [2]. Brewka and Eiter did not consider it as a principle. On the other hand [13] did.

8   Comparison with Existing Approaches

In this section we investigate the connection of preferred answer sets as defined in Definition 10 to existing approaches. We focus our attention to the selective approaches that stay in the NP complexity class.

In [12] Schaub and Wang have shown that the approaches $\mathcal{PAS}_{DST}$ [4], $\mathcal{PAS}_{WZL}$ [17] and $\mathcal{PAS}_{BE}$ [2] form a hierarchy. We will use $\mathcal{PAS}_{DST}(\mathcal{P})$, $\mathcal{PAS}_{WZL}(\mathcal{P})$ and $\mathcal{PAS}_{BE}(\mathcal{P})$ to denote the set of all the preferred answer sets of a program according to the respective semantics.

Theorem 12 ([12]). *Let $\mathcal{P} = (P, <)$ be a logic program with preferences. Then $\mathcal{PAS}_{DST}(\mathcal{P}) \subseteq \mathcal{PAS}_{WZL}(\mathcal{P}) \subseteq \mathcal{PAS}_{BE}(\mathcal{P}) \subseteq \mathcal{AS}(P)$.*

We show that our semantics continues to be in this hierarchy. We start by an alternative definition of our semantics.

Definition 12. *Let $\mathcal{P} = (P, <)$ be a logic program with preferences. An answer set X of P is called $<$-satisfying iff for each $r \in P \setminus \Gamma_P(X)$ we have that:*

- $body^+(r) \not\subseteq X$, or
- $body^-(r) \cap \{head(t) : t \in \Gamma_P(X) \text{ and } r \text{ does not override } t\} \neq \emptyset$.

Lemma 1. *Let $\mathcal{P} = (P, <)$ be a logic program with preferences. A consistent set of literals X is a preferred answer set iff X is a $<$-satisfying answer set of P.*

Proof. ($\Rightarrow$) There is a set of rules $R = \mathcal{Q}(\mathcal{P}^R)$ such that $X = head(R)$.

Let $r \in P \setminus \Gamma_P(X)$. Since R is a generating set, we have $R = \Gamma_P(X)$. Hence $r \notin R = \mathcal{Q}(\mathcal{P}^R)$. From that $body^+(r) \not\subseteq X = head(R)$ or $r \notin \mathcal{P}^R$. If $r \notin \mathcal{P}^R$, then there must be a rule $t \in R$ such that $head(t) \in body^-(r)$ and r does not override t.

($\Leftarrow$) Let $R = \Gamma_P(X)$ and consider $\mathcal{P}^R$. Since $R = \Gamma_P(X)$, we have $R \subseteq \mathcal{P}^R$ and $R \subseteq \mathcal{Q}(\mathcal{P}^R)$.

Assume that $\mathcal{Q}(\mathcal{P}^R) \not\subseteq R$, i.e., there is a rule $r \in \mathcal{Q}(\mathcal{P}^R)$ such that $r \notin R$.

Since $r \in \mathcal{Q}(\mathcal{P}^R)$, we have that $body^+(r) \subseteq head(\mathcal{Q}(\mathcal{P}^R))$, i.e., every literal in $body^+(r)$ is supported by a rule in $\mathcal{Q}(\mathcal{P}^R)$.

As $r \notin R$, we have that $body^+(r) \not\subseteq X = head(R)$ or $body^-(r) \cap head(R) \neq \emptyset$.

Assume there is $t \in R$ such that $head(t) \in body^-(r)$ and r does not override t. Then $r \notin \mathcal{P}^R$. A contradiction. Hence no such t exists. Since $r \in P \setminus \Gamma_P(X)$, and X is $<$-satisfying, we get $body^+(r) \not\subseteq X$.

We have shown that $body^+(r) \not\subseteq X = head(R)$. Then there is a literal in $body^+(r)$ that is not supported by a rule from R. Hence there is a literal in $body^+(r)$ that is supported solely by a rule from $\mathcal{Q}(\mathcal{P}^R) \setminus R$. Hence each rule in $\mathcal{Q}(\mathcal{P}^R) \setminus R$ positively depends on a literal that can be derived only by a rule from $\mathcal{Q}(\mathcal{P}^R) \setminus R$. Then from minimality of $\mathcal{Q}(\mathcal{P}^R)$ we get that $r \notin \mathcal{Q}(\mathcal{P}^R)$. A contradiction. Therefore $\mathcal{Q}(\mathcal{P}^R) \subseteq R$.

Finally, $\mathcal{Q}(\mathcal{P}^R) = R$, and $X = head(R)$ is a preferred answer set of $\mathcal{P}$.

Definition 13 (Alternative definition of $\mathcal{PAS}_{BE}$ [12]). *Let $\mathcal{P} = (P, <)$ be a logic program with preferences. An answer set X of P is a BE preferred answer set of $\mathcal{P}$ iff there is an enumeration $\langle r_i \rangle$ of $\Gamma_P(X)$ such that for each i, j:*

- *if $r_i < r_j$, then $j < i$, and*
- *if $r_i < r$ and $r \in P \setminus \Gamma_P(X)$, then*
  - $body^+(r) \not\subseteq X$ or
  - $body^-(r) \cap \{head(r_j) : j < i\} \neq \emptyset$ or
  - $head(r) \in X$.

The difference between our semantics and $\mathcal{PAS}_{BE}$ can be seen directly from Definitions 12 and 13. One of the main differences is that Definition 12 completely drops the condition for enumeration of the rules, i.e., *preferences are not interpreted as an order, in which the rules are applied*. As the result, the second condition requiring how a preferred rule is defeated changes. A preferred rule can be defeated only by a rule that is not less preferred and conflicting. Hence the second difference: an *explicit definition of conflict* is used, and *preferences between non-conflicting rules are ignored*. The condition $head(r) \in X$ is also completely dropped.

Very similar differences hold for $\mathcal{PAS}_{DST}$ and $\mathcal{PAS}_{WZL}$ (Definitions of $\mathcal{PAS}_{DST}$ and $\mathcal{PAS}_{WZL}$, similar to Definition 13, can be found in [12]).

Theorem 13. *Let $\mathcal{P}$ be a logic program with preferences. Then $\mathcal{PAS}_{BE}(\mathcal{P}) \subseteq \mathcal{PAS}(\mathcal{P})$.*

Proof. Let $\mathcal{P}$ be a logic program with preferences and $X \in \mathcal{PAS}_{BE}(\mathcal{P})$. Then there is an enumeration $\langle r_i \rangle$ of $\Gamma_P(X)$ satisfying the conditions from Definition 13.

Assume there is $r \in P \setminus \Gamma_P(X)$ such (i) $body^+(r) \subseteq X$, and (ii) for each rule $t \in \Gamma_P(X)$ such that $head(t) \in body^-(r)$ it holds that r overrides t.

Since $body^+(r) \subseteq X$ and $r \notin \Gamma_P(X)$, we have that $body^-(r) \cap X \neq \emptyset$. There is a rule $r_i \in \Gamma_P(X)$ with $head(r_i) \in body^-(r)$. Then r_i and r are conflicting, i.e., $head(r) \in body^-(r_i)$. Since $r_i \in \Gamma_P(X)$ we have that $head(r) \notin X$. We also have that $r_i < r$.

We have shown that $r_i \in \Gamma_P(X)$, $r \in P \setminus \Gamma_P(X)$, $r_i < r$, $body^+(r) \subseteq X$ and $head(r) \notin X$.

Since $X \in \mathcal{PAS}_{BE}(\mathcal{P})$ we have that there is a rule $r_k \in \Gamma_P(X)$ with $head(r_k) \in body^-(r)$ for $k < i$. From the conditions above, we have that r_k and r are conflicting and $r_k < r$. By the same argument as before, there is a rule $r_l \in \Gamma_P(X)$ with $head(r_l) \in body^-(r)$ and $r_l < r$ for some $l < k$, and so on, until we reach the beginning of the enumeration and no such rule can be found. A contradiction. Hence (i) $body^+(r) \not\subseteq X$, or (ii) $body^-(r) \cap \{head(t) : t \in R$ and r does not override$t\} \neq \emptyset$.

Therefore X is $<$-satisfying, and $X \in \mathcal{PAS}(\mathcal{P})$.

Theorem 14. *It does not hold $\mathcal{PAS}(\mathcal{P}) \subseteq \mathcal{PAS}_{BE}(\mathcal{P})$ for each logic program with preferences $\mathcal{P}$.*

Proof. Consider the program $\mathcal{P}$ from Example 5.5 from [2]

r_1: c $\leftarrow$ $not\ b$
r_2: b $\leftarrow$ $not\ a$ $r_2 < r_1$

The program is stratified. $\mathcal{PAS}_{BE}(\mathcal{P}) = \emptyset$. On the other hand $\mathcal{PAS}(\mathcal{P}) = \{\{b\}\}$.

Theorem 15. *Let $\mathcal{P} = (P, <)$ be a logic program with preferences.*
  Then $\mathcal{PAS}_{DST}(\mathcal{P}) \subseteq \mathcal{PAS}_{WZL}(\mathcal{P}) \subseteq \mathcal{PAS}_{BE}(\mathcal{P}) \subseteq \mathcal{PAS}(\mathcal{P}) \subseteq \mathcal{AS}(P).$

Proof. It follows directly from Theorems 4, 12 and 13.   □

Theorems 14 and 15 can be interpreted as follows. Our semantics continues to be in the hierarchy of approaches $\mathcal{PAS}_{DST}$, $\mathcal{PAS}_{WZL}$ and $\mathcal{PAS}_{BE}$. It preserves the preferred answer sets of these semantics and admits additional ones, which were rejected because of preferences between non-conflicting rules.

Theorem 6 is another distinguishing feature of our semantics. None of the approaches $\mathcal{PAS}_{DST}$, $\mathcal{PAS}_{WZL}$ and $\mathcal{PAS}_{BE}$ satisfies it. We consider Theorem 6 to be important as a stratified program contains no conflicting rules and its meaning is given by a unique answer set. Theorem 6 also allows us to resolve the problematic program shown in the proof of Theorem 14.

9   Conclusion and Future Work

In this paper we have developed a descriptive semantics for logic programs with preferences on rules. The main idea is to add an additional condition to the Gelfond-Lifschitz reduction: a rule cannot be removed because of a conflicting less preferred rule. As a result, the approach uses an explicit definition of conflicting rules and ignores preferences between non-conflicting rules. This feature, not present in other approaches, is important for scenarios where preferences between rules are automatically induced from preferences between modules, as we do not want such preferences to cause any side effects.

The semantics continues to be in the hierarchy of approaches [2, 4, 17]. It preserves the preferred answer sets of these semantics and admits additional ones, which were rejected because of preferences between non-conflicting rules. The semantics satisfies both principles for preferential reasoning proposed in [2]. In contrast to [2, 4, 17], it guarantees existence of a preferred answer set for the class of call-consistent head-consistent extended logic programs. The semantics can be also characterized by a transformation from logic programs with preferences to logic programs without preferences such that the preferred answer sets of an original program correspond to the answer sets of the transformed program. The transformation is based on a simple idea: we remove a default negated literal from a rule's body if it is derived by a conflicting less preferred rule. We have also developed a prototypical solver for preferred answer sets using meta-interpretation technique from [6]. A description of the solver can be found in the technical report [15] and an implementation can be downloaded from [16].

In this paper we have only considered the most common type of conflict – the heads of two conflicting rules are in each others negative bodies. An important extension, which we are dealing with in the ongoing research, are indirect conflicts – literals in negative bodies are not derived directly by conflicting rules, but via other rules. Preliminary results in this direction can be found in a technical report [15].

Acknowledgement. We would like to thank the anonymous reviewers for detailed and useful comments. This work was supported by the grant UK/276/2013 of Comenius University in Bratislava and 1/1333/12 of VEGA.

References

1. Apt, K.R., Blair, H.A., Walker, A.: Towards a theory of declarative knowledge. In: Foundations of Deductive Databases and Logic Programming, pp. 89–148. Morgan Kaufmann, New York (1988)
2. Brewka, G., Eiter, T.: Preferred answer sets for extended logic programs. Artif. Intell. **109**(1–2), 297–356 (1999)
3. Chandra, A.K., Harel, D.: Horn clauses queries and generalizations. J. Log. Program. **2**(1), 1–15 (1985)
4. Delgrande, J.P., Schaub, T., Tompits, H.: A framework for compiling preferences in logic programs. Theory Pract. Log. Program. **3**(2), 129–187 (2003)
5. Dung, P.M.: On the relations between stable and well-founded semantics of logic programs. Theoret. Comput. Sci. **105**(1), 7–25 (1992)
6. Eiter, T., Faber, W., Leone, N., Pfeifer, G.: Computing preferred answer sets by meta-interpretation in answer set programming. Theory Pract. Log. Program. **3**(4–5), 463–498 (2003)
7. Gelfond, M., Lifschitz, V.: Classical negation in logic programs and disjunctive databases. New Gener. Comput. **9**(3/4), 365–386 (1991)
8. Illic, M., Leite, J., Slota, M.: ERASP - a system for enhancing recommendations using answer-set programming. Int. J. Reasoning-based Intell. Syst. **1**, 147–163 (2009)
9. Kunen, K.: Signed data dependencies in logic programs. J. Log. Program. **7**(3), 231–245 (1989)
10. Sakama, C., Inoue, K.: Prioritized logic programming and its application to commonsense reasoning. Artif. Intell. **123**(1–2), 185–222 (2000)
11. Sato, T.: On consistency of first-order logic programs. Technical report TR 87–12, ETL (1987)
12. Schaub, T., Wang, K.: A semantic framework for preference handling in answer set programming. Theory Pract. Log. Program. **3**(4–5), 569–607 (2003)
13. Šefránek, J.: Preferred answer sets supported by arguments. In: Proceedings of 12th International Workshop on Non-Monotonic Reasoning (NMR 2008), pp. 232–240 (2008)
14. Turner, H.: Signed logic programs. In: Logic Programming: Proceedngs of the 1994 International Symposium (ILPS'94), pp. 61–75 (1994)
15. Šimko, A.: Logic programming with preferences on rules. Technical report TR-2013-035, Comenius University in Bratislava (2013). http://kedrigern.dcs.fmph.uniba.sk/reports/display.php?id=50

16. Šimko, A.: Meta-interpreter for logic programs with preferences on rules, April 2013. http://dai.fmph.uniba.sk/~simko/lpp/
17. Wang, K., Zhou, L., Lin, F.: Alternating fixpoint theory for logic programs with priority. In: Palamidessi, C., et al. (eds.) CL 2000. LNCS (LNAI), vol. 1861, pp. 164–178. Springer, Heidelberg (2000)
18. Zhang, Y., Foo, N.Y.: Answer sets for prioritized logic programs. In: Proceedings of the 1997 International Logic Programming Symposium (ILPS'97), pp. 69–83 (1997)
19. Zhang, Y., Foo, N.Y.: Towards generalized rule-based updates. In: Proceedings of the Fifteenth International Joint Conference on Artificial Intelligence, IJCAI 97 (1997)

Some Experiments on Light-Weight Object-Functional-Logic Programming in Java with Paisley

Baltasar Trancón y Widemann[1,2]([⊠]) and Markus Lepper[2]

[1] Programming Languages and Compilers,
Technische Universität Ilmenau, Ilmenau, Germany
baltasar.trancon@tu-ilmenau.de
[2] <semantics/> GmbH, Berlin, Germany

Abstract. The Paisley library and embedded domain-specific language provides light-weight nondeterministic pattern matching on the Java platform. It fully respects the imperative semantics and data abstraction of the object-oriented paradigm, while leveraging the declarative styles of pattern-based querying and searching of complex object models. Previous papers on Paisley have focused on the functional paradigm and data flow issues. Here, we illustrate its use under the logic paradigm. We discuss the expressiveness and evaluate the performance of Paisley in terms of the well-known combinatorial search problem "send more money" and its generalizations.

1 Introduction

We describe one link in a chain of efforts to bring the object-oriented programming paradigm closer to the more declarative functional and logic paradigms. Historically, there have been many attempts to reconstruct or reinvent objects on top of a logic platform, for instance the early [1,2], with later implementations sharing the same basic design considerations; however, our basic approach is exactly opposite. Our starting point is a full commitment to mainstream, statically typed object-orientation, undoubtedly the dominant paradigm of our times, with unparalleled tool and library support for real-world programming. We develop "prosthetic" tools and programming techniques that amend well-known weaknesses in the expressiveness of plain object-orientation, without sacrificing broadness of scope or forcing programmers to leave their comfort zone. See the homepage at [3].

The present paper presents first results on the use of our Paisley library and language, designed for object-oriented pattern matching, as a toolkit for logic programming on an object-oriented platform.

1.1 Outline

The remainder of this paper is structured as follows: Sect. 2 summarizes the design of Paisley and its practical consequences under the various paradigms,

M. Hanus and R. Rocha (Eds.): KDPD 2013, LNAI 8439, pp. 218–233, 2014.
DOI: 10.1007/978-3-319-08909-6_14, © Springer International Publishing Switzerland 2014

as far as needed for understanding the following case studies. Technical details, further examples and comparison to related work can be found in [4,5].[1] Section 3 demonstrates logic programming in the Paisley style by means of cryptarithmetic puzzles and their most famous instance, "send more money". Section 4 presents comparative performance measurements. Section 5 summarizes the experiences gained so far, and gives some outlook into future work.

2   Paisley

2.1   Design Considerations

The Paisley library and programming style [4,5] provide sorely missed pattern matching capabilities to the Java platform. For both theoretical and practical reasons, it does so in the form of a light-weight embedded domain-specific language (DSL). The following paragraphs discuss the conceptual implications of the approach concerning style and software engineering. The philosophically unconcerned reader is welcome to skip ahead for more technical matters.

The qualifier "embedded" means that absolutely no extension of the language or associated tools such as compilers or virtual machines is required. Extending an evolving language such as Java, although academically attractive, is fraught with great practical problems, mostly of maintenance and support: History shows that language extensions either get adopted into the main branch of development quickly, or die as academic prototypes. Instead, an embedded language shares the syntax, type system and first-class citizens of its host language, that is in the Java case, objects. Ideally, it is also "reified", meaning that elements expose their DSL-level properties at host-level expressive public interfaces, and can be constructed, queried and manipulated freely and compositionally by the user.

The qualifier "light-weight" means that there is no technical distinction between "source" and "executable" forms of the DSL. No global pre-processing or compilation procedure is required, and no central interpreter engine exists. The capabilities of the DSL are distributed modularly over the implementation of DSL elements in the host language.

The two qualifiers together have wide-ranging implications for programming: They ensure that the embedded language is open and can easily be extended and customized by the user. They also guarantee tight integration and "impedance match" of interfaces, with fast and precise transfer of control and without data marshaling, between domain-specific and host-level computations. The price for this freedom is that a compositional structure precludes some global optimizations and refactorings of the DSL implementation.

For illustration purposes, consider parser combinators in a functional language as a prime example of reified light-weight embedded DSLs. Subparsers are ordinary (monadic) functions and can be defined and used directly as such,

---

[1] Online documentation and demonstration package with Paisley library binaries and multiple examples, including source code quoted here, available from [3] at http://bandm.eu/metatools/paisley/.

including definitions of whole multi-level context-free grammars as recursive functional programs. On the downside, global syntax analysis such as performed routinely by monolithic parser generators is poorly supported in a combinatorial setting.

By contrast, consider regular expression notations for string matching as a prime example of DSLs that are neither reified, embedded nor light-weight. Typical implementations involve compilation to nondeterministic finite automata (NFA), and hide their implementation behind terse global interfaces. User control over control features, most importantly nondeterminism, is indirect and awkward (in the form of a plethora of analogous combinators with subtly different amounts of greediness). On the upside, the NFA implementation is well-known to improve global performance greatly in appropriate cases. The XPath language for XML document navigation is another prominent example of the same kind.

Non-reified DSLs are typically used in a monolithic fashion: whole DSL programs are passed textually at the platform interface, and the computational means for *programming* the DSL are conceptually and technically separate from the means for *implementing* it. Conversely, reified DSLs lend themselves to compositional programming: In the simplest case a DSL program is a statically nested constructor expression, that is the abstract syntax analog of a literal non-reified DSL program. But the real power comes from more complex uses, where the structure of the program under construction is either *abstracted* into host-level functions, or *dynamized* by host-level control flow. (Contrast the construction code depicted in Figs. 2, 3, 4, 5, 6, and 7 with the resulting DSL programs depicted in Fig. 10.)

Besides language embedding strategies, there are two important features to consider, which we have done in detail in [4,5]: Firstly, whether the static type system of the host language (if any) is reflected in patterns. A positive answer rules out dynamic host languages like Python or Ruby, but brings the obvious great clarity and safety benefits. Secondly, whether nondeterminism is supported as a first-class aspect orthogonal to data projection. Pure paradigms are divided about this issue: it is a characteristic feature of logic programming, but not present in relevant functional or object-oriented languages.

2.2 Patterns, Object-Orientedly

Several theoretically well-founded paradigms for pattern matching exist: regular expressions, inverse algebraic semantics (in functional programming), term unification (in logic programming). However, for a tool to be practically useful in an object-oriented environment, it is of crucial importance not to impose any of the axioms of such theories, since they are typically not warranted for realistic object data models, and pretending otherwise causes impedance mismatch and is a source of much trouble and subtle bugs.

What, then, is object-oriented pattern matching proper? Starting from the rough approximation that (mainstream) object-orientation is imperative programming with data abstraction (encapsulation), patterns are a declarative specialist notation for data *queries*: They are applicable to object data models solely in terms of their public interface, which may not safely be assumed to have sound mathematical properties, such as statelessness, invertibility, completeness or extensionality. Object-oriented pattern matching organizes actual (getter) method calls, not meta-level semantic case distinctions.

Four generic aspects of querying can be discerned: data are subjected to *tests* for acceptability; matching may proceed to other accessible data by *projections*; information may flow back to the user in the form of side-effect variable *bindings*; control flow of matching links different patterns according to the outcome of tests with logical *combinators*. The absence of compositional programming constructs for these aspects in Java leads to awkward idioms, discussed in detail in [5].

In Paisley, a pattern that can process data of some object type A is an object of type Pattern⟨A⟩. A match is attempted by invoking method **boolean** match(A target), with the return value indicating success (determined by the test aspect of the pattern). Nondeterminism is generally allowed, in the sense that a pattern may match the same target data in more than one way. These multiple solutions can be explored by repeatedly invoking method **boolean** matchAgain() until it fails. See Fig. 9 for a typical loop-based usage example.

Extracted information (determined by the projection and binding aspects of the pattern) is not available from the pattern root object, nor reflected in its type. Instead, references to the variables occurring in the pattern must be retained by the user. Variables are objects of type Variable⟨A⟩ **extends** Pattern⟨A⟩. They match any target data deterministically and store a reference by side-effect to represent the binding, which can then be extracted with method A getValue(). Again, see Fig. 9 for the extraction of binding values in the loop body. Variables are imperative, in the sense that they have no discernible unbound state, and may be reused (sequentially) at no cost.

Elementary patterns performing particular test and projection duties are predefined in the Paisley library and may be extended freely by the user. They are combined by two universal control flow combinators for *conjunction* (all) and *disjunction* (some). These are fully aware of nondeterminism (analogous to the Prolog operators , and ;), and also guarantee strict sequentiality of side-effects and have very efficient implementations for deterministic operands (analogous to the C-family operators && and | |). As an immediate consequence, subpatterns may observe bindings of variables effected in earlier branches of a conjunction.

2.3   Patterns, Functionally

The core concept of functional pattern matching, namely that initial algebra semantics can be imposed on data and inverted for querying, is valid only for degenerate cases of object-oriented programming. Real-world interface contracts are more subtle; while effective query strategies can be devised for particular problems, automatic "optimizations" such as transparent pattern restructuring for compilation of pattern-based definitions [6, 7] are generally out of the question. Nevertheless, several functional principles can be used to good effect in the design of a powerfully abstract pattern object library:

Pattern building blocks effecting projections often correspond directly to a getter method of the object data model. Getters of class C with result type D can be conceived as functions from C to D; patterns of type Pattern⟨A⟩ can be conceived as functions from A to some complex solution/effect type. Hence each getter induces a contravariant lifting from Pattern⟨D⟩ to Pattern⟨C⟩ by mere function composition (the Hom-functor for the categorically-minded).

Functions between pattern types, both patterns as ad-hoc functions of a distinguished variable, and encapsulated pattern factories, are a very powerful abstraction, and a prerequisite for higher-order pattern operations. In Paisley they are represented by the interface Motif⟨A, B⟩ with a method Pattern⟨B⟩ apply(Pattern⟨A⟩).

2.4   Patterns, Logically

The main contribution of the logic paradigm to the design of Paisley is ubiquitous and transparent nondeterminism. It integrates with the operational semantics of patterns by having a fixed and precise resolution strategy, namely backtracking with strictly ordered choices. The other key idea of patterns in logic programming, namely unification, does not carry over soundly to the object-oriented paradigm, because of the lack of a stable global notion of equality for objects.

Nondeterminism can be introduced ad-hoc using explicit disjunctive combinators. But a natural kind of more abstract and useful sources of nondeterminism is the *imprecise* lifting of parameterized getter functions, abstracting from their qualifying parameters. For instance, the method A get (int index) of the Java collection interface List⟨A⟩ gives rise not only to a deterministic lifting from (Pattern⟨A⟩, int) to Pattern⟨List⟨A⟩⟩, but also to a nondeterministic variant from Pattern⟨A⟩ to Pattern⟨List⟨A⟩⟩ that tries each element of the target list in order. It is implemented in Paisley as the factory method CollectionPatterns.anyElement.

Nondeterminism, once introduced, is operationalized by the logical pattern combinators. A highly portable backtracking implementation is realized by eliminating the choice stack from the call stack (to which the programmer has limited access on the Java platform), and storing choice points on the heap, directly in the objects that instantiate pattern conjunction and disjunction. This has the notable effect that dynamic backtracking state is reified alongside static pattern structure, and can be deferred indefinitely, canceled abruptly, cloned and reused, committed to persistent storage etc., without interfering with normal control flow. The price for this flexibility is that the call stack needs to be reconstructed for backtracking (by iterated recursive descent), and that some caveats regarding pattern sharing and reentrance apply.

In the functional-logic spirit, matches of a pattern p as nondeterministic function of a distinguished variable x applied to target data t can be exhaustively explored (encapsulated search) by writing x.eagerBindings(p, t) or x.lazyBindings (p, t), with immediate or on-demand backtracking, respectively.

Combinatorial search problems can be encoded in the Paisley style as follows: Nondeterministic *generator* patterns for the involved variables are combined conjunctively (spanning the Cartesian product of solution candidates) and combined with *constraints*. Constraints are represented as patterns that take no target data, may observe previously bound variables (by earlier branches of a conjunction) and succeed at most once. They are implemented in Paisley by the class Constraint extends Pattern⟨Object⟩ with the method **boolean** test().

There is a strong trade-off between the effort to determine that a constraint is safe to test because all concerned variables have been bound, and the associated gain due to early pruning of the search tree. The following case study discusses a prominent combinatorial search problem, its generic object-oriented implementation in the Paisley style, and various strategies spread along the axis of the trade-off, from brute force to complex scheduling. It demonstrates how well Paisley handles logic-intensive and massively nondeterministic problems. We expect the positive results to scale to fairly substantial problems in an object-oriented scenario, where nondeterminism is usually much milder and limited to local searching queries.

3   Case Study

The arithmetical puzzle "send more money" [8] is a well-known combinatorial problem that has been used ubiquitously to exemplify notations and implementations of logic programming. It specifies an assignment of decimal digits to variables $\{D, E, M, N, O, R, S, Y\}$, such that $SEND + MORE = MONEY$ in usual decimal notation. This equation, together with the implicit assumptions that the assignment is *injective* (the variable values are all different) and the numbers are *normal* (leading digits are nonzero), has a unique solution, namely $\{D = 7, E = 5, M = 1, N = 6, O = 0, R = 8, S = 9, Y = 2\}$.

Figure 1 shows a straightforward implementation in the (functional-)logical paradigm.[2] This implementation is typical in the sense that it distinguished several subproblems, indicated by the flush right comments, which shall reappear in our object-oriented model. However, it is has more heavyweight platform requirements than our implementation: In particular, a finite domain constraint solver and corresponding driver for the CLPFD frontend must be available, thus restricting the range of compatible Curry implementations. By contrast, our approach makes do with any standard Java platform and elementary Paisley logics.

The particular "send more money" problem easily suggests a number of generalizations, and has indeed not been the first of its kind. As a fairly broad class of similar problems, we consider the *cryptarithmetic puzzles* with arbitrary number of digits in each term, arbitrary number of terms in the sum, and arbitrary choice of base.

The following sections present different solution strategies with increasing performance and implementation complexity. The meta-level discussion is complemented with corresponding fragments of the actual application code, written in Java 7 using the Paisley libraries and style. Only basic knowledge of the Java

---

[2] Available from the Curry homepage at http://www.informatik.uni-kiel.de/~curry/examples/CLP/smm.curry.

```
import CLPFD

smm l =
    l =:= [ s,e,n,d,m,o,r,y] &                                        -- variables
    domain l 0 9 &                                                    -- domain
    s >#0 & m >#0 &                                                   -- no leading zeroes
    allDifferent  l  &                                               -- all different
                        1000 *# s +# 100 *# e +# 10 *# n +# d
    +#                  1000 *# m +# 100 *# o +# 10 *# r +# e
    =# 10000 *# m +# 1000 *# o +# 100 *# n +# 10 *# e +# y &          -- sum
    labeling []  l
    where s,e,n,d,m,o,r,y free
```

Fig. 1. "Send more money" in Curry with finite domain constraints

syntax and collection framework is necessary to fully appreciate the code fragments; no particularly advanced or obscure coding techniques are employed. As a truly embedded language, Paisley can be interspersed finely with host code, and consequently hard to spot. For the reader's convenience, all Paisley-specific types and operations are underlined. Note that the purpose of most of the Java code is to *construct* an embedded Paisley program for combinatorial search; for some possible results see Fig. 10 below.

3.1 Basic Model

The basic model of cryptarithmetic puzzles is depicted in Fig. 2. It is parameterized at construction time with the chosen base, and terms encoded as strings. For instance, the original puzzle can be specified concisely as:

```
new CryptArith(10, "SEND", "MORE", "MONEY")
```

Also at construction time, Paisley objects for computations independent of a particular solution strategy are allocated internally: A variable is assigned to each character occurring in the terms; computed by a method vars(String...) not shown. A local non-zero constraint is assigned to each character occurring in leading position; computed by method noLeadingZeroes(String...). A global constraint expressing the sum equation in terms of the created variables is formed from the terms; computed by method sum(String...).

The constraint sum created by the latter adds the values of all terms but the last, and compares the sum to the value of the last term. For term evaluation it resorts to the auxiliary method number that computes a number from its b-adic representation by the currently bound values of the sequence of digit variables. Hence the constraints sum, as well as the constraints noLeadingZeroes depend on all or one variable, respectively, and must be tested only after the concerned variables have been bound successfully.

```java
public class CryptArith {
  private final int base;
  private final Map⟨Character, Variable⟨Integer⟩⟩ vars;
  private final Map⟨Variable⟨Integer⟩, Constraint⟩ noLeadingZeroes;
  private final Constraint sum;

  public CryptArith(int base, String... args) {
    if (base < 2 || args.length < 1)
      throw new IllegalArgumentException();
    this.base = base;
    this.vars = vars(args);
    this.noLeadingZeroes = noLeadingZeroes(args);
    this.sum = sum(args);
  }

  private Constraint sum(String... args) {
    final int n = args.length;
    final List⟨List⟨Variable⟨Integer⟩⟩⟩ rows = new ArrayList⟨⟩(n);
    for (String s : args) {
      // add characterwise list of variables to rows
    }
    return new Constraint() {
      public boolean test() {
        int s = 0;
        for (List⟨Variable⟨Integer⟩⟩ r : rows.subList(0, n − 1))
          s += number(r);
        return s == number(rows.get(n − 1)) ;
      }
    };
  }

  private Map⟨Variable⟨Integer⟩, Constraint⟩ noLeadingZeroes(String... args) {
    final Map⟨Variable⟨Integer⟩, Constraint⟩ result = new HashMap⟨⟩() ;
    for (String s : args) {
      final Variable⟨Integer⟩ v = vars.get(s.charAt(0)) ;
      result.put(v, Constraints.neq(v, 0)) ;
    }
    return result ;
  }

  private int number(List⟨? extends Variable⟨Integer⟩⟩ vs) {
    int n = 0;
    for (Variable⟨Integer⟩ v : vs)
      n = n * base + v.getValue();
    return n;
  }
  // ... see Figs. 3–5, 7–9
}
```

Fig. 2. Basic model of cryptarithmetic puzzles

Any solution strategy for the cryptarithmetic puzzles consists of nondeterministic matches of all variables against valid digit values (ranging from zero, inclusive, to the given base, exclusive), and constraints equivalent to injectiveness (pairwise difference of all variables), absence of leading zeroes and the sum equation. Strategies differ in, and draw their varying efficiency from, the early use of constraints to prune the search tree.

3.2 Brute-Force Generate and Test

The programmatically simplest, least efficient strategy is to defer all constraints until after all variables have been bound. This is of course the infamous *generate and test* pattern for combinatorial search. The implementation is depicted in Fig. 3. It uses a generic auxiliary method generate to produce a generator pattern by mapping a nondeterministic Motif pattern function over the collection of variables, and another generic auxiliary method allDifferent to constrain them. The latter traverses a triangle matrix of all variables in order to produce pairwise inequality constraints.

This strategy refers to auxiliary methods, shared by the other strategies, depicted in Fig. 4. The method domain() produces the nondeterministic motif

```
public Pattern⟨Iterable⟨? extends Integer⟩⟩ strategy1() {
    return Pattern.all(generate(domain(), vars.values()),      // generate
                       allDifferent(vars.values()),            // and test,
                       Pattern.all(noLeadingZeroes.values()),  // test,
                       sum);                                    // test.
}

private ⟨A, B⟩ Pattern⟨B⟩ generate(Motif⟨A, B⟩ m,
                                  Collection⟨Variable⟨A⟩⟩ vars) {
    final List⟨Pattern⟨B⟩⟩ ps = new ArrayList⟨⟩();
    for (Variable⟨A⟩ v : vars)
        ps.add(m.apply(v));                     // search space: single variable,
    return Pattern.all(ps);                     // and Cartesian product
}

private ⟨A⟩ Constraint allDifferent(Collection⟨Variable⟨A⟩⟩ vars) {
    final List⟨Variable⟨A⟩⟩ done = new ArrayList⟨⟩(vars.size());
    final List⟨Constraint⟩ neqs = new ArrayList⟨⟩();
    for (Variable⟨A⟩ v : vars) {
        for (Variable⟨A⟩ u : done)
            neqs.add(neq(u, v));
        done.add(v);
    }
    return Constraint.all(neqs);
}
```

Fig. 3. Strategy 1: naïve generate and test

```
private Motif⟨Integer, Iterable⟨? extends Integer⟩⟩ domain() {
  return CollectionPatterns.anyElement();
}

private ⟨A⟩ Constraint neq(final Variable⟨A⟩ v, final Variable⟨A⟩ w) {
  return new Constraint() {
    public boolean test() {
      return !v.getValue().equals(w.getValue());
    }
  };
}
```

Fig. 4. All strategies: generic utilities

used to generate candidate values for variables from a given enumeration, by simply instantiating a generic motif for nondeterministic element selection from the Paisley collection framework. We envisage the "abuse" of the pattern matching target to represent the search space as a general style pattern for logic programming in Paisley. The method neq(Variable, Variable) produces a single inequality constraint between the current values of two variables.

3.3 Early Checking of Simple Constraints

The preceding brute-force strategy 1 has the disadvantage of actually generating all possible variable assignments, that is b^n combinations for n variables over base b. But most of the constraints that prune the search tree (in fact all of them except the sum equation proper) concern at most two variables. Hence it is easy to predict the earliest point in the search plan where they can be checked.

This observation gives rise to an improved strategy 2 depicted in Fig. 5. It works by splicing together the first three phases of strategy 1, each of which has a loop over the variables, into a single loop. Only the global constraint of the sum equation, which concerns all variables and must necessarily come last, is left behind. This straightforward refactoring reduces the number of generated assignments greatly, to less than $n! \cdot \binom{b}{n}$.

3.4 Exploiting Partial Sums

It is known that early pruning of the search tree can be improved further by approximations to the sum using modular arithmetics. Each partial sum of the k least significant digits must be satisfied up to carry, which can be expressed as a congruence modulo b^k. While these partial sum relations are implied by the exact sum equation (a congruence modulo infinity), and hence logically redundant, they have the practically advantageous property of concerning fewer variables. Hence they can be checked earlier in the search plan. Figure 6 depicts the partial sum congruences for "send more money" with $k = 1, \ldots, 5$, each together with the set of variables concerned for that k at the earliest.

```
public Pattern⟨Iterable⟨? extends Integer⟩⟩ strategy2() {
  return Pattern.all(generateAndTestEarly(domain(), vars.values()), sum);  // ... test.
}

private ⟨A, B⟩ Pattern⟨B⟩ generateAndTestEarly(Motif⟨A, B⟩ m,
                                   Collection⟨Variable⟨A⟩⟩ vars) {
  final List⟨Variable⟨A⟩⟩ done = new ArrayList⟨⟩ (vars.size());
  final List⟨Pattern⟨? super B⟩⟩ pats = new ArrayList⟨⟩ ();
  for (Variable⟨A⟩ v : vars) {
    pats.add(m.apply(v));                    // generate
    if (noLeadingZeroes.containsKey(v))
      pats.add(noLeadingZeroes.get(v));      // and test,
    for (Variable⟨A⟩ u : done)
      pats.add(neq(u, v));                    // test, ...
    done.add(v);
  }
  return Pattern.all(pats);
}
```

Fig. 5. Strategy 2: generate with early checks

$D +$	$E \equiv$	Y	mod	10	$\{D, E, Y\}$
$ND +$	$RE \equiv$	EY	mod	100	$\{N, R\}$
$END +$	$ORE \equiv$	NEY	mod	1\,000	$\{O\}$
$SEND + MORE \equiv$		$ONEY$	mod	10\,000	$\{M, S\}$
$SEND + MORE \equiv MONEY$			mod	∞	$\{\}$

Fig. 6. Partial sum modular congruences for "send more money"

The strategic information discussed above is reified in the model extension depicted in Fig. 8 below. The inner class PartialSum encapsulates both the set of concerned variables and the associated congruence as a constraint. It refers to an extended version of the auxiliary method sum (compare Fig. 2), parameterized with the number of digits under consideration. The sequence of partial sums is precomputed at model construction time. The full strategy 3 can then be generated by a loop over this sequence, issuing generators for newly introduced variables, inequality and nonzero checks, and partial sum congruences in turn, as depicted in Fig. 7.

3.5 Dynamic Constraint Scheduling

If precise constraint scheduling, as in the preceding strategy, is deemed unfeasible, one can still resort to dynamic scheduling techniques. Constraints can be implemented such that their evaluation is suspended if some concerned variable is not yet bound, and resumed when that condition changes. Then, a trivial strategy simply places the generators last. However, tracking the state of variables has

```
public Pattern⟨Iterable⟨? extends Integer⟩⟩ strategy3() {
  return generatePartialSums(domain());
}

⟨B⟩ Pattern⟨B⟩ generatePartialSums(Motif⟨Integer, B⟩ m) {
  final List⟨Variable⟨Integer⟩⟩ done = new ArrayList⟨⟩();
  final List⟨Pattern⟨? super B⟩⟩ pats = new ArrayList⟨⟩();
  for (PartialSum s : partialSums) {
    for (Variable⟨Integer⟩ v : s.getDependencies())
      if (!done.contains(v)) {
        pats.add(m.apply(v));                    // generate
        if (noLeadingZeroes.containsKey(v))
          pats.add(noLeadingZeroes.get(v));      // and test,
        for (Variable⟨Integer⟩ u : done)
          pats.add(neq(v, u));                   // test,
        done.add(v);
      }
    pats.add(s.getConstraint());                 // test.
  }
  return Pattern.all(pats);
}
```

Fig. 7. Strategy 3: generate with partial sums (operation)

a significant run-time overhead, especially in Paisley where all components are as light-weight as possible, and nondeterminism is expected to incur no appreciable cost if not actually used.

We have added a prototype implementation of suspendable constraints to our case study, implemented using the well-known *observer* pattern of object-oriented programming. Strategies 2' and 3' are the analogs of 2 and 3, respectively, but with suspendable constraints preceding the generators they depend on.

4   Evaluation

All experiments have been performed on a single MacBook Air containing an Intel Core i5-3317U CPU with 4 cores at 1.7 GHz and 8 GiB of RAM running the OpenJDK 7-21 Java environment on Ubuntu 12.10. Reported times are wall-clock time intervals, measured with System.nanoTime() to the highest available precision. All experiments have been repeated 10 times, without restarting the Java machine or interfering with automatic memory management.

Table 1 summarizes the findings for all strategies, obtained by exhaustive search in a loop, as depicted in Fig. 9. Measurements are fairly consistent, with little random variation. Memory management appears to have negligible impact, as expected for a combinatorial computation with tiny data footprint.

```
private final List ⟨PartialSum⟩ partialSums = new ArrayList ⟨⟩ () ;

public CryptArith (int base, String... args) {
  // ...
  this.partialSums = partialSums (args);        // analogous to noLeadingZeroes
}

class PartialSum {

  private final int length;
  private final Set ⟨Variable ⟨Integer⟩ ⟩ dependencies ;        // cf. Fig. 6 right
  private final String[] args;

  PartialSum (int length,
              Set ⟨? extends Variable ⟨Integer⟩ ⟩ dependencies,
              String... args) {
    // initialize fields
  }

  public Set ⟨Variable ⟨Integer⟩ ⟩ getDependencies () {
    return dependencies ;
  }

  public Constraint getConstraint () {
    return sum (length, args);
  }
}

private Constraint sum (int length, String... args) {
  final int n = args.length;
  final List ⟨List ⟨Variable ⟨Integer⟩ ⟩ ⟩ rows = new ArrayList ⟨⟩ (n);
  for (String s : args) {
    // add characterwise list of last length variables to rows
  }
  final int m = power (base, length);
  return new Constraint () {
    public boolean test () {
      int s = 0;
      for (List ⟨Variable ⟨Integer⟩ ⟩ r : rows.subList (0, n − 1))
        s += number (r);
      return s % m == number (rows.get (n − 1)) % m ;        // congruence
    }
  };
}
```

Fig. 8. Strategy 3: generate with partial sums (model extension)

```
public void run(final Pattern⟨Iterable⟨? extends Integer⟩⟩ p) {    // p produced by a strategy
  if (p.match(digits())) do {                                       // confer Fig. 1 (domain)
    for (Map.Entry⟨Character, Variable⟨Integer⟩⟩ e : vars.entrySet())
      System.out.println(e.getKey() + "␣=␣" + e.getValue().getValue());
    System.out.println();
  } while (p.matchAgain());
}

private Collection⟨Integer⟩ digits() {
  final List⟨Integer⟩ result = new ArrayList⟨Integer⟩(base);
  for (int i = 0; i < base; i++)
    result.add(i);
  return result;
}
```

Fig. 9. Running a strategy

Table 1. Experimental evaluation: send more money, $N = 10$

Strategy	Time (ms)		
	Min	Median	Max
1	5 396.93	5 470.24	5 775.04
2	737.56	770.25	809.48
3	2.34	2.37	3.60
2'	761.37	771.93	797.53
3'	850.30	863.21	881.72

$D < b, E < b, S < b, R < b, N < b, O < b, M < b, Y < b,$
$D \neq E, D \neq S, E \neq S, D \neq R, E \neq R, S \neq R, D \neq N, E \neq N, S \neq N, R \neq N,$
$D \neq O, E \neq O, S \neq O, R \neq O, N \neq O, D \neq M, E \neq M, S \neq M, R \neq M, N \neq M,$
$O \neq M, D \neq Y, E \neq Y, S \neq Y, R \neq Y, N \neq Y, O \neq Y, M \neq Y,$
$M \neq 0, S \neq 0,$
$SEND + MORE \equiv MONEY \bmod \infty$

$D < b, E < b, D \neq E, S < b, S \neq 0, D \neq S, E \neq S, R < b, D \neq R, E \neq R, S \neq R,$
$N < b, D \neq N, E \neq N, S \neq N, R \neq N, O < b, D \neq O, E \neq O, S \neq O, R \neq O, N \neq O,$
$M < b, M \neq 0, D \neq M, E \neq M, S \neq M, R \neq M, N \neq M, O \neq M,$
$Y < b, D \neq Y, E \neq Y, S \neq Y, R \neq Y, N \neq Y, O \neq Y, M \neq Y,$
$SEND + MORE \equiv MONEY \bmod \infty$

$Y < b, D < b, D \neq Y, E < b, E \neq Y, E \neq D, D + E \equiv Y \bmod 10, R < b, R \neq Y,$
$R \neq D, R \neq E, N < b, N \neq Y, N \neq D, N \neq E, N \neq R, ND + RE \equiv EY \bmod 100,$
$O < b, O \neq Y, O \neq D, O \neq E, O \neq R, O \neq N, END + ORE \equiv NEY \bmod 1\,000,$
$S < b, S \neq 0, S \neq Y, S \neq D, S \neq E, S \neq R, S \neq N, S \neq O,$
$M < b, M \neq 0, M \neq Y, M \neq D, M \neq E, M \neq R, M \neq N, M \neq O, M \neq S,$
$SEND + MORE \equiv ONEY \bmod 10\,000, SEND + MORE \equiv MONEY \bmod \infty$

Fig. 10. Unfolded search plans generated by strategies 1, 2, 3, respectively

5   Conclusion

Figure 10 gives a synopsis of the inner structure of patterns produced by the strategies 1–3 for the "send more money" example. In each case elementary generator and constraint patterns are composed associatively into a global conjunction. For the more advanced strategies, more powerful constraints appear earlier in the sequence. The patterns can be used immediately as depicted in Fig. 9, or used in every other conceivable way as ordinary Java objects. As such, our implementation of the problem domain on top of Paisley acts technically as a domain-specific compiler to a threaded code back-end, given by the Paisley operations.

The Paisley approach leads to a style that has some of the best of both worlds: The object-oriented paradigm has excellent support for data abstraction and encapsulation. Object-oriented models of the problem domain have expressive interfaces close to the programmer's intentions and intuitions, and high documentation value.

On the other hand, the declarative style of the logic paradigm allows for abstraction from the complex control flow of searching by composition of simple nondeterministic fragments. Note that all explicit control flow in the given code samples is exclusively for the *construction* of a particular instance of the generic puzzle model. The actual control flow of searching is hidden entirely in the invocations of pattern combinators, most notably Pattern.all, consequently reified in a complex Pattern object that both represents and implements the search, and finally effected using Pattern.match and Pattern.matchAgain, as depicted in Fig. 9.

The influence of the functional paradigm is evidently the weakest in the examples discussed here: They make a single use of the Motif class. But there is considerable, unfulfilled potential: Virtually all of the loops in the example code express *comprehensions*, and could be rephrased in terms of the higher-order functions *map*, *reduce* and friends. These are conspicuously absent from the traditional Java collection framework; but there is hope that the rise of anonymous functions in Java 8 [9] will improve the situation. The Paisley approach is expected to profit greatly from equally high expressiveness in all three paradigms. Alternatively, Paisley could be ported to Scala, in order to reap the benefits of decent functional programming immediately.

The relative performance of more intelligent strategies within the Paisley framework is encouraging: A measured speedup of over three orders of magnitude by means of a moderately complex model extension that captures only well-understood heuristics about the problem domain is certainly worth the effort.

The absolute performance of Paisley implementations is of course no match for low-level optimized solver code. For instance, the C program obtainable from [10] takes approximately 0.17 ms to solve "send more money" on our test machine (compiled with gcc -O), about an order of magnitude less than our best effort with strategy 3. On the other hand, we have tested a simple, portable implementation in the functional-logic programming language Curry.[3] It differs

---

[3] Kindly contributed by F. Reck, co-developer of the KiCS2 compiler [11].

from the implementation depicted in Fig. 1 mainly by being self-contained pure Curry and not referring to any constraints requiring additional solver modules. Hence it follows a similar strategy as our strategy 2. Evaluation takes about 7.49 s on the same test machine (compiled with KiCS2 0.2.4 +optimize), an order of magnitude more than its most direct Paisley competitor, and even more than our brute-force strategy 1. Note that this result is not representative of the language at large; more sophisticated Curry implementations of cryptarithmetic puzzles such as the one described in [12] can be fairly competitive. The same holds to some degree for finite domain implementations as depicted in Fig. 1, depending on the strength of the available backend solver.

Considering the costs of portable backtracking and object-oriented data abstraction, Paisley appears to be well on the way. The only disappointment so far is the performance of the dynamically scheduled constraints in strategies 2' and 3', although the current implementation is merely a proof-of-concept prototype. Here the scheduling overhead clearly dominates the actual computation. More research into efficient implementations is needed.

References

1. Conery, J.S.: Logical objects. In: Kowalski, R., Bowen, K. (eds.) Proceedings of the 5th International Conference on Logic Programming, pp. 420–434. MIT Press (1988)
2. Kahn, K., Tribble, E.D., Miller, M.S., Bobrow, D.G.: Vulcan: logical concurrent objects. In: Shapiro, E. (ed.) Concurrent Prolog: Collected Papers, pp. 274–303. MIT Press, Cambridge (1987)
3. Lepper, M., Trancón y Widemann, B.: Metatools homepage (2013)
4. Trancón y Widemann, B., Lepper, M.: Paisley: pattern matching à la carte. In: Hu, Z., de Lara, J. (eds.) ICMT 2012. LNCS, vol. 7307, pp. 240–247. Springer, Heidelberg (2012)
5. Trancón y Widemann, B., Lepper, M.: Paisley: a pattern matching library for arbitrary object models. In: Proceedings der 6. Arbeitstagung Programmiersprachen (ATPS 2013). LNI, vol. 215, pp. 171–186. Gesellschaft fr Informatik (2013)
6. Augustsson, L.: Compiling pattern-matching. In: Jouannaud, J.-P. (ed.) FPCA 1985. LNCS, vol. 201, pp. 368–381. Springer, Heidelberg (1985)
7. Pettersson, M.: A term pattern-match compiler inspired by finite automata theory. In: Pfahler, P., Kastens, U. (eds.) CC 1992. LNCS, vol. 641, pp. 258–270. Springer, Heidelberg (1992)
8. Dudeney, H.E.: Strand Magazine **68**, p. 97 (1924)
9. Goetz, B.: Lambda expressions for the Java programming language (draft review 3). Java Specification Request 335, Oracle (2013)
10. Tamura, N.: Cryptarithmetic puzzle solver (2004). Accessed 1 June 2013
11. Braßel, B., Hanus, Michael, Peemöller, B., Reck, F.: KiCS2: a new compiler from curry. In: Kuchen, Herbert (ed.) WFLP 2011. LNCS, vol. 6816, pp. 1–18. Springer, Heidelberg (2011)
12. Antoy, S., Hanus, M.: Concurrent distinct choices. J. Funct. Program. **14**, 657–668 (2004)

Heuristic Search over Program Transformations

Claus Zinn[✉]

Department of Computer Science, University of Konstanz, Konstanz, Germany
claus.zinn@uni-konstanz.de

Abstract. In prior work, we have developed a method for the automatic reconstruction of buggy Prolog programs from correct programs to model learners' incorrect reasoning in a tutoring context. The method combines an innovative variant of algorithmic debugging with program transformations. Algorithmic debugging is used to indicate a learner's error and its type; this informs a program transformation that "repairs" the expert program into a buggy variant that is closer at replicating a learner's behaviour. In this paper, we improve our method by using heuristic search. To search the space of program transformations, we estimate the distance between programs. Instead of only returning the first irreducible disagreement between program and Oracle, the algorithmic debugger now traverses the entire program. In the process, all irreducible agreements and disagreements are counted to compute the distance metrics, which also includes the cost of transformations. Overall, the heuristic approach offers a significant improvement to our existing blind method.

1 Introduction

Typically, programs have bugs. We are interested in runtime bugs where the program terminates with output that the programmer judges incorrect. In these cases, Shapiro's algorithmic debugging technique can be used to pinpoint the location of the error. A dialogue between the debugger and the programmer unfolds until the meta-interpretation of the program reaches a statement that captures the cause of disagreement between the program's actual behaviour and the programmer's intent of how the program should behave. Once the bug has been located, it is the programmer's task to repair the program, and then, to start another test-debugging-repair cycle. Let us make the following assumption: there exists an Oracle that relieves the programmer from answering any of the questions during the debugging cycle; the Oracle "knows" the programmer's intent for each and every piece of code. With the mechanisation of the Oracle to locate the program's bugs, we now seek to automate the programmer's task to repair the bug, and thus, to fully automate the test-debug-repair cycle.

In the tutoring context, Oracles can be mechanised: for a given domain of instruction, there is always a reference model that defines expert problem solving behaviour. Moreover, a learner's problem solving behaviour is judged with regard to this model; a learner commits a mistake whenever the learner deviates

M. Hanus and R. Rocha (Eds.): KDPD 2013, LNAI 8439, pp. 234–249, 2014.
DOI: 10.1007/978-3-319-08909-6_15, © Springer International Publishing Switzerland 2014

from the expert problem solving path. Algorithmic debugging can be used to identify the location of learners' erroneous behaviour. For this, we have to turn Shapiro's method on its head: we take the expert program to take the role of the buggy program, and the learner to take the role of the programmer, that is, the Oracle. As in the traditional method, any disagreement between the two parties indicates the location of the bug. Moreover, we can relieve the learner from answering Oracle questions. Answers to all questions can be reconstructed from the learner's answer to a given problem, using the expert model [11].

With the ability to locate a learner's error, we now seek to "repair" the expert program (assumed buggy) in such a way that it reproduces the learner's erroneous (assumed expert) behaviour. The resulting program acts as symbolic artifact of a deep diagnosis of a learner's problem solving process; it can be used to inform effective remediation, helping learners to realize and correct their mistakes. Ideally, repair operators shall mirror typical learner errors. This is feasible indeed. There is a small set of error types, and many of them can be formally described in a domain-independent manner.

With the identification of an error's location, and a small, effective set of mutation operators for program repair, we strive to fully automate the test-debug-repair cycle in the tutoring context. Our approach is applicable for a wider context, given the specification of an ideal program and a theory of error.

Main contributions. To address an important issue in intelligent tutoring, the deep diagnosis of learner input, we cast the problem of automatically deriving one (erroneous) program from another (expert) program as a heuristic search problem. We define a metric that quantifies the distance of two given programs with regard to an input/output pair. We define a number of domain-independent code perturbation operators whose execution transforms a given program into its mutated variant. Most mutation operators encode typical actions that learners perform when encountering an impasse during problem solving. We show the effectiveness of our approach for the most frequent learner bugs in the domain of multi-column subtraction. Erroneous procedures are automatically derived to reproduce these errors. This work extends and generalises our previous work in this area [11,12] with regard to the heuristic search approach, which is novel.

Overview. Section 2 gives a very brief review on student errors in tutoring. It presents multi-column subtraction as domain of instruction and gives an encoding of the expert model in Prolog. For each of the top-eight learner errors in this domain, we demonstrate how the expert model needs to be perturbated to reproduce them. We show that most perturbations are based on a small but effective set of mutation operators. Also, we briefly review our existing method of error diagnosis in the tutoring context. In Sect. 3, we improve and generalise our method. The problem of deriving one program from another is cast in terms of a heuristic search problem. We introduce a distance metrics between programs that is based on algorithmic debugging, and use a best-first search algorithm to illustrate and evaluate the effectiveness of our approach. Section 4 discusses our approach and relates it to existing work. Section 5 concludes with future work.

2   Background

2.1   Human Error in Tutoring

When learning something new, one is bound to make mistakes. Effective teaching depends on deep cognitive analyses to diagnose learners' problem solving paths, and subsequently to repair the incorrect parts. Good teachers are thus capable to reconstruct students' erroneous procedures and use this information to inform their remediation. In the area of elementary school mathematics, our chosen tutoring domain, the seminal works of Brown and Burton [2,3], O'Shea and Young [10], and VanLehn [9], among others, extensively studied the subtraction errors of large populations of pupils. Their research included a computational account of errors by manually constructing cognitive models that reproduced learners' most frequent errors. The main insight of this research is that student errors are seldom random. There are two main causes. The first cause is that student errors may result from *correctly* executing an erroneous procedure; for some reasons, the erroneous rather than the expert procedure has been acquired. The second cause is based on VanLehn's theory of *impasses* and *repairs*. Following VanLehn, learners "know" the correct procedure, but face difficulties executing it. They "treat the impasse as a problem, solve it, and continue executing the procedure" [9, p. 42]. The repair strategies to address an impasse are known to be common across student populations and domains. Typical repairs include executing only the steps known to the learner and to skip all other steps, or to adapt the situation to prevent the impasse from happening.

2.2   Expert Model for Multi-column Subtraction

Figure 1 depicts the entire cognitive model for multi-column subtraction using the decomposition method. The Prolog code represents a subtraction problem as a list of column terms (M, S, R) consisting of a minuend M, a subtrahend S, and a result cell R. The main predicate subtract/2 determines the number of columns and passes its arguments to mc_subtract/3.[1] This predicate processes columns from right to left until all columns have been processed and the recursion terminates. The predicate process_column/3 receives a partial sum, and processes its right-most column (extracted by last/2). There are two cases. Either the column's subtrahend is larger than its minuend, when a borrowing operation is required, or the subtrahend is not larger than the minuend, in which case we can subtract the former from the latter (calling take_difference/4). In the first case, we add ten to the minuend (add_ten_to_minuend/3) by borrowing from the left (calling decrement/3). The decrement operation also consists of two clauses, with the second clause being the easier case. Here, the minuend of the column left to the current column is not zero, so we simply reduce the minuend by one. If the minuend is zero, we need to borrow again, and hence decrement/3 is called recursively. When we return from recursion, we first add ten to the minuend, and then reduce it by one.

---

[1] The argument CurrentColumn is passed onto most other predicates; it is only used to help automating the Oracle.

```
01 :  subtract(PartialSum, Sum) ←
02 :           length(PartialSum, LSum),
03 :           mc_subtract(LSum, PartialSum, Sum).

04 :  mc_subtract(_, [], []).
05 :  mc_subtract(CurrentColumn, Sum, NewSum) ←
06 :           process_column(CurrentColumn, Sum, Sum1),
07 :           shift_left(CurrentColumn, Sum1, Sum2, ProcessedColumn),
08 :           CurrentColumn1 is CurrentColumn − 1,
09 :           mc_subtract(CurrentColumn1, Sum2, SumFinal),
10 :           append(SumFinal, [ProcessedColumn], NewSum).

11 :  process_column(CurrentColumn, Sum, NewSum) ←
12 :           last(Sum, LastColumn), allbutlast(Sum, RestSum),
13 :           minuend(LastColumn, M), subtrahend(LastColumn, S),
14 :           S > M, !,
15 :           add_ten_to_minuend(CurrentColumn, M, M10),
16 :           CurrentColumn1 is CurrentColumn − 1,
17 :           decrement(CurrentColumn1, RestSum, NewRestSum),
18 :           take_difference(CurrentColumn, M10, S, R),
19 :           append(NewRestSum, [(M10, S, R)], NewSum).

20 :  process_column(CurrentColumn, Sum, NewSum) ←
21 :           last(Sum, LastColumn), allbutlast(Sum, RestSum),
22 :           minuend(LastColumn, M), subtrahend(LastColumn, S),
23 :           % S =< M,
24 :           take_difference(CurrentColumn, M, S, R),
25 :           append(RestSum, [(M, S, R)], NewSum).

26 :  shift_left( _CurrentColumn, SumList, RestSumList, Item ) ←
27 :           allbutlast(SumList, RestSumList), last(SumList, Item).

28 :  decrement(CurrentColumn, Sum, NewSum ) ←
29 :           irreducible,
30 :           last( Sum, (M, S, R) ), allbutlast( Sum, RestSum),
31 :           M == 0, !,
32 :           CurrentColumn1 is CurrentColumn − 1,
33 :           decrement(CurrentColumn1, RestSum, NewRestSum ),
34 :           NM is M + 10,
35 :           NM1 is NM − 1,
36 :           append( NewRestSum, [(NM1, S, R)], NewSum),

37 :  decrement(CurrentColumn, Sum, NewSum) ←
38 :           irreducible,
39 :           last( Sum, (M, S, R) ), allbutlast( Sum, RestSum),
40 :           % \+ (M == 0),
41 :           M1 is M − 1,
42 :           append( RestSum, [(M1, S, R)], NewSum ).

43 :  add_ten_to_minuend( _CC, M, M10) ← irreducible, M10 is M + 10.
44 :  take_difference(_CC, M, S, R) ← irreducible, R is M − S.

45 :  minuend( (M, _S, _R), M).
46 :  subtrahend( (_M, S, _R), S).

47 :  allbutlast([], []).
48 :  allbutlast([_H], []).
49 :  allbutlast([H1|[H2|T]], [H1|T1]) ← allbutlast([H2|T], T1).

50 :  irreducible.
```

Fig. 1. The decomposition method for subtraction in prolog

```
              9
      3      10     11
      4       0      1
  -   1       9      9
  =   2       0      2
   (a) correct solution
```
```
      4       0      1
  -   1       9      9
  =   3       9      8
   (b) smaller-from-larger
```
```
      3      10     11
      4       0      1
  -   1       9      9
  =   2       1      2
   (c) stops-borrow-at-zero
```

```
      2
      3      10     11
      4       0      1
  -   1       9      9
  =   1       1      2
   (d) borrow-across-zero
```
```
              9     11
      4       0      1
  -   1       9      9
  =   3       0      2
   (e) borrow-from-zero
```
```
             10     11
      4       0      1
  -   1       9      9
  =   3       1      2
   (f) borrow-no-decrement
```

```
                    11
      4       0      1
  -   1       9      9
  =   3       9      2
   (g)  stops-borrow-at-
   zero diff-0-N=N
```
```
      2
      3      11     11
      4       1      1
  -   1       9      9
  =   1       2      2
   (h) always-borrow-left
```
```
      3              11
      4       0      1
  -   1       9      9
  =   2       9      2
   (i)  borrow-across-zero
   diff-0-N=N
```

Fig. 2. A correct solution, and the top-eight bugs sets, see [9, p. 195].

2.3 Buggy Sets in Multi-column Subtraction

Figure 2(a) depicts the correct solution to the subtraction problem $401 - 199$, the Fig. 2(b)–(i) show how the top-eight bug sets from the DEBUGGY study [9, p. 195, p. 235] manifest themselves in the same task. All erroneous answers are rooted in learners' difficulty to borrow: the errors in Fig. 2(b) and (f) result from the learners' more general impasse "does not know how to borrow", and the errors in Fig. 2(c)–(e) results from the learners' more specific impasse "does not know how to borrow from zero". All other errors, but Fig. 2(h), are variations of the previous error types. Figure 2(h) is better explained by the incorrect acquisition of knowledge rather than within the impasse-repair theory.

We now describe how the expert procedure given in Fig. 1 needs to be "repaired" to reproduce each of the top-eight bugs.

smaller-from-larger: *the student does not borrow, but in each column subtracts the smaller digit from the larger one* [9, p. 228]. The impasse "learner does not know how to borrow" is overcome by not letting borrowing to happen. The expert model is perturbated at the level of `process_column/3`. In its first clause, we delete the calls to `add_ten_to_minuend/3` (line 15) and `decrement/3` (line 17). As a consequence, we replace all remaining occurrences of `M10` and `NewRestSum` with `M` and `RestSum`, respectively. Moreover, we swap the arguments for `M` and `S` when taking differences (line 18).

borrow-no-decrement: *when borrowing, the student adds ten correctly, but does not change any column to the left* [9, p. 223]. The learner addresses the impasse "does not know how to borrow" with a partial skipping of steps. In the first clause of process_column/3, the subgoal decrement/3 (line 17) is deleted; the remaining occurrence of NewRestSum is then replaced by RestSum (line 19).

stops-borrow-at-zero: *instead of borrowing across a zero, the student adds ten to the column he is doing, but does not change any column to the left* [9, p. 229]. The impasse "learner does not know how to borrow from zero" is overcome by not performing *complete* borrowing when the minuend in question is zero. The recursive call to decrement/3 (line 33) and the goals producing NM1 and NM (lines 34, 35) are removed, and the remaining occurrence of NM1 replaced by M (line 36).

borrow-across-zero: *when borrowing across a 0, the student skips over the 0 to borrow from the next column. If this causes him to have to borrow twice, he decrements the same number both times* [9, p. 114, p. 221]. Same impasse, different repair. The clauses that produce NM1 and NM (lines 34, 35) are removed; the remaining occurrence of NM1 in append/3 replaced by M (line 36).

borrow-from-zero: *instead of borrowing across a zero, the student changes the zero to nine, but does not continue borrowing from the column to the left* [9, p. 223]. Same impasse, yet another repair: the assignments NM and NM1 stay in place, but the recursive call to decrement/3 (line 33) is deleted; the occurrence of NewRestSum is replaced by RestSum (line 36).

stops-borrow-at-zero diff0-N=N: *when the student encounters a column of the form $0 - N$, he does not borrow, but instead writes N as the answer, possibly combined with* stops-borrow-at-zero. For diff-0-N=N, we shadow the existing clause for taking differences with take_difference(M, S, R):- M == 0, R = S. To ensure that no borrowing operation is performed in case the minuend is zero, the first clause of process_column/3 is modified. The constraint S>M (line 14) is complemented with \+ (M == 0); line 23 is changed to (S =< M) ; (M == 0).

always-borrow-left: *the student borrows from the left-most digit instead of borrowing from the digit immediately to the left* [9, p. 225]. This error is best explained by the incorrect acquisition of knowledge rather than within the impasse-repair theory. To reproduce it, we shadow the existing clauses for decrement/3 with decrement([(M,S,R)|OtherC], [(M1,S,R)|OtherC]) :- !, M1 is M - 1.

borrow-across-zero diff-0-N=N: *see above.* With both errors already been dealt with, we combine the respective perturbations to reproduce this error.

Summary. All error types except always-borrow-left require the deletion of one or more subgoals, with a tidying-up phase for their input and output arguments. For smaller-than-larger, the swapping of arguments was necessary. For always-borrow-left, we shadowed the existing clauses for decrement/3 with a new clause. While the top five errors can be reproduced by syntactic means, the last three errors seem to require elements whose construction will be hard to mechanise.

```
1: function RECONSTRUCTERRONEOUSPROCEDURE(Program, Problem, Solution)
2:     (Disagr, Cause) ← AlgorithmicDebugging(Program, Problem, Solution)
3:     if Disagr = nil then
4:         return Program
5:     else
6:         NewProgram ← PERTURBATION(Program, Disagr, Cause)
7:         RECONSTRUCTERRONEOUSPROCEDURE(NewProgram, Problem, Solution)
8:     end if
9: end function

10: function PERTURBATION(Program, Clause, Cause)
11:     return chooseOneOf(Cause)
12:         DELETECALLTOCLAUSE(Program, Clause)
13:         DELETESUBGOALSOFCLAUSE(Program, Clause)
14:         SWAPCLAUSEARGUMENTS(Program, Clause)
15:         SHADOWCLAUSE(Program, Clause)
16: end function
```

Fig. 3. Pseudo-code: compute variant of *Program* to reproduce a learner's *Solution*.

2.4   Existing Method

In [12], we have presented a method that interleaves algorithmic debugging with program transformations for the automatic reconstruction of learners' erroneous procedure, see Fig. 3. The function `ReconstructErroneousProcedure/3` is recursively called until a program is obtained that reproduces learner behaviour, in which case there are no further disagreements. Note that multiple perturbations may be required to reproduce single bugs, and that multiple bugs are tackled by iterative applications of algorithmic debugging and code perturbation.

The irreducible disagreement resulting from the algorithmic debugging phase locates the code pieces where perturbations must take place; its cause determines the kind of perturbation. The function `Perturbation/3` can invoke various kinds of transformations: the deletion of a call to the clause in question, or the deletion of one of its subgoals, or the shadowing of the clause in question by a more specialized instance, or the swapping of the clause' arguments. These perturbations reflect the repair strategies learners use when encountering an impasse.

Our algorithm for clause call deletion, e.g., traverses a given program until it identifies a clause whose body contains a call to the clause `Clause` in question; once identified, it removes `Clause` from the body and replaces all occurrences of its output argument by its input argument in the adjacent subgoals as well as in the clause's head, if present. Then, the modified program is returned.

There are many choice points as an action can materialise in many different ways. Our original method uses Prolog's built-in depth-first mechanism to *blindly* search the space of program transformations. Our new method uses a heuristics to make *informed* decisions during search.

3   Heuristic Search over Program Transformations

The problem of automatically reconstructing a Prolog program to model a learner's incorrect reasoning can be cast as a heuristic search problem. The initial state holds a Prolog program that solves arbitrary multi-column subtraction tasks in an expert manner. The goal state holds the program's perturbated variant whose execution reproduces the learner's erroneous behaviour. For each state s, a successor state s' can be obtained by the application of a single perturbation operator op_i. We seek a sequence of perturbation actions $op_1, op_2, ...op_n$ to define a path between start and goal state, with minimal costs.

3.1   Heuristic Function

Best-first search depends on a heuristic function to evaluate a node's distance to the goal node. For this, we extend our variant of algorithmic debugging. A heuristic score could be obtained, e.g., by counting the number of agreements until the first irreducible disagreement is found; however, when errors occur early in the problem solving process, this simple scoring performs poorly. Modifying the algorithmic debugger to always traverse the entire program and count all irreducible agreements and disagreements during traversal yields a better score.

Figure 4 depicts the algorithmic debugger in pseudo-code; it extends a simple meta-interpreter. Before start, both counters are initialised, and the references set for Goal, Problem, Solution to hold the top-level goal, the task to be solved and the learner's Solution to the task, respectively. There are four main cases. The meta-interpreter encounters either (i) a conjunction of goals, (ii) a goal that is a system predicate, (iii) a goal that does not need to be inspected, or (iv) a goal that needs to be inspected. For (i), algorithmic debugging is called recursively on each of the goals of the conjunctions; for (ii), the goal is called; and for (iii), we obtain the goal's body and ask the meta-interpreter to inspect it. The interesting aspect is case (iv) for goals marked relevant. Here, the goal is evaluated by both the expert program (using call/1) and the Oracle. The Oracle retrieves the learner's solution for the given Problem and reconstructs from it the learner's answer to the goal under discussion. Now, there are two cases. If system and learner agree on the goal's result, then the goal's weight is determined and added to the number of agreements; if they disagree, the goal must be inspected further to identify the exact location of the disagreement. If the goal is a leaf node, the irreducible disagreement has been identified and the disagreement counter is incremented by one; otherwise, the goal's body is retrieved and subjected to algorithmic debugging. The heuristic score is obtained by subtracting the number of agreements from the number of disagreements.

3.2   Best-First Search: Guiding Program Transformations with A^*

Typically, a search method maintains two lists of states: an *open list* of all states still to be investigated for the goal property, and a *closed list* for all states that were already checked for the goal property but where the check failed. Among

```
1:  NumberAgreements ← 0, NumberDisagreements ← 0
2:  Problem ← current task to be solved, Solution ← learner input to task
3:  Goal ← top-clause of routine, with input Problem and output Solution
4:  procedure ALGORITHMICDEBUGGING(Goal)
5:      if Goal is conjunction of goals (Goal1, Goal2) then
6:          ← algorithmicDebugging(Goal1)
7:          ← algorithmicDebugging(Goal2)
8:      end if
9:      if Goal is system predicate then
10:          ← call(Goal)
11:     end if
12:     if Goal is not on the list of goals to be discussed with learners then
13:         Body ← getClauseSubgoals(Goal)
14:             ← algorithmicDebugging(Body)
15:     end if
16:     if Goal is on the list of goals to be discussed with learners then
17:         SystemResult ← call(Goal)
18:         OracleResult ← oracle(Goal)
19:         if results agree on Goal then
20:             Weight ← computeWeight(Goal)    ▷ compute # of skills in proof tree
21:             NumberAgreements ← NumberAgreements + Weight
22:         else
23:             if Goal is leaf node (or marked as irreducible) then
24:                 NumberDisagreements ← NumberDisagreements + 1
25:             else
26:                 Body ← getClauseSubgoals(Goal)
27:                     ← algorithmicDebugging(Body)
28:             end if
29:         end if
30:     end if
31: end procedure
32: Score ← NumberDisagreements − NumberAgreements
```

Fig. 4. Pseudo-code: top-down traversal, keeping track of (dis-)agreements.

all the open states, *greedy* best-first search always selects the most promising candidate, i.e., the candidate that is most likely the closest to a given goal state. Our approach also takes into account the cost of program transformations. With the heuristic function defined as $f(n) = g(n) + h(n)$, we thus implement the A^*-algorithm. The cost function $g(n)$ returns the cost of producing state n. The function $h(n)$ estimates the distance between the program in state n and the goal state; it is described as `Score` in Fig. 4.

Representation. Each state in the search tree is represented by the term (`Algorithm, IrreducibleDisagreement, Path`), encoding a reified version of a Prolog program, the *first* irreducible agreement between program and learner behavior, and the path of prior perturbation actions to reach the current state. Each state n is also associated with a numerical value $f(n)$ that quantifies its production cost as well as the `Algorithm`'s distance to the algorithm of the goal state. A successor state of a given state results from applying a perturbation action. The action obtains a Prolog program, performs some sort of mutation, and returns a modified program. – We discuss our approach by example.

Initialisation. The start node holds the expert program (see Fig. 1) that produces the correct solution. Sought is a mutated variant of the expert program to produce the learner's erroneous solution, here the error `smaller-from-larger`:

$$
\begin{array}{c}
\begin{array}{r}
9 \\
3\ \cancel{10}\ 11 \\
4\ \cancel{0}\ \cancel{1} \\
-1\ 9\ 9 \\
\hline
=2\ 0\ 2
\end{array}
\quad
\xleftarrow[\text{producing}]{\text{expert}}
\boxed{\text{Start Node}}
\xleftarrow[\text{search}]{\text{heuristic}}
\boxed{\text{Goal Node}}
\xrightarrow[\text{producing}]{\text{learner}}
\begin{array}{r}
4\ 0\ 1 \\
-1\ 9\ 9 \\
\hline
=3\ 9\ 8
\end{array}
\end{array}
$$

Best-first search starts with initialising a heap data structure. For this, the start node's distance to the goal node is estimated, using the algorithm given in Fig. 4. There is no single agreement between expert program solving behaviour and learner behaviour, i.e., no single subtraction cell has been filled out the same way. There are six disagreements, yielding a heuristic score of $6 - 0 = 6$.

To inform the generation of the node's children, the first of the six irreducible disagreements – `add_ten_to_minuend(3, 1, 1)` – (1 instead of 11) is attached to the node's second component. The third component is initialised with the empty path `[]` (cost 0). The node and its estimate is then added to the empty heap.

Checking for Goal State. A state is a goal state when its associated program passes algorithmic debugging with zero disagreements. In this case, best-first search terminates with the goal state, returning the node's algorithm and its path, i.e., a list of actions that were applied to reach the goal state. Here, the initial node, with a non-zero number of disagreements is not the goal node.

Generation of Successor Nodes. If a given state is not the goal state, the state's successors are computed. Given the state's algorithm and the first irreducible disagreement that indicates the location of the "error", Prolog is asked to `findall` applicable perturbation actions, see Fig. 3. For the initial state, we obtain:

n_1 `DeleteCallToClause/2`: deletion of the call to `add_ten_to_minuend/3` in the first program clause `process_column/3` (line 15).

n_2 `ShadowClause/2`: addition of the irreducible disagreement (learner's view) `add_ten_to_minuend(3, 1, 1) :- irreducible.` to the program.

n_3 `DeleteSubgoalsOfClause/2`: deletion of subgoals from the definition of the predicate `add_ten_to_minuend/3`. As the goal `irreducible/0` cannot be deleted as it is needed by the Oracle, the only permissible action is to delete the subgoal `M10 is M + 10`, and to replace `M10` by `M` in the clause' header.

To add a successor node to the heap, the existing path is extended with the respective action taken. Also, for each node's algorithm, its first irreducible disagreement with the learner must be identified, and the distance to the goal node must be determined. For all successor nodes, we get the irreducible disagreement `decrement(2,[(4,1,S1), (0,9,S2)],[(3,1,S1), (9,9,S2)])`.

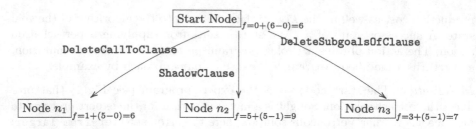

Fig. 5. Best-first search over program transformations: first expansion.

Now, consider the nodes' (dis-) agreement scores. Each of the nodes has five disagreements, one less than in the parent node; the second and third child now also feature one agreement. In node n_2, there is an agreement with the new clause add_ten_to_minuend/3 added; in node n_3, there is an agreement with the perturbated clause add_ten_to_minuend/3.

Action Cost. Some program transformations are better than others. Consider the ShadowClause action, yielding mutations that are specific to a given input/output pair. The resulting program will, thus, reproduce the learner's error only for the given subtraction task, not for other input. The action's lack of generality is acknowledged by giving it a high cost, namely 5. Hence, the action will only be used when more general and less costly actions are not applicable.

The action DeleteSubgoalsOfClause can delete more than a single subgoal from the predicate indicated by the disagreement. This adds a notion of focus to the perturbation and mirrors the fact that learners often address an impasse with skipping one or more steps of the skill in question. Its cost is defined by the number of subgoals deleted; a penalty is added, however, when the resulting body is left with the single subgoal **irreducible**. The mutations performed by DeleteCallToClause and SwapClauseArguments have a cost of 1.

Score and Continued Search. Figure 5 depicts the scores obtained. Best-first search selects the child with the lowest valuation, n_1. Its irreducible disagreement on the decrement/3 operation in column 2 can be addressed by eight different repairs: the deletion of the call to decrement/3 in the first clause of process_column/3 (line 17), the addition of the disagreement clause (the learner's view) to the program, and the removal of one or more subgoals in any of the two clause definitions for decrement/3 (6 possible repairs). It shows that the deletion of line 17 yields the algorithm with the lowest overall estimate; the algorithm's irreducible disagreement at take_difference(3,1,9,8) (8 *vs.* −8) lets best-first search determine the final perturbation action, which is SwapClauseArguments/2 to swap the arguments of take_difference/4 in the first clause of process_column/3.

3.3   Evaluation

We have tested our new approach against the eight most frequent bugs of Van-Lehn's study (Fig. 2). For this, we have implemented the following three search

Table 1. Evaluation: blind-search *vs.* cost-based search *vs.* A^*

Error	Blind	Cost-based	A^*
smaller-from-larger	33.882K-N	2.385K-Y	659K-Y
borrow-no-decrement	425K-N	425K-Y	425K-Y
stops-borrow-at-zero	406K-N	640K-N	407K-Y
borrow-across-zero	444K-N	1.126K-Y	446K-Y
borrow-from-zero	457K-N	510K-Y	457K-Y
stops-borrow-at-zero. diff0-N=N	5.054K-N	108.832K-C	2.786K-C
always-borrow-left	111K-N	2.831K-C	111K-C
borrow-across-zero. diff0-N=N	5.053K-N	203.125K-C	1.874K-C

methods: blind (depth-first) search where each node n is associated with the value $f(n) = 1$, cost-based search where each node is associated with its construction cost $f(n) = g(n)$, and A^*-search where each node is associated with its construction cost and its estimated distance to the goal node: $f(n) = g(n)+h(n)$.

Table 1 compares the three search methods using the two metrics *performance* and *quality of solution*. Performance is measured in terms of inferences required to obtain the first goal node (as computed by SWI-Prolog's `time/1` predicate). The quality of solution is measured using the perturbations described in Sect. 2.3 as gold standard. Inference numbers are either annotated with "Y" (the gold standard has been reproduced automatically), "C" (the reproduction is close to the gold standard), and "N" (no reproduction).

All three search methods have access to the same arsenal of actions, which includes the action `ShadowClause`. This perturbation acts as a fallback mechanism and ensures that all search terminates with a program mutation whose execution reproduces the learner's erroneous answer to a given subtraction task. If a `ShadowClause` action has been applied, the resulting mutation is task-specific; it usually fails to reproduce a learner's consistent erroneous behaviour across other tasks. The explanatory power of the resulting mutation is rather limited.

In blind search, all perturbation actions have equal cost. Therefore, blind search often yields programs that result from applying `ShadowClause` perturbations. As Table 1 shows, blind search often terminates with less inferences than the other two methods, but at the cost of low-quality solutions. None of the typical errors were reproduced faithfully.

Cost-based search and A^*-search offer a vast improvement to blind-search. Here, `ShadowClause` transformations are only chosen when no other transformations are available. While cost-based search reproduces four of the top-five errors, A^* manages to get all five reproductions right. For `stops-borrow-at-zero`, cost-based search constructs a buggy variant of the expert program by deleting only a single subgoal in the first definition of `decrement/3` (line 35); A^*-search performs three deletions in this clause, effectively rendering it into a null operation. While both variants reproduce the learner error, their dynamics is different: the first program forces `process_column/3` to enter its second clause for processing the middle column, while A^* forces `process_column/3` into its first clause. Clearly, A^* returns a more faithful reproduction of the given error.

In terms of inferences, A^* has equal or better performance than the other two methods, while returning equal or better solutions. The benefit of A^* is dramatic for `smaller-from-larger`, where blind search delivers a low-quality path of length 5, and cost-based search and A^* a high-quality path of length 3.

For the last three errors, we can only obtain solutions that are close to our gold standard. This is due to the current lack of inductive capabilities in the test framework. The perturbations to reproduce the error `stops-borrow-at-zero`, `diff0-N=N` follow, by and large, the perturbations performed for `stops -borrow-at-zero`. The presence of the error `diff0-N=N`, however, implies that no decrement operation is necessary for the given task $401 - 199$. Rather than just making `decrement/3` a null operation (as in `stops-borrow-at-zero`), the perturbations for `stops-borrow-at-zero`, `diff0-N=N` delete the call to `decrement/3` in `process_column/3` altogether. In addition, two task-specific clauses are added for the `diff0-N=N` case. Similar remarks apply to `borrow_across_zero_diff0_N_eq_N`. We find the performance gain in both of the cases significant.

4 Related Work

4.1 Program Testing

Our research has an interesting link to program testing and the design and reliability of test data [4]. The theory of program testing rests on the *competent programmer hypothesis*, which states that programmers "create programs that are *close* to being correct" [4]. In other words, if a program is buggy, then it differs from the correct program only by a combination of simple errors. Moreover, programmers have a rough idea of the kind of error that are likely to occur, and they have the ability to examine their programs in detail. Program testing is also thought to be aided by the *coupling effect*: test cases that detect simple types of faults are sensitive enough to detect more complex types of faults. The analogy to VanLehn's theory of impasses and repairs is striking. When learners encounter an impasse in executing a correct procedure, they address the impasse by a local repair, which often can be explained in terms of simple errors. Also, teachers have a rough idea of the kind of errors learners are likely to make (and learners might be aware of their repairs, too). Good teachers are able to reconstruct the erroneous procedure a learner is executing, and learners are able to correct their mistakes either themselves or under teacher supervision.

In program testing, the technique of *mutation testing* aims at identifying deficiencies in test suites, and to increase the programmer's confidence in the tests' fault detection power. A mutated variant p' of a program p is created only to evaluate the test suite designed for p on p'. If the behaviour between p and p' on test t is different, then the mutant p' is said to be dead, and the test suite "good enough" wrt. the mutation. If they are equal, then p and p' are equivalent, or the test set is not good enough. In this case, the programs' equivalence must be examined by the programmer; if they are not equivalent, the test suite must be extended to cover the critical test. This relates to our approach. When a given program is unable to reproduce a learner's solution, we create a set of perturbated variants, or mutants. If one of them reproduces the learner's solution, it passes the test, and we are done. Otherwise, we choose the best mutant, given the heuristic function f, and continue with the perturbations. The originality of our approach is due to our systematic search for mutations and the use of f to measure the distance between mutants wrt. a given input/output.

In [6], Kilperäinen & Mannila describe a general method for producing complete sets of test data for simple Prolog programs. Their method is based on the competent programmer hypothesis, and works by mutating list processing programs with a small class of suitable modifications. In [8], the authors give a wide range of mutation operators for Prolog. At the clause level, they have operators for the removal of subgoals, for changing the order of subgoals, and for the insertion, removal, or permutation of cuts. At the operator level, they propose mutations that change one arithmetic or relational operator by another one. Moreover, they propose mutations that act on Prolog variables or constants, e.g., the changing of one variable into another variable, an anonymous variable, or a constant, or the changing of one constant into another one. All mutations are syntactic, and aim at capturing typical programmer errors. So far, our approach makes use of a subset of the aforementioned mutation operators. It is surprising that the top-five bugs, accounting for nearly 50 % of all learner errors, can be explained by learners skipping steps, i.e., mostly in terms of clause deletions.

4.2 Intelligent Tutoring Systems

In the intelligent tutoring community, most system designers follow a rule-based approach to implement *interactive exercises* that help students learn. The ACT* architecture is both theoretical embedding and practical implementation basis for ITSs such as the LISP Tutor or the PAT algebra tutor, see the overview [1]. In the ACT* approach, a set of production rules models the skills to be acquired by the learner. To capture erroneous learner behaviour, expert rules are complemented by buggy rules. The rule engine's step-wise interpretation of the rule system allows the tracing of learner actions in terms of the model. Learner actions are on-path when reproducible by the execution of expert rules, or off-path when explainable in terms of buggy production rules, or when no sequence of rules can be found. Positive and remedial feedback, which is attached to rules, can be generated to support learners' problem solving.

Tutors built upon production rule systems have two major drawbacks: they have high authoring costs, and they need to keep learners close to the correct solution path to tame the combinatorial explosion of the (forward reasoning) rule engine. We focus on the first aspect. While rule-based systems offer an adequate formalism to represent the *logic* of a given domain using a *set* of rules, it seems to be much harder to encode a domain's *control* aspect. The hierarchical aspect of the domain algorithm can only be modeled in terms of goal structures that reside in the rule system's *working memory* and which must be explicitly maintained and manipulated using the rules' pre- and postconditions. The ACT*-like encoding of control creates rules that depend on each other, and hence, makes the authoring and managing of large rule bases a costly undertaking.

In our approach, logic and control are encoded using Prolog, where goal structures are automatically taken care of. Moreover, our tracing of learner actions does not require an *a priori* encoding of buggy rules; buggy program variants are generated on the fly, using a clever variant of algorithmic debugging, which compares expert with learner behaviour, and program transformation techniques that are based on well-defined perturbation operators. In rule-based systems, there is no representation that encodes the difference between an expert and a buggy rule, and hence, little support for modeling learners' repair strategies.

The Icarus cognitive architecture [7] addresses some of the drawbacks of rule-based systems. Inspired by Prolog, it allows rules (*skills*) to explicitly mention sub-skills (i.e., other rules) without making indirect references to them through the working memory. Nevertheless, Icarus retains the overall flavour of a production system by following a recognize-act-cycle. A more radical approach to separate logic (*domain-specific rules*) from control (*strategic guidance*) is proposed by Heeren *et al.* [5]. They separate: (i) information about the domain (e.g., the subtraction matrix and its place-value system), (ii) rules for manipulating expressions in this domain (e.g., performing a complete borrow-payback operation, or taking the difference in a column), (iii) a strategy for solving the exercise (e.g., performing subtraction from right to the left), and (iv) buggy knowledge for modeling both incorrect expression manipulations and incorrect strategies. In their approach, a strategy language is defined that has rules as smallest building blocks, and which controls their combination using rule sequencing, rule choice, and a recursion mechanism. Using the language, a strategy can be defined as a context-free grammar. With the tracing of learner actions reduced to a parsing problem, Heeren *et al.* define a strategy recognizer that is able to compute several types of feedback to support learners when incrementally solving interactive exercises. In this approach, the strategy recognizer should be capable of coping with learner errors that result from strategic (skipping the step) repairs.

5 Conclusion and Future Work

In this paper, we propose a method to automatically transform an initial Prolog program into another program capable of producing a given input/output behaviour. The method depends on a heuristic function that estimates the distance between programs. An experimental evaluation demonstrated the benefits of using heuristic search when compared to the blind (depth-first) search we have used in [12]. It shows that the test-debug-repair cycle can be mechanised in the tutoring context. Here, there is always a reference model to encode ideal behaviour; moreover, many learner errors can be captured and reproduced by a combination of simple, syntactically-driven program transformation actions.

In the near future, we would like to include more mutation operators (see [8]), investigate their interaction with our existing ones, fine-tune the cost function, and study whether erroneous procedures can be obtained that better reflect learners' incorrect reasoning. Ideally, the new operators can be used to "un-employ" the costly `ShadowClause` operator, whose primary purpose is to serve as a fall-back action when all other actions fail. Moreover, we are currently working on a web-based interface for multi-column subtraction tasks that we want to give to learners, and where we plan an evaluation in terms of pedagogical benefits.

In the long term, we would like to take on another domain of instruction to underline the generality of our approach. The domain of learning programming in Prolog is particularly interesting. In the subtraction domain discussed in this paper, we are systematically modifying an expert program into a buggy program to model a learner's erroneous behaviour. In the "learning Prolog domain", we can re-use our program distance measure in a more traditional sense. When learners do specify an executable Prolog program, we compare its behaviour with the prescribed expert program, identify their (dis-)agreement score, and then repair the learner's program, step by step, to become the expert program.

Acknowledgments. The research was funded by the DFG (ZI 1322/2-1).

References

1. Anderson, J.R., Corbett, A.T., Koedinger, K.R., Pelletier, R.: Cognitive tutors: lessons learned. J. Learn. Sci. **4**(2), 167–207 (1995)
2. Brown, J.S., Burton, R.R.: Diagnostic models for procedural bugs in basic mathematical skills. Cogn. Sci. **2**, 155–192 (1978)
3. Burton, R.R.: Debuggy: diagnosis of errors in basic mathematical skills. In: Sherman, D., Brown, J.S. (eds.) Intelligent Tutoring Systems. Academic Press, London (1982)
4. DeMillo, R.A., Lipton, R.J., Sayward, F.G.: Hints on test data selection: help for the practicing programmer. Computer **11**(4), 34–41 (1978)
5. Heeren, B., Jeuring, J., Gerdes, A.: Specifying rewrite strategies for interactive exercises. Math. Comput. Sci. **3**(3), 349–370 (2010)
6. Kilperäinen, P., Mannila, H.: Generation of test cases for simple prolog programs. Acta Cybern. **9**(3), 235–246 (1990)
7. Langley, P., Cummings, K.: Hierarchical skills and cognitive architectures. In: 26th Annual Conference of the Cognitive Science Society, pp. 779–784 (2004)
8. Toaldo, J.R., Vergilio, S.R.: Applying mutation testing in prolog programs. http://www.lbd.dcc.ufmg.br/colecoes/wtf/2006/st2_1.pdf
9. VanLehn, K.: Mind bugs: The Origins of Procedural Misconceptions. MIT Press, Cambridge (1990)
10. Young, R.M., O'Shea, T.: Errors in children's subtraction. Cogn. Sci. **5**(2), 153–177 (1981)
11. Zinn, C.: Algorithmic debugging to support cognitive diagnosis in tutoring systems. In: Bach, J., Edelkamp, S. (eds.) KI 2011. LNCS (LNAI), vol. 7006, pp. 357–368. Springer, Heidelberg (2011)
12. Zinn, C.: Program analysis and manipulation to reproduce learners' erroneous reasoning. In: Albert, E. (ed.) LOPSTR 2012. LNCS, vol. 7844, pp. 228–243. Springer, Heidelberg (2013)

Author Index

Printed in the United States
By Bookmasters

Printed in the United States
By Bookmasters